SOIL AND CONSERVATION

Second Edition

SOIL EROSION AND CONSERVATION

Second Edition

R. P. C. Morgan
Silsoe College,
Cranfield University

 LONGMAN

Addison Wesley Longman Limited
Edinburgh Gate, Harlow
Essex CM20 2JE, England
and Associated Companies throughout the world.

First published 1986
Second edition 1995
Reprinted 1996 (twice)

British Library Cataloguing in Publication Data
A catalogue entry for this title is available from the British Library.

ISBN 0–582–24492–7

Library of Congress Cataloging-in-Publication data
A catalog entry for this title is available from the Library of Congress.

Set by 3 in Plantin
Produced through Longman Malaysia, TCP

Contents

Preface

Soil erosion is a hazard traditionally associated with agriculture in tropical and semi-arid areas and is important for its long-term effects on soil productivity and sustainable agriculture. It is, however, a problem of far wider significance occurring additionally on land devoted to forestry, transport and recreation. Since the first edition of *Soil Erosion and Conservation* was published in 1986, erosion has become a global issue because of its environmental consequences such as sedimentation and pollution. Major pollution problems can arise from quite moderate and frequent erosion events in both temperate and tropical climates. Erosion control is now a necessity in almost every country of the world under virtually every type of land use.

Whilst it has been tempting to address the environmental issues in detail, many other texts fulfil this role. The philosophy of this book therefore remains unchanged, namely to provide a text which concentrates on the processes of erosion and how to combat them. Whatever the environmental context, there is still a need to control erosion and the design of control measures requires a thorough understanding of the processes of erosion and their controlling factors. Despite a substantial publication output in the field over the last ten years, there is still no other text which covers soil conservation from a substantive treatment of erosion and also incorporates a global research perspective.

The mechanics of erosion are reviewed in the first part of the book with emphasis on the extent and deficiencies in current knowledge. Techniques of classifying land with respect to erosion risk are also examined and a simple working method is presented. A discussion of various approaches to modelling soil erosion focuses on the value of models for predicting rates of soil loss and planning conservation work. The second part of the book is concerned with soil conservation practices. Strategies for erosion control are first examined with the emphasis on the integration of agronomic measures, engineering structures and soil management. The following chapters discuss the range of measures available under each of these headings, paying particular attention to their design and, through a review of research findings, their effectiveness.

Since 1986, major advances have been made in the study of erosion, particularly on the initiation of rills, the importance of cohesion as a component of soil erodibility and the role of vegetation. These advances are forming the basis for a new generation of erosion models which are being developed to address the environmental issues for which empirical models such as the Universal Soil Loss Equation are no longer appropriate. Chapters 2, 3 and 5 have been substantially expanded and rewritten to reflect these developments.

The increase in research output from Africa, Asia and Latin America, referred to briefly in the Preface to the 1986 edition, has continued. As a result, greater selectivity has been possible in choosing examples to demonstrate the success or otherwise of different soil management and conservation techniques. Generally, preference has been given to studies that have been carried out over several years. At the same time, the temptation to replace all the references with more recent examples has been resisted. Reference to what may be considered seminal researches of the 1940s to 1970s has been retained, partly to give an important historical context but also to maintain awareness of what has been achieved in the past in an attempt to prevent mere repetition of previous work.

The resolution of soil erosion problems will not be achieved by technical solutions alone; there is an equally important socio-economic context. Although this was implicit in the 1986 edition, it received a somewhat scant treatment. The opportunity has now been taken to expand on the social, economic, political and institutional issues not in a substantive way, since this would require a separate text in its own right, but to set the background within which technical solutions have to be designed and implemented. As a result, Chapters 7 and 11 have been considerably expanded.

Other changes since 1986 take the form of additions to and amendments of the text, either to up-date the material, or to broaden the context. For example, the use of caesium-137 as a method for measuring rates of erosion is now included in Chapter 6, along with new work on the use of splash cups and sand collectors; the sections on agroforestry, contour grass strips, soil conditioners and geotextiles have been rewritten; and more attention is given to problems of erosion control on road banks and footpaths.

The text continues to be based on courses which I give at Silsoe College and again contains material from research carried out by myself, my colleagues and students. The contributions of the last two groups are referenced and much appreciated. The text remains intended for undergraduate and postgraduate students studying soil erosion and conservation as part of their courses in geography, environmental science, agriculture, agricultural engineering, hydrology, soil science, ecology and civil engineering. With the expansion in the socio-economic content, the text is even more integrative of disciplines than before. In addition, it provides an introduction to the subject for those concerned with resources survey and development, agricultural, recreational and rural planning, and for those working on soil erosion and conservation, either in an advisory capacity or at research and experimental stations.

Again, I am grateful to my wife, Gillian, and my sons, Richard and Gerald, for accepting the disruptions caused by the preparation of the book.

R. P. C. Morgan
Silsoe
July 1994

Foreword

Over the last decade a widely stated objective in land resource management has been the adoption of strategies to ensure the sustainable use of land. Such an objective has been central to soil conservation ever since its emergence as an applied science in the 1930s. The aims of any policy dealing with sustainable use of soils are to maintain soil quality, properties, processes and diversity. The formulation of policies on sustainable development is thus on the agendas of national governments and international agencies. At the same time soil erosion continues to degrade the global land resource base with approximately 30 per cent of the present cultivated area having been substantially affected. Soil erosion is a major global environmental issue. The need for greater as well as more reliable agricultural yields is a continuing and ever-increasing global challenge. A wide range of economic, social, environmental, demographic and political factors interact to cause the dire situation that exists in many countries. Sometimes it is thought that erosion is confined to tropical, arid or semi-arid regions but it is increasingly realized that a range of insidious processes also exists in cool temperate regions. Current concerns associated with the erosion of soils are the long-term impacts on such topics as soil quality, agricultural productivity, movement of pollutants, ecological diversity in streams and wetlands, river channel change and effects on flooding.

Despite a research record stretching over 60 years and the fact that many governments are striving towards policies to ensure sustainable land use systems, there is still a dearth of knowledge on the nature and incidence of soil erosion. The tendency has been to approach soil conservation issues as engineering problems for which technological solutions have to be determined. The key point which emerges from this book by Roy Morgan is that a soil conservation strategy must evolve from detailed knowledge and understanding of actual erosional processes; furthermore, the evolution of such a strategy must accord with local economic, social, demographic and political circumstances. Discussion of strategies for erosion control, crop and vegetation management and soil management are preceded by examination of erosion processes and associated factors, erosion hazard assessment, and modelling and measurement of soil erosion. The book concludes with a thoughtful discussion of issues associated with the implementation of soil conservation strategies; particular emphasis is given to the adoption of a bottom-up approach involving the participation of local land users. In essence the message is the need for a blend of good applied science with local practical knowledge and experience. Such a discussion implies that in many countries new ways have to be found for the implementation at the local level of soil conservation strategies.

The first version of this book was published in 1979 in the series *Topics in Applied Geography*. It was a short volume which provided readers with a succinct yet comprehensive overview of soil erosion. The success of this initial volume led to the publication in 1986 of the First Edition of *Soil Erosion and Conservation*. This book gave Roy Morgan the opportunity to update his text, but also to write a much longer second part dealing with conservation strategies. This approach is retained in this new edition which has been written in order to include recent advances on soil erosion processes and modelling in particular. The reader will be impressed by the extent to which Roy Morgan's own research experience is evident. His early work was in Malaysia, but in more recent years he has been much associated with monitoring of erosion at sites in Bedfordshire as well as with collaborative projects within the European Union. His final point is a plea for people with broad perspectives of such disciplines as geomorphology, agriculture, engineering, economics and sociology as well as a specialist skill in one area to become involved with soil erosion questions. His book is an excellent demonstration of such involvement and it is hoped that it is widely read by all concerned with this important subject.

Donald A. Davidson

Acknowledgements

We are grateful to the following for permission to reproduce copyright material:

British Trust for Conservation Volunteers for Fig. 10.7 (Agate, 1983); Department of Agricultural Engineering at the University of Nairobi for Fig. 8.1 (Mututho, 1989); Division de Recherches sur les Systèmes for Fig. 7.3 (Hijkoop, van der Poel and Kaya, 1991); Elsevier Science for Fig. 2.3 (Savat, 1982); John Wiley & Sons Ltd. for Fig. 6.6 (Walling and Quine, 1990. Use of caesium-137 to investigate patterns and rates of soil erosion on arable fields. In J. Boardman, I. D. L. Foster and J. A. Dearing (eds), *Soil erosion in agricultural land*. Copyright © 1990. Reprinted by permission of John Wiley & Sons, Ltd.); UCL Press Ltd. for Fig. 2.2 (Poesen, 1992).

Whilst every effort has been made to trace the owners of copyright material, in a few cases this has proved impossible and we take this opportunity to offer our apologies to any copyright holders whose rights we may have unwittingly infringed.

CHAPTER I

Distribution of Soil Erosion

The rapid erosion of soil by wind and water has been a problem ever since land was first cultivated. The consequences of soil erosion occur both on- and off-site. On-site effects are particularly important on agricultural land where the redistribution of soil within a field, the loss of soil from a field, the breakdown of soil structure and the decline in organic matter and nutrient result in a reduction of cultivable soil depth and a decline in soil fertility. Erosion also reduces available soil moisture, resulting in more drought-prone conditions. The net effect is a loss of productivity which, at first, restricts what can be grown and results in increased expenditure on fertilizers to maintain yields, but later threatens food production and leads, ultimately, to land abandonment. It also leads to a decline in the value of the land as it changes from productive farmland to wasteland. Off-site problems result from sedimentation downstream or downwind which reduces the capacity of rivers and drainage ditches, enhances the risk of flooding, blocks irrigation canals and shortens the design life of reservoirs. Many hydro-electricity and irrigation projects have been ruined as a consequence of erosion. Sediment is also a pollutant in its own right and, through the chemicals adsorbed to it, can increase the levels of nitrogen and phosphorus in water bodies and result in eutrophication.

The on-site costs of erosion are necessarily borne by the farmer although they may be passed on in part to the community in terms of higher food prices as yields decline or land goes out of production. The farmer bears little of the off-site costs which fall on local authorities for road clearance and maintenance, insurance companies and all the land holders in the local community affected by sedimentation and flooding. Only rarely does the central government have to consider the true costs of erosion.

The causes of soil erosion are still poorly understood. Whilst considerable research, particularly since 1940, has resulted in better knowledge of the mechanics of erosion processes and their relationship with the physical environment, only recently has systematic research been undertaken on the social, economic, political and institutional factors that influence where and when erosion occurs. Conventional wisdom favours explaining erosion as a response to increasing pressure on the land brought about by a growing world population and the abandonment of large areas of formerly productive land as a result of erosion, salinization or alkalinization. In the loess plateau region of China for example, annual soil loss has increased exponentially since about 220 BC in a simple relationship with the total population (Wen 1993). Population pressure forces people to farm more marginal land, often unwisely, especially in the Himalayas, the Andes and many mountainous areas of the humid tropics.

In other parts of the world, however, erosion can be seen as a direct response to abandonment of land associated with rural depopulation. A dramatic example comes from the terraced mountain slopes of the Haraz in the Yemen where land abandonment began in a drought early in this century, was repeated in other droughts in the 1940s and between 1967 and 1973, and then increased markedly in the 1970s as people migrated to Saudi Arabia and the Gulf States. All types of water erosion are now visible, following collapse of the terrace walls, and erosion is reducing the depth of the already shallow soil by 1 to 3 cm/y (Vogel 1990). In much of Mediterranean Europe, policies to reduce the number of people employed in agriculture and to increase farm size and the level of mechanization have had a two-fold effect. First, fewer people remain on the land to maintain traditional terracing which is left to decay. Second, the increase in farm size is often accompanied by large-scale earth moving and land levelling operations which make the soil more erodible. Almost everywhere that land consolidation programmes have been carried out, rates of soil erosion have increased.

The prevention of soil erosion, which means reducing the rate of soil loss to approximately that which would occur under natural conditions, relies on selecting appropriate strategies for soil conservation and this, in turn, requires a thorough understanding of the processes of erosion. The factors which influence the rate of erosion are rainfall, runoff, wind, soil, slope, plant cover and the presence or absence of conservation measures. These and other related factors may be considered under three headings: energy, resistance and protection. The energy group includes the potential ability of rainfall, runoff and wind to cause erosion. This ability is termed erosivity. Also included are those factors which directly affect the power of the erosive agents such as the reduction in the length of runoff or wind blow through the construction of terraces and wind breaks respectively. Fundamental to the resistance group is the erodibility of the soil which depends upon its mechanical and chemical properties. Factors which encourage the infiltration of water into the soil and thereby reduce runoff decrease erodibility whilst any activity that pulverizes the soil increases it. Thus cultivation may decrease the erodibility of clay soils but increase that of sandy soils. The protection group focuses on factors relating to the plant cover. By intercepting rainfall and reducing the velocity of runoff and wind, a plant cover can protect the soil from erosion. Different plant covers afford different degrees of protection so that human influences, by determining land use, can control the rate of erosion to a considerable degree.

The rate of soil loss is normally expressed in units of mass or volume per unit area per unit of time. In a review of erosion under natural conditions, Young (1969) quotes annual rates of the order of 0.0045 t/ha for areas of moderate relief and 0.45 t/ha for steep relief. For comparison, rates from agricultural land are in the range of 45 to 450 t/ha. Theoretically, whether or not a rate of soil loss is severe is judged relative to the rate of soil formation. If soil properties such as nutrient status, texture and thickness remain unchanged through time, it is assumed that the rate of erosion balances the rate of soil formation.

1.1 Spatial Variations

On a world scale, investigations of the relationship between soil loss and climate (Langbein and Schumm 1958) show that erosion reaches a maximum in areas with an effective mean annual precipitation of 300 mm. Effective precipitation is that required to produce a known quantity of runoff under specified temperature conditions. At precipitation totals below 300 mm, erosion increases as precipitation increases. But, as precipitation increases so does the vegetation cover, resulting in better protection of the soil surface. At precipitation totals above 300 mm, the protection effect counteracts the erosive effect of greater rainfall, so that soil loss decreases as precipitation increases. There is some evidence (Douglas 1967), however, to indicate that soil loss may increase again with precipitation as the latter and therefore the runoff reach still higher levels (Fig. 1.1). The availability of information

on sediment loads in rivers has increased enormously over the last 30 years. Although the data relate only to the material carried in suspension because of the technical difficulties in measuring bed load and the relative paucity of observations on solute concentrations, they can be used to provide a reasonably reliable assessment of the global pattern of water erosion (Fig. 1.2; Walling and Webb 1983).

The major feature revealed by this assessment is the vulnerability to erosion of the semi-arid and semi-humid areas of the world, especially in China, India, the western USA, central Russia and the Mediterranean lands. The problem of soil erosion in these areas is compounded by the need for water conservation and the ecological sensitivity of the environment, so that removal of the vegetation cover for cropping or grazing results in rapid declines in organic content of the soil, followed by soil exhaustion and the risk of desertification. Other areas of high erosion rates include mountainous terrain, such as much of the Andes, the Himalayas, the Karakoram, parts of the

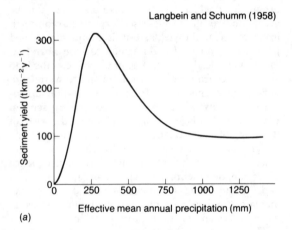

(a)

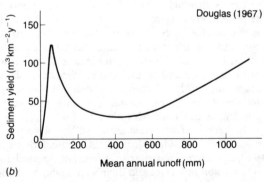

(b)

Figure 1.1 Proposed relationships between sediment yield and (a) effective mean annual precipitation; (b) mean annual runoff (after Gregory and Walling 1973).

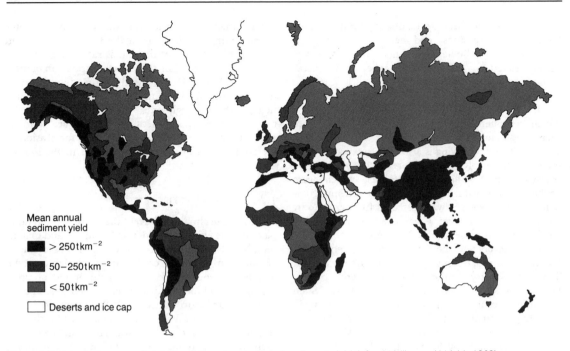

Figure 1.2 A tentative map of global variations in suspended sediment yield (after Walling and Webb 1983).

Legend:

Mean annual sediment yield

- ■ > 250 t km^{-2}
- ■ 50–250 t km^{-2}
- ■ < 50 t km^{-2}
- □ Deserts and ice cap

Rocky Mountains and the African Rift Valley; and areas of volcanic soils, for example Java, South Island of New Zealand, Papua New Guinea and parts of Central America.

Within any particular area there will be a considerable variation in erosion rates but if the rates are grouped into those related to natural vegetation, cultivated land and bare soil, each group follows a broadly

Table 1.1 Annual rates of erosion in selected countries (t/ha)

	Natural	Cultivated	Bare soil
China	0.1–2	150–200	280–360
USA	0.03–3	5–170	4–9
Australia	0.0–64	0.1–150	44–87
Ivory Coast	0.03–0.2	0.1–90	10–750
Nigeria	0.5–1	0.1–35	3–150
India	0.5–5	0.3–40	10–185
Ethiopia	1–5	8–42	5–70
Belgium	0.1–0.5	3–30	7–82
UK	0.1–0.5	0.1–20	10–200

Sources: Boardman 1990; Bollinne 1978; Browning *et al.* 1948; Edwards 1993; Fournier 1972; Hurni 1993; Jiang, Qi and Tan 1981; Lal 1976; Morgan 1985a; Roose 1971; Singh, Babu and Chandra 1981.

similar pattern of global variation (Table 1.1). One exception to this is the humid tropics. Measurements

of soil loss from hillslopes in West Africa (Roose 1971) ranging in steepness from 0.3° to 4° yield mean annual rates of erosion of 0.15, 0.20 and 0.03 t/ha under natural conditions of open savanna grassland, dense savanna grassland and tropical rain forest respectively. Clearance of the land for agriculture increases the rates to 8, 26 and 90 t/ha whilst leaving the land as bare soil produces rates of 20, 30 and 170 t/ha respectively. Thus removal of the rain forest results in much greater rises in erosion rates than does removal of the savanna grassland. Where the land is cleared, erosion is highest in the rain forest areas whereas under natural conditions it is highest in the dense savanna areas. These measurements emphasize the high degree of protection afforded to the soil by the rain forest but also reflect the erosive capacity of the high rainfalls of the humid tropics when the protection is destroyed. They indicate the high sensitivity of the rain forest ecosystem to disturbance. Given the present rate of removal of the tropical rain forest, a second area of high vulnerability to erosion is being created which is likely to prove as severe a problem as that of the semi-arid and semi-humid areas of the world.

A third area of high risk of erosion, not discernible from Fig. 1.2, relates to regions where the landforms and associated soils are relics of a previous climate. Over much of southern Africa stratigraphical evidence shows sequences of periods of comparative stability in the landscape, indicated by the development of humic layers and stone lines, and periods of instability, represented by colluvial sediments, often up to 5 m thick, related to gully and sheet erosion (Price-Williams, Watson and Goudie 1982). Throughout much of Swaziland and Zimbabwe, present-day gully erosion is particularly severe on the colluvial deposits which are often fine sandy or silty in nature and, therefore, inherently highly erodible (Shakesby and Whitlow 1991). Gullying is also extensive world-wide on areas of deeply-weathered regoliths or saprolite overlying granites and granodiorites, which were probably formed during a more humid tropical climate when the surface was protected from erosion by a dense vegetation cover. Clearance of the vegetation has led to an increase in runoff and erosion. Once the upper soil layers have been removed, the underlying, highly-weathered, often very fine substrate is exposed. This offers very limited resistance and rapidly becomes deeply dissected (Rohdenburg 1989). Such conditions occur not only in southern Africa, again including much of Zimbabwe and Swaziland, but also on the margins of the savanna lands in West Africa, and in Brazil and southern China. These examples indicate that the response of the land to changes in vegetation cover brought about by human influences can depend greatly on geomorphological inheritance from previous climatic conditions.

Care should be taken in interpreting data on rates of erosion, like those given in Table 1.1, because the rate varies with the size of the area being considered. Part of the sediment removed from hillsides, embankments and cuttings finds it way into rivers but part is deposited on footslopes and in flood plains where it remains in temporary storage, sometimes until the next storm, or, at other times, as in the case of the colluvial material mentioned above, for millions of years. Since larger drainage basins tend to have larger proportions of these sediment sinks, erosion rates expressed per unit area are generally higher for small basins and decrease as the catchment becomes bigger. The proportion of the sediment eroded from the land surface which discharges into the river is known as the sediment delivery ratio. Surprisingly few studies have been made of this, considering it is fundamental to assessing sediment yield, but it varies from about 3 to 90 per cent, decreasing with greater basin area and lower average slope (Walling 1983).

Within relatively small areas, rainfall characteristics are relatively uniform and the distribution of sediment sources and sinks varies with soils, slopes and land use. Boardman (1990) found that most erosion between 1982 and 1987 in the area between Brighton and Lewes in the South Downs, England, occurred on the sides of the major dry valleys where the relief was

greater than 100 m and the land was under winter cereals. Out of 300 fields in the study area, erosion was limited to 114, with annual rates ranging from 0.01 to 200 m^3/ha. The frequency distribution of rates was strongly positively skewed with median values in the range of 0.6 to 5.0 m^3/ha in any one year. Thus, fields with very high rates of erosion which are the major sources of sediment production are spatially rather limited. In 1987, the worst year for erosion in the study period, erosion exceeded 10 m^3/ha in only 28 fields.

1.2 Temporal Variations

Most geomorphologists accept that most erosion takes place during events of moderate frequency and magnitude simply because extreme or catastrophic events are too infrequent to contribute appreciably to the quantity of soil eroded over a long period of time. Magnitude-frequency ideas were applied to studies of sediment transport in rivers by Wolman and Miller (1960). They found that the dominant event, the one responsible for most work, was larger in magnitude than the most frequent event but was by no means extreme, corresponding to the stage of bankfull discharge, an event with a return period of between 1.33 and 2 years.

A similar concept may be applied to erosion on hillsides. Experimental studies of Roose (1967) in Senegal show that between 1959 and 1963, 68 per cent of the soil loss took place in rain storms of 15 to 60 mm; these storms have a frequency of about ten times per year. Studies of erosion in mid-Bedfordshire, England (Morgan, Martin and Noble 1986) indicate that in the period 1973 to 1979, 80 per cent of the erosion occurred in 13 storms, the greatest soil loss, comprising 21 per cent of the erosion, resulting from a storm of 57.2 mm. These storms have a frequency of between two and four times a year. Measurements of erosion near Ibadan, Nigeria, showed that half the annual soil loss can occur with between two and seven storms (Lal 1976). In contrast to this evidence supporting a somewhat frequent dominant event for hillslope erosion, Hudson (1981) emphasizes the role of the more dramatic event. Quoting from research in Zimbabwe, he states that 50 per cent of the annual soil loss occurs in only two storms and that, in one year, 75 per cent of the erosion took place in ten minutes.

The more dramatic events may become important where erosion is not a function of climate alone but depends on the frequency at which potentially erosive events coincide with ground conditions that favour erosion. Analysis of 28 years of data for nine small catchments under a four-year rotation of maize-wheat–grass–grass at Coshocton, Ohio (Edwards and Owens 1991) showed that the three largest storms, all with return periods of 100 years or more, accounted for 52 per cent of the erosion and that 92 per cent of the soil loss occurred in the years when the land was under maize.

A comparison of the studies of sediment transport on hillslopes with that in rivers indicates a more frequent dominant event for the former and suggests that the dominant event frequency may vary for different erosion processes. Further evidence for this is provided by research on shallow debris slides and mudflows on cultivated fields and grassland in the Mgeta area of Tanzania for which the dominant event has a return period of once in five years (Temple and Rapp 1972). Moderate events also account for most of the erosion carried out by wind. Studies on coastal dunes at Cape Moreton, New South Wales, show that most sand transport occurs in strong winds of about 14 m/s with relatively little being carried out by winds of gale force or above because their greater competence is compensated for by their rarity (Chapman 1990).

Even though most erosion of soil from hillsides and transport of sediment in rivers occurs in dominant events of moderate magnitude, the effects of an extreme event can be very dramatic. Extreme events can have effects which are different in kind as well as degree and may well be longer lasting. For example, rainstorms with return periods greater than once in a hundred years may initiate headward erosion of gullies and the formation of new channel heads. A slow-moving equatorial storm disturbance within the northeast monsoon deposited 631 mm of rain on 28 December 1926 and 1194 mm between 26 and 29 December in the Kuantan area of Malaysia, resulting in extensive gully erosion and numerous landslides. The scars produced in the landscape were still visible 35 years later (Nossin 1964).

In addition to the variations in erosion associated with the frequency and magnitude of single storms, rates of erosion often follow a seasonal pattern. This is best illustrated with reference to a rainfall regime with a wet and a dry season (Fig. 1.3). The vegetation growth follows a similar pattern but peaks later in the season than the rainfall. The most vulnerable time for erosion is the early part of the wet season when the rainfall is high but the vegetation has not grown sufficiently to protect the soil. Thus the erosion peak precedes the rainfall peak.

Somewhat more complex seasonal patterns occur with less simple rainfall regimes or where the land is used for arable farming. Generally, the period between ploughing and the growth of the crop beyond the seedling stage contains an erosion risk if it coincides with heavy rainfall or strong winds. Thus spring is often a peak time for erosion in western Europe, including the UK.

Also, superimposed on the frequency-magnitude patterns of erosion are changes brought about by alteration of the plant cover. A typical sequence of events is described by Wolman (1967) for Maryland where soil erosion rates increased with the conversion of woodland to cropland after AD 1700 (Fig. 1.4). They declined as the urban fringe extended across the area in the 1950s and the land reverted to scrub when the farmers sold out to speculators, before accelerating rapidly, reaching annual rates of 7,000 t/ha, when the area was laid bare during housing construction. With the completion of urban development, runoff from concrete surfaces is concentrated into gutters and sewers, and annual soil loss falls to below 4 t/ha.

Based on stratigraphical and archaeological evidence of valley floor deposits and archival material, Bork (1989) has reconstructed the history of soil erosion in Niedersachsen, Germany. From the early Holocene, when soils developed under the natural woodlands, up to the early Middle Ages, erosion rates were extremely low. With the clearance of forest for agriculture between AD 940 and 1340, erosion increased and reached annual rates of about 10 t/ha.

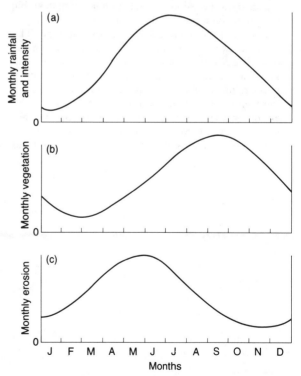

Figure 1.3 Seasonal cycles of rainfall, vegetation cover and erosion in a semi-humid climate (after Kirkby 1980a).

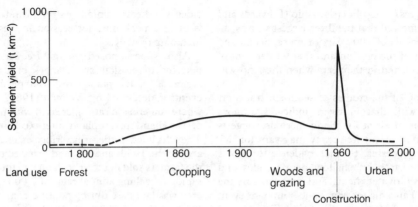

Figure 1.4 Relationship between sediment yield and changing land use in the Piedmont region of Maryland, USA (after Wolman 1967).

Between 1340 and 1350, annual erosion rates rose dramatically to 2,250 t/ha as a result of gully erosion induced by extreme climatic events. Erosion declined afterwards, partly as a result of a decrease in the area under arable as land was abandoned due to impoverishment by erosion. The rate of erosion did not return to early medieval levels, however, but remained at an average annual rate of around 25 t/ha. The higher rate reflected sheet erosion on the land remaining in arable as production went over to the three-field system with one-third of the land in fallow at any one time. The period between 1750 and 1800 saw a second episode of gullying, with an average annual soil loss around 160 t/ha, in response to an increase in the frequency of heavy rainfall events. The soil loss did not reach mid-fourteenth century levels, however, because of the establishment of terraces, the use of contour plough-ing and grass strips, and a higher proportion of the land under grass and trees. Since 1800 annual soil loss

has averaged 20 t/ha but has increased in recent years following land consolidation which has resulted in larger fields, removal of terraces and grass strips, and land levelling.

Erosion and land use change are very strongly related. Rates of soil loss accelerate quickly to unacceptably high levels whenever land is misused. Under these conditions, the effects of geomorphological events of moderate and high magnitude in particular are exaggerated. Only in extreme events is the land cover influence overshadowed and extensive areas of land damaged regardless of how they are managed. A combination of historical and geomorphological analysis emphasizes that erosion is a natural process but that its rate and spatial and temporal distribution depend on the interaction of physical and human circumstances. Soil erosion is therefore an integral part of both the natural and cultural environment.

CHAPTER 2

Processes and Mechanics of Erosion

Soil erosion is a two-phase process consisting of the detachment of individual particles from the soil mass and their transport by erosive agents such as running water and wind. When sufficient energy is no longer available to transport the particles, a third phase, deposition, occurs.

Rainsplash is the most important detaching agent. As a result of raindrops striking a bare soil surface, soil particles may be thrown through the air over distances of several centimetres. Continuous exposure to intense rainstorms considerably weakens the soil. The soil is also broken up by weathering processes, both mechanical, by alternate wetting and drying, freezing and thawing and frost action, and biochemical. Soil is disturbed by tillage operations and by the trampling of people and livestock. Running water and wind are further contributors to the detachment of soil particles. All these processes loosen the soil so that it is easily removed by the agents of transport.

The transporting agents comprise those which act areally and contribute to the removal of a relatively uniform thickness of soil and those which concentrate their action in channels. The first group consists of rainsplash, surface runoff in the form of shallow flows of infinite width, sometimes termed sheet flow but more correctly called overland flow, and wind. The second group covers water flow in small channels, known as rills, which can be obliterated by weathering and ploughing, or in the larger more permanent features of gullies and rivers. A distinction is sometimes made for water erosion between rill erosion and erosion on the land between the rills by the action of raindrop impact and overland flow combined. This is termed interrill erosion. To these agents which act externally, picking up material from and carrying it over the ground surface, should be added transport by mass movements such as soil flows, slides and creep, in which water affects the soil internally, altering its strength.

The severity of erosion depends upon the quantity of material supplied by detachment and the capacity of the eroding agents to transport it. Where the agents have the capacity to transport more material than is supplied by detachment, the erosion is described as detachment-limited. Where more material is supplied than can be transported, the erosion is transport-limited.

The energy available for erosion takes two forms: potential and kinetic. Potential energy (*PE*) results from the difference in height of one body with respect to another. It is the product of mass (*m*), height difference (*h*) and acceleration due to gravity (*g*), so that

$$PE = mhg \qquad (2.1)$$

which, in units of kg, m and ms^{-2} respectively, yields a value in Joules. The potential energy for erosion is converted into kinetic energy (*KE*), the energy of motion. This is related to the mass and the velocity (*v*) of the eroding agent in the expression

$$KE = \frac{1}{2}mv^2 \qquad (2.2)$$

which, in units of kg and $(ms^{-1})^2$, also gives a value in Joules. Most of this energy is dissipated in friction with the surface over which the agent moves so that only 3 to 4 per cent of the energy of running water and 0.2 per cent of that of falling raindrops is expended in erosion (Pearce 1976). An indication of the relative efficiencies of the processes of water erosion can be obtained by applying these figures to calculations of kinetic energy, using Eq. 2.2, based on typical velocities (Table 2.1). The concentration of running water in rills affords the most powerful erosive agent but raindrops are potentially more erosive than overland flow. Most of the raindrop energy is used in detachment, however, so that the amount available for transport is less than that from overland flow. This is illustrated by measurements of soil loss in a field in mid-Bedfordshire, England. Over a 900-day period on an 11° slope on a sandy soil, transport across a centimetre width of slope amounted to 19,000 g of sediment by rills, 400 g by overland flow and only 20 g by rainsplash (Morgan, Martin and Noble 1986).

7

Table 2.1 Efficiency of forms of water erosion

Form	Mass*	Typical velocity (m/s)	Kinetic energy[†]	Energy for erosion[‡]	Observed sediment transport[§] (g/cm)
Raindrops	R	9	$40.5R$	$0.081R$	20
Overland flow	$0.5R$	0.01	$2.5 \times 10^{-5}R$	$7.5 \times 10^{-7}R$	400
Rill flow	$0.5R$	4[¶]	$4R$	$0.12R$	19 000

* Assumes rainfall of mass R of which 50 per cent contributes to runoff.
† Based on $\frac{1}{2}\,m\,v^2$.
‡ Assumes that 0.2 per cent of the kinetic energy of raindrops and 3 per cent of the kinetic energy of runoff is utilized in erosion.
§ Totals observed in mid-Bedfordshire on an 11° slope, on sandy soil, over 900 days. Most of the energy of raindrops contributes to detachment rather than transport.
¶ Estimated using the Manning equation of flow velocity for a rill, 0.3 m wide and 0.2 m deep, on a slope of 11°, at bankfull, assuming a roughness coefficient of 0.02.

2.1 Hydrological Basis of Erosion

The processes of water erosion are closely related to the routes taken by water in its passage through the vegetation cover and its movement over the ground surface. Their base thus lies in the hydrological cycle. During a rainstorm, part of the water falls directly on the land, either because there is no vegetation or because it passes through gaps in the plant canopy. This component of the rainfall is known as direct throughfall. Part of the rain is intercepted by the canopy from where it is either returned to the atmosphere by evaporation or finds its way to the ground by dripping from the leaves, a component termed leaf drainage, or by running down the stem as stemflow. The action of direct throughfall and leaf drainage produces rainsplash erosion. The rain which reaches the ground may be stored in small depressions or hollows on the surface or it may infiltrate the soil, contributing either to soil moisture storage or, by percolating deeper, to groundwater. When the soil is unable to take in more water, the excess moves laterally downslope within the soil as subsurface or interflow or it contributes to runoff on the surface, resulting in erosion by overland flow or by rills and gullies.

The rate at which water passes into the soil is known as the infiltration rate and this exerts a major control over the generation of surface runoff. Water is drawn into the soil by gravity and by capillary forces whereby it is attracted to and held as a thin molecular film around the soil particles. During a rainstorm, the spaces between the soil particles become filled with water and the capillary forces decrease so that the infiltration rate starts high at the beginning of a storm and declines to a level which represents the maximum sustained rate at which water can pass through the soil (Fig. 2.1). This level, the infiltration capacity or terminal infiltration rate, corresponds theoretically to the saturated hydraulic conductivity of the soil. In practice, however, the infiltration capacity is often lower than the saturated hydraulic conductivity because of air entrapped in the soil pores as the wetting front passes downwards through the soil.

Various attempts have been made to describe the change in infiltration rate over time mathematically. One of the most widely-used equations is the mod-

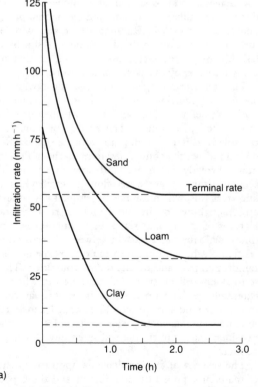

(a)

Figure 2.1 Typical infiltration rates for various soils (after Withers and Vipond 1974).

ification of the Green and Ampt (1911) equation proposed by Mein and Larson (1973):

$$i = A + \frac{B}{t} \qquad (2.3)$$

where i is the instantaneous rate of infiltration, A is the transmission constant or saturated hydraulic conductivity of the soil, B is the sorptivity, defined by Talsma (1969) as the slope of the line when i is plotted against t, and t is the time elapsed since the onset of the rain. This equation has been found to describe well the infiltration behaviour of soils in southern Spain (Scoging and Thornes 1979) and Arizona (Scoging, Parsons and Abrahams 1992) but Bork and Rohdenburg (1981), also working in the south of Spain, obtained better results with the equation proposed by Philip (1957):

$$i = A + \frac{B}{\sqrt{t}} \qquad (2.4)$$

whilst Gifford (1976) found neither equation satisfactory for semi-arid rangelands in northern Australia and Utah.

Infiltration rates depend upon the characteristics of the soil. Generally, coarse-textured soils such as sands and sandy loams have higher infiltration rates than clay soils because of the larger spaces between the soil particles. Infiltration capacities may range from more than 200 mm/h for sands to less than 5 mm/h for tight clays (Fig. 2.1). In addition to the role played by the inter-particle spacing or micropores, the larger cracks or macropores exert an important influence over infiltration. They can transmit considerable quantities of water so that clays with well-defined structures can have infiltration rates that are much higher than would be expected from their texture alone. Infiltration behaviour on many soils is also rather complex because the soil profiles are characterised by two or more layers of differing hydraulic conductivities; most agricultural soils, for example, consist of a disturbed plough layer and an undisturbed subsoil. Local variability in infiltration rates can be quite high because of differences in the structure, compaction, initial moisture content and profile form of the soil and in vegetation density. Field determinations of average infiltration capacity using infiltrometers may have coefficients of variation of 70 to 75 per cent. Eyles (1967) measured infiltration capacity on soils of the Melaka Series near Temerloh, Malaysia, and obtained values ranging from 15 to 420 mm/h with a mean of 147 mm/h.

According to Horton (1945), if rainfall intensity is less than the infiltration capacity of the soil, no surface runoff occurs and the infiltration rate equals the rainfall intensity. If the rainfall intensity exceeds the infiltration capacity, the infiltration rate equals the infiltration capacity and the excess rainfall forms surface runoff. As a mechanism for generating runoff, however, this comparison of rainfall intensity and infiltration capacity does not always hold. Studies in Bedfordshire, England (Morgan, Martin and Noble 1986) on a sandy soil show that infiltration capacity is greater than 400 mm/h and that rainfall intensities rarely exceed 40 mm/h. Thus no surface runoff would be expected whereas, in fact, the mean annual runoff is about 55 mm from a mean annual rainfall of 550 mm.

The important control for runoff production on these soils is not infiltration capacity but a limiting soil moisture content. When the actual moisture content is below this value, pore water pressure in the soil is less than atmospheric pressure and water is held in capillary form under tensile stress or suction. When the limiting moisture content is reached and all the pores are full of water, pore water pressure equates to atmospheric pressure, suction reduces to zero and surface ponding occurs. This explains why sands which have low levels of capillary storage produce runoff very quickly even though their infiltration capacity is not exceeded by the rainfall intensity. Since hydraulic conductivity is a flux partly controlled by rainfall intensity, increases in intensity can cause conductivity to rise so that, although runoff may have formed rapidly at a relatively low rainfall intensity, higher rainfall intensities do not always produce greater runoff. This mechanism explains why infiltration rates sometimes increase with rainfall intensity (Nassif and Wilson 1975). Bowyer-Bower (1993) found that, for a given soil, infiltration capacity was higher with higher rainfall intensities because of their ability to disrupt surface seals and crusts which would otherwise keep the infiltration rate low. The importance of crusting and sealing is also emphasized by Poesen (1984) who found that infiltration rates were higher on steeper slopes where the higher erosion rate prevented the seal from forming.

Once water starts to pond on the surface, it is held in depressions or hollows and runoff does not begin until the storage capacity of these is satisfied. On agricultural land, depression storage varies seasonally depending upon the type of cultivation which has been carried out and the time since cultivation for the roughness to be reduced by weathering and raindrop impact. Reid (1979) shows that clay soils near Aylesbury, England, have a depression storage capacity of 5 to 7 mm following autumn ploughing but that by the following spring this has been reduced to about 3 mm. Clay soils have 1.6 to 2.3 times more storage volume than sandy clay loams after ploughing, but following drilling and rolling of the seed bed the difference is reduced to 1.1 times (Evans 1980). For sandy soils depression storage can range from about 2.3 mm on a ploughed surface to 0.02 mm on a freshly-prepared seed bed (Auerswald personal communication).

2.2 Rainsplash Erosion

The action of raindrops on soil particles is most easily understood by considering the momentum of a single raindrop falling on a sloping surface. The downslope component of this momentum is transferred in full to the soil surface but only a small proportion of the component normal to the surface is transferred, the remainder being reflected. The transfer of momentum to the soil particles has two effects. First, it provides a consolidating force, compacting the soil, and second, it produces a disruptive force as the water rapidly disperses from and returns to the point of impact in laterally flowing jets. The local velocities in these jets are nearly double those of the raindrop impact (Huang, Bradford and Cushman 1982) and are sufficient to impart a velocity to some of the soil particles, launching them into the air, entrained within water droplets which are themselves formed by the break-up of the raindrop on contact with the ground (Mutchler and Young 1975). Thus, raindrops are agents of both consolidation and dispersion.

The consolidation effect is best seen in the formation of a surface crust, usually only a few millimetres thick, which results from clogging of the pores by soil compaction. It has been suggested that this is associated with the dispersal of fine particles from soil aggregates or clods which then fill in the pores (Poesen 1992). Studies of crust development under simulated rainfall show that crusts have a dense surface skin or seal, about 0.1 mm thick, with well-oriented clay particles. Beneath this is a layer, 1 to 3 mm thick, where the larger pore spaces are filled by finer washed-in material (Tackett and Pearson 1965). Hillel (1960) explained crusting by the collapse of the soil aggregates on saturation but Farres (1978) shows that raindrop impact is the critical process. After a rainstorm, most aggregates on the soil surface have been destroyed but those in the lower layer of the crust remain intact even though completely saturated. A tap of these aggregates, however, causes their instant breakdown. From this evidence, it seems that although saturation reduces the internal strength of the aggregate, it does not disintegrate until struck by raindrops.

A distinction is often made between a crust and a seal. Sealing refers to the reorganization of the surface soil layer during a rainstorm and crusting is the hardening of the surface seal as the soil dries out. Over a number of rainstorms, the result is, first, the devel-opment of a structural crust, formed *in situ* and, second, the formation of a depositional crust by deposition of fine particles in microdepressions where puddles form during the rain (Boiffin 1985). The most important effect of a surface crust is to reduce the infiltration capacity and thereby promote greater surface runoff. On loamy soils in north-east France, crusting can reduce the infiltration capacity from about 45 mm/h on an uncrusted soil to about 6 mm/h with a structural crust and 1 mm/h with a depositional crust (Boiffin and Monnier 1985). Measurements in Israel show that the infiltration capacity declines as a result of crusting from 100 mm/h to 8 mm/h on a sandy soil and from 45 mm/h to 5 mm/h on a loess soil (Morin, Benyamini and Michaeli 1981). The infiltration capacity on sandy soils in Mali ranges from 100 to 200 mm/h, but where crusting has developed it reduces to about 10 mm/h. Only a few storms are needed to bring about this change; a 50 per cent reduction can occur in one storm (Hoogmoed and Stroosnijder 1984). Such a rapid initial response to rainfall is in line with the observations of Govers and Poesen (1985) which show that the percentage area of the soil surface affected by crust development increases exponentially with cumulative rainfall and those of Boiffin and Monnier (1985) indicating an exponential decrease in infiltration capacity with increasing crusted area.

The actual response of a soil to a given rainfall depends upon the moisture content and, therefore, the structural state of the soil, and the intensity of the rain. Le Bissonnais (1990) describes three possible responses.

1. If the soil is dry and the rainfall intensity is high, the soil aggregates break down quickly by slaking. This is the breakdown by compression of air ahead of the wetting front. Infiltration capacity reduces rapidly and on very smooth surfaces runoff can be generated after only a few millimetres of rain. With rougher surfaces, depression storage is greater and runoff takes longer to form.
2. If the aggregates are initially partially wetted or the rainfall intensity is low, microcracking occurs and the aggregates break down into smaller aggregates. Surface roughness thus decreases but infiltration remains high because of the large pore spaces between the microaggregates.
3. If the aggregates are initially saturated, infiltration capacity depends on the saturated hydraulic conductivity of the soil and large quantities of rain are required to seal the surface. Nevertheless, soils with less than 15 per cent clay content are vulnerable to sealing if the intensity of the rain is high.

The susceptibility of soils to crusting has been related

by De Ploey (1981a) to a consistency index (C_{5-10}). The index is defined as $w_5 - w_{10}$ where w_5 and w_{10} are the water contents as a percentage dry weight at which two sections of a pat of soil placed in a Casagrande cup touch each other over a distance of 10 mm, after five and ten blows respectively. Based on studies of soils in Brabant and Flanders, Belgium, values of the index > 3 denote stable soils and those < 2.5 indicate crust-prone soils. Crustability decreases with increasing contents of clay and organic matter since these provide greater strength to the soil. Thus loams and sandy loams are the most vulnerable to crust formation.

Studies of the kinetic energy required to detach one kilogram of sediment by raindrop impact show that minimal energy is needed for particles of 0.125 mm and that particles between 0.063 mm and 0.250 mm

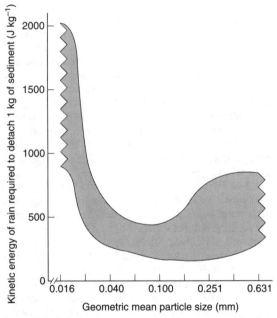

Figure 2.2 Relation between geometric mean particle size of the soil and rainfall energy required to detach 1 kg of sediment. Shaded area shows range of experimental values (after Poesen 1992).

are the most vulnerable to detachment (Fig. 2.2; Poesen 1985). Coarser particles are resistant to detachment because of their weight and clay particles are resistant because the raindrop energy has to overcome the adhesive or chemical bonding forces which link the minerals comprising the clay particles (Yariv 1976). This means that soils with high percentages of particles within the most vulnerable range, for example silt loams, loams, fine sands and sandy loams, are the most detachable. Selective removal of the soil

particles by rainsplash can cause variations in soil texture downslope. Splash erosion on stony loamy soils in the Luxembourg Ardennes has resulted in the soils on the valley sides becoming deficient in clay and silt-size particles and high in gravel and stone content whereas the colluvial soils at the base of the slopes are enriched by the splashed-out material (Kwaad 1977). Selective erosion can affect soil aggregates as well as primary particles. Rainfall simulation experiments on clay soils from Italy show that splashed-out material is enriched in soil aggregates of 0.063 to 0.50 mm in size (Torri and Sfalanga 1986).

The detachability of soil depends not only on its texture but also on top soil shear strength (Cruse and Larson 1977), a finding which has prompted attempts to understand splash erosion in terms of shear. The detachment of soil particles represents a failure of the soil by the combined mechanism of compression and shear under raindrop impact, an event which is most likely to occur under saturated conditions when the shear strength of the soil is lowest (Al-Durrah and Bradford 1982). Generally, detachment decreases exponentially with increasing shear strength. Broadly linear relationships have been obtained, however, between the quantity of soil particles detached by raindrop impact and the ratio of the kinetic energy of the rainfall to soil shear strength (Al-Durrah and Bradford 1981; Torri, Sfalanga and Del Sette 1987; Bradford, Truman and Huang 1992), and also the ratio of the rate of soil aggregate breakdown to shear strength (Ekwue and Ohu 1990).

Rain does not always fall on to a dry surface. During a storm it may fall on surface water in the form of puddles or overland flow. Studies by Palmer (1964) show that as the thickness of the surface water layer increases, so does splash erosion. This is believed to be due to the turbulence which impacting raindrops impart to the water. There is, however, a critical water depth beyond which erosion decreases again. Once this critical depth has been reached, detachment decreases exponentially with increasing water depth because more of the rainfall energy is dissipated in the water and does not affect the soil surface. Laboratory experiments have shown that the critical depth is approximately equal to the diameter of the raindrops (Palmer 1964) or to one-fifth (Torri and Sfalanga 1986) or one-third (Mutchler and Young 1975) of the diameter. These studies were made with a variety of soils ranging from clays to silt loams, loams and sandy loams. No increase in splash erosion with water depth has been observed, however, on sandy soils (Ghadiri and Payne 1979; Poesen 1981).

Experimental studies show that the rate of detachment of soil particles with rainsplash varies with the 1.0 power of the instantaneous kinetic energy of the rain (Free 1960; Quansah 1981) or with the square of the instantaneous rainfall intensity (Meyer 1981). The detachment rate (D_r) on bare soil can be expressed by equations of the form:

$$D_r \propto I^a S^c \qquad (2.5)$$

$$D_r \propto KE^b . S^c . e^{-dh} \qquad (2.6)$$

where I is the rainfall intensity (mm/h), S is the slope expressed in m/m or as a sine of the slope angle, KE is the kinetic energy of the rain (J/m^2), and h is the depth of surface water (m). Although 2.0 is a convenient value for a, the value may be adjusted to allow for variations in soil texture using the term $a = 2.0 - (0.01 \times \% \text{ clay})$ (Meyer 1981). Similarly, the value of 1.0 for b may be varied from about 0.8 for sandy soils to 1.8 for clays (Bubenzer and Jones 1971). Values for c are in the range of 0.2 to 0.3 (Quansah 1981; Torri and Sfalanga 1986), also varying with the texture of the soil (Torri and Poesen 1992). It should be remembered that the slope term in this equation refers to the local slope for a distance equivalent to only a few drop diameters from the point of raindrop impact, for example that on the side of a soil clod, and not the average ground slope. Thus, for practical purposes, the slope term is often omitted from calculations of soil particle detachment. The value of 2.0 is convenient for d as representative of a range of values between 0.9 and 3.1 for different soil textures (Torri, Sfalanga and Del Sette 1987).

In contrast, average ground slope is important when considering the overall transport of splashed particles. On a sloping surface more particles are thrown downslope than upslope so that there is a net downslope movement of material. Splash transport per unit width of slope (T_r) can be expressed by the relationship:

$$T_r \propto I^j S^f \qquad (2.7)$$

where $j = 1.0$ (Meyer and Wischmeier 1969) and $f = 1.0$ (Savat 1981; Quansah 1981). There is some evidence to suggest that the value of f decreases on steeper slopes; Mosley (1973) gives a value of 0.8 and Moeyersons and De Ploey (1976) a value of 0.75 for data where slope angles rise to 20° and 25° respectively. Foster and Martin (1969) and Bryan (1979) found splash transport increases with slope angle to reach a maximum at about 18° and that on steeper slopes f becomes negative.

These relationships for detachment and transport of soil particles by rainsplash ignore the role of wind. Windspeed imparts a horizontal force to a falling raindrop until its horizontal velocity component equals the velocity of the wind. As a result, the kinetic energy of the raindrop is increased. Not surprisingly, detachment of soil particles by impacting wind-driven raindrops can be some 1.5 to 3 times greater than that resulting from rains of the same intensity without wind (Disrud and Krauss 1971; Lyles, Dickerson and Schmeidler 1974). Wind also causes raindrops to strike the surface at an angle from the vertical. This affects the relative proportions of upslope versus downslope splash. Moeyersons (1983) shows that where the angle between the falling raindrop and the vertical is 20°, net splash transport is reduced to zero for slopes of 17–19° and has a net upslope component on gentler slopes. Where the angle between the falling raindrop and the vertical is 5°, zero splash transport occurs on a slope of 3°.

Since splash erosion acts uniformly over the land surface its effects are seen only where stones or tree roots selectively protect the underlying soil and splash pedestals or soil pillars are formed. Such features frequently indicate the severity of erosion. Splash erosion is most important for detaching the soil particles which are subsequently eroded by running water. However, on the upper parts of hillslopes, particularly on those of convex form, splash transport may be the dominant erosion process. In Calabria, southern Italy, under forest and also scattered herb and shrub vegetation, splash transport accounts for 30 to 95 per cent of the total transport of material by water erosion (van Asch 1983). Studies in Bedfordshire, England, show that splash transport accounts for 15 to 52 per cent of the total transport on land under cereals and grass but only 3 to 10 per cent on bare ground (Morgan, Martin and Noble 1986). As runoff and soil loss increase, the importance of splash transport declines, although very low contributions of splash to total transport were also measured in Bedfordshire under woodland because of the protective effect of a dense litter layer. Govers and Poesen (1988) found that although raindrop impact detached 152 t/ha of soil over one year on a bare loam soil on a 14° slope in Belgium, splash transport accounted for only 0.2 t/ha of the soil loss. By far the most important aspect of splash erosion was its contribution of sediment to overland flow which was the main agent of sediment transport in the interrill areas.

2.3 Overland Flow

Overland flow occurs on hillsides during a rainstorm when surface depression storage and either, in the case of prolonged rain, soil moisture storage or, with intense rain, the infiltration capacity of the soil are exceeded. The flow is rarely in the form of a sheet of water of uniform depth and more commonly is a mass of anastomosing or braided water courses with no pronounced channels. The flow is broken up by large stones and cobbles and by the vegetation cover, often swirling around tufts of grass and small shrubs.

Hydraulic characteristics

The hydraulic characteristics of the flow are described by its Reynolds number (Re) and its Froude number (F), defined as follows:

$$Re = \frac{vr}{\nu} \qquad (2.8)$$

$$F = \frac{v}{\sqrt{gr}} \qquad (2.9)$$

where r is the hydraulic radius which, for overland flow, is taken as equal to the flow depth and ν is the kinematic viscosity of water. The Reynolds number is an index of the turbulence of the flow. The greater the turbulence, the greater is the erosive power generated by the flow. At numbers less than 500, laminar flow prevails and at values above 2,000 flow is fully turbulent. In laminar flow, each fluid layer moves in a straight line with uniform velocity and there is no mixing between the layers, whereas turbulent flow has a complicated pattern of eddies, producing considerable localized fluctuations in velocity, and a continuous interchange of water between the layers. Intermediate values are indicative of transitional or disturbed flow, often a result of turbulence being imparted to laminar flow by raindrop impact (Emmett 1970). The Froude number is an index of whether or not gravity waves form in the flow. When the Froude number is less than 1.0, gravity waves do not form and the flow, being relatively smooth, is described as tranquil or subcritical. Froude numbers greater than 1.0 denote rapid or supercritical flow, characterized by gravity waves, which is more erosive. Field studies of overland flow in Bedfordshire, however, reveal Reynolds numbers less than 75 and Froude numbers less than 0.5 (Morgan 1980a). Flows with Reynolds numbers less than 40 and Froude numbers less than 0.13 were observed by Pearce (1976) in the field near Sudbury, Ontario. In various field experiments on semi-arid hillsides in the Walnut Gulch Experimental Watershed, Arizona, Froude numbers in overland flow were consistently less than 0.5 even though Reynolds numbers ranged from about 100 to 1,200 (Parsons, Abrahams and Luk 1990).

Detachment of soil particles by flow

The important factor in these hydraulic relationships is the flow velocity. Because of an inherent resistance of the soil, velocity must attain a threshold value before erosion commences. Basically, the detachment of an individual soil particle from the soil mass occurs when the forces exerted by the flow exceed the forces keeping the particle at rest. Shields (1936) made a fundamental analysis of the processes involved and the forces at work to determine the critical conditions for initiating particle movement over relatively gentle slopes in rivers in terms of the dimensionless shear stress (Θ) of the flow and the particle roughness Reynolds number ($Re^{\star}$), defined respectively by:

$$\Theta = \frac{\rho_w u_{\star}^2}{g(\rho_s - \rho_w)D} \qquad (2.10)$$

$$Re^{\star} = \frac{u_{\star}D}{\nu} \qquad (2.11)$$

where Θ is known as the Shields number, ρ_w is the density of water, $u_{\star}$ is the shear velocity of the flow, g is the acceleration due to gravity, ρ_σ is the density of the sediment and D is the diameter of the particle. The shear velocity ($u_{\star}$) is expressed as:

$$u_{\star} = \sqrt{grs} \qquad (2.12)$$

When the value of $Re^{\star}$ is greater than 40 (turbulent flow), the critical value of $\Theta(\Theta_c)$ for particle movement assumes a constant value of 0.05. Unfortunately, this value does not hold when, as is the case with overland flow, the particles are not fully submerged or the flow has Reynolds numbers in the laminar range. Studies with rock fragments in shallow flows suggest that Θ_c is about 0.01 in value (Poesen 1987; Torri and Poesen 1988). For laminar flow, Yalin and Karahan (1979) showed that the value of Θ_c was not constant but depended on the value of $Re^{\star}$. Other research on the initiation of particle movement by overland flow (Govers 1987; Guy and Dickinson 1990; Torri and Borselli 1991) indicates that the Shields number consistently overpredicts the hydraulic requirements for particle movement. This implies that the initiation of movement is not solely a fluid shear stress phenomenon but is enhanced by other factors. Among the additional factors not accounted for by the Shields number are the effects of raindrop impact on the flow, the angle of repose of the particle in relation to the ground slope, the strong influence of gravity as the slope steepness increases, the cohesion of the soil, changes in the density of the fluid as sediment concentration in the flow increases and abrasion between particles moving in the flow and the soil beneath.

Since the above approach has not proved entirely satisfactory, a more empirical procedure has been adopted based on a critical value of the flow's shear velocity for initiating particle movement. As can be seen in Fig. 2.3 (Savat 1982), for particles larger than 0.2 mm in diameter, the critical shear velocity

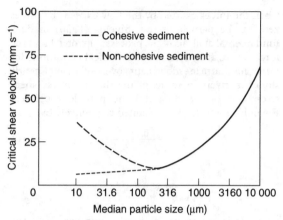

Figure 2.3 Critical shear velocity in turbulent water flow for soil particle detachment as a function of particle size (after Savat 1982).

increases with particle size. A larger force is required to move larger particles. For particles smaller than 0.2 mm, the critical shear velocity increases with decreasing particle size. The finer particles are harder to erode because of the cohesiveness of the clay minerals of which they are comprised unless they have been previously detached and, as a result, lost their cohesion. They can then be moved at very low shear velocities. In practice, the critical shear velocities required to erode soil may differ from those shown in Fig. 2.3 because the latter are derived for surfaces of uniform particle size. With mixed particle sizes, the finer particles are protected by the coarser ones so that they are not removed until the shear velocity is great enough to pick up the larger particles. Counteracting this effect, however, is the action of rainsplash which may detach soil particles and throw them into the flow.

Once the critical conditions for particle movement are exceeded, soil particles may be detached from the soil mass at a rate which is dependent on the shear velocity of the flow and the unit discharge (Govers and Rauws 1986). The direct application of this relationship is only valid, however, if the shear velocity is exerted solely on the soil particles, which implies that the resistance to the flow is entirely due to grain resistance. This situation is only true for completely smooth bare soil surfaces. In practice, resistance due to the microtopographic form of the soil surface and the plant cover is usually more important and grain resistance may be as little as 5 per cent of the total resistance offered to the flow (Abrahams, Parsons and Hirsch 1992). Since it is difficult to determine the level of grain resistance, only very generalized relationships can be developed for describing detachment rate

(D_f). These depend on a simple relationship between detachment and flow velocity.

The velocity of flow is dependent upon the flow depth or hydraulic radius (r), the roughness of the surface and the slope (s). This relationship is commonly expressed by the Manning equation:

$$v = \frac{r^{2/3} s^{1/2}}{n} \tag{2.13}$$

where n is the Manning coefficient of roughness. Equation 2.13 assumes fully turbulent flow moving over a rough surface. Using the continuity and Manning's velocity equations, Meyer (1965) shows that:

$$v \propto s^{1/3} Q^{1/3} \tag{2.14}$$

for constant roughness conditions, where Q is discharge or flow rate. Assuming that the detachment rate varies with the square of velocity (Meyer and Wischmeier 1969):

$$D_f \propto Q^{2/3} s^{2/3} \tag{2.15}$$

Quansah (1985) obtained experimentally higher exponent values, however, for a range of soil types from clay to sand:

$$D_f \propto Q^{1.50} s^{1.44} \tag{2.16}$$

Equations 2.15 and 2.16 both relate only to the action of water flow over the soil surface. Quansah (1985) found that the exponents decreased in value when the flow was accompanied by rainfall to give:

$$D_f \propto Q^{1.12} s^{0.64} \tag{2.17}$$

indicating that raindrop impact inhibits the ability of flow to detach soil particles.

Based on the findings from laboratory experiments by Meyer and Monke (1965) who observed that the rate of detachment depended on the amount of sediment already in the flow, Foster and Meyer (1972) proposed that the term, D_f, in the above equations applies to the detachment capacity rate which occurs only when the flow is clear. Under other conditions, D_f depends upon the difference between the actual sediment concentration in the flow (C) and the maximum concentration that the flow can hold (C_{max}):

$$D_f \propto (C_{max} - C) \tag{2.18}$$

This implies that the detachment rate declines as sediment concentration increases and that when the maximum sediment concentration is reached, the detachment rate is zero. Recent laboratory studies on sandy loam soils by Kamalu (1993), however, showed that the detachment rate for flow without rainfall remained at the capacity rate up to the time that maximum sediment concentration was reached, at which point detachment ceased. Thus there is still considerable uncertainty about the nature of the mechanisms involved in soil particle detachment by flow.

Transport of soil particles by flow

Once sediment has been entrained within the flow, it will be transported until such time as deposition occurs. Meyer and Wischmeier (1969) proposed that the transporting capacity of the flow (T_f) varies with the fifth power of the velocity so that:

$$T_f \propto Q^{5/3} s^{5/3} \qquad (2.19)$$

This equation compares closely with those below derived respectively by Carson and Kirkby (1972) and Morgan (1980a) from a consideration of the hydraulics of sediment transport:

$$T_f = 0.0085\, Q^{1.75} s^{1.625} D_{84}^{-1.11} \qquad (2.20)$$

$$T_f = 0.0061\, Q^{1.8} s^{1.13} n^{-0.15} D_{35}^{-1} \qquad (2.21)$$

where D_{84} and D_{35} define the particle size of the surface material at which respectively 84 and 35 per cent of the grains are finer. All these equations relate to the action of overland flow on its own whereas, in practice, the flow is usually accompanied by rainfall. The interaction with raindrop impact causes a slight rise in the value of the exponents for discharge and slope. Laboratory experiments by Quansah (1982) with overland flow and rainfall combined gave the relationship:

$$T_f \propto Q^{2.13} s^{2.27} \qquad (2.22)$$

Thus whilst, as seen by Eqs. 2.16 and 2.17, the detachment capacity of flow is reduced by raindrop impact, its transporting capacity is enhanced (Savat 1979; Guy and Dickinson 1990; Proffitt and Rose 1992). The degree of enhancement depends on the resistance of the soil, the diameter of the raindrops and the depth and velocity of the flow. According to Kinnell (1991), the enhancement is greatest when the depth of flow is less than 1.5 times the diameter of the raindrops. At greater flow depths, the contribution from splash is much less (Section 2.2) and most of the sediment load comes from detachment by flow (Kinnell 1990a). Although the enhancement is linearly related to flow velocity (Moss 1988; Kinnell 1990a) the relationship is complicated because an increase in sediment concentration may, in turn, have an effect on flow velocity. Govers (1989) found that high sediment concentrations could increase velocity by up to 40 per cent, especially at low discharges and flow depths. His experiments, however, were carried out for flow without rain whereas Guy et al. (1990) found that the impact of rain decreased flow velocity by about 12 per cent.

Govers (1990) has investigated three different types

of equations, based on grain shear stress, effective stream power and unit stream power for describing the transport capacity of overland flow, defined as the maximum sediment concentration that can be carried. For ease of use, the relationships based on unit stream power are preferred since this is simply the product of slope and flow velocity. He found that:

$$C_{max} = a(sv - 0.4)^b \qquad (2.23)$$

where a and b are empirical coefficients depending on grain size.

Instead of trying to define transport capacity only in terms of flow properties, some researchers have attempted to relate transport capacity to the maximum sediment concentration that the flow can carry when a balanced condition exists between detachment and deposition (Rose et al. 1983; Styczen and Nielsen 1989). The rate of deposition (D_p) is:

$$D_p = v_s \,.\, C \qquad (2.24)$$

where v_s is the settling velocity of the particles. Torri and Borselli (1991) took data from the experiments of Govers (1990) and obtained a good agreement between the transport capacity of the flow estimated from a balance-based approach and that estimated from Eq. 2.23, indicating that this equation is a reasonable expression of the transport capacity of flow.

Given the rather shallow depths of overland flow, the considerable role played by surface roughness and the generally low Reynolds and Froude numbers, it can be proposed that most of the sediment being transported is derived from detachment by raindrop impact and that, except on steep slopes or on smooth bare soil surfaces, grain shear velocity rarely attains the level necessary to detach soil particles. Since, as seen earlier, particles between 0.063 and 0.250 mm in size are the most detachable by raindrop impact and, from Fig. 2.3, it can be seen that the most detachable particles by flow are within the 0.1 and 0.3 mm range, the sediment carried in overland flow is deficient in particles larger than 1 mm and enriched in finer material. Thus, over time, areas of erosion on a hillside will become progressively sandier and areas of deposition, particularly in valley floors, will be enriched with clay particles (Alberts, Moldenhauer and Foster 1980).

The trend towards increasing sandiness in eroded areas is also brought about by another mechanism. Most of the sediment splashed into the flow is moved only relatively short distances before being deposited. Since deposition is a particle-size selective process with the coarser particles being deposited first, the deposited layer becomes progressively coarser (Proffitt, Rose and Hairsine 1991) and, as seen in

Section 2.2, may develop into a depositional crust. Less of the finer material is then exposed to erosion. This mechanism can take place even within an individual storm so that detachment is highest at the beginning of the storm and transport capacity is reached very quickly.

Plots of the relationship between sediment transport by overland flow and discharge, as measured in the field, do not always conform with that expected from the research described above. Work in Bedfordshire (Jackson 1984; Morgan, Martin and Noble 1986) and in southern Italy (van Asch 1983) shows that sediment transport varies with discharge raised by a power of 0.6 to 0.8. The similarity of this value to that in Eq. 2.15 has led to an interpretation of the erosion being detachment-limited. An alternative interpretation is that the transport process is dominantly one of rolling of the particles over the soil surface as bed load. Equations for bed load show that sediment transport is related to discharge raised by a power that is generally between 0.6 and 1.0 in value. Studies of soil particle transport in shallow unchannelled flows indicate that whilst some suspended load is present, coarser particles and a high proportion of the soil aggregates are moved as bed load. Kinnell (1990b) considers that the sediment component contributed to overland flow by raindrop impact is moved as bed load and the component contributed by detachment by the flow itself is moved as suspended load. This implies that sediment transport may be better expressed by Eq. 2.15 on low slopes where soil particle detachment is solely by rainsplash but by Eq. 2.22 on steeper slopes or with higher flow velocities when particle detachment by flow also takes place.

Spatial distribution

The effectiveness of overland flow as an eroding agent depends on its spatial extent and distribution over a hillside. Horton (1945) describes overland flow as covering two-thirds or more of the hillslopes in a drainage basin during the peak period of a storm. The flow results from the rainfall intensity being greater than the infiltration capacity of the soil and is distributed over the land surface in the following pattern. At the top of a slope is a zone without flow which forms a belt of no erosion. At a critical distance from the crest sufficient water has accumulated on the surface for flow to begin. Moving further downslope, the depth of flow increases with distance from the crest until, at a further critical distance, the flow becomes concentrated into fewer and deeper flow paths which occupy a progressively smaller proportion of the hillslope (Parsons, Abrahams and Luk 1990). Hydraulic efficiency improves allowing the increased discharge

to be accommodated by a higher flow velocity. Nevertheless, the hydraulic characteristics of the flow vary greatly over very short distances because of the influence of bed roughness associated with vegetation and stones. As a result, erosion is often localized and after a rainstorm the surface of a hillside displays a pattern of alternating scours and sediment fans (Moss and Walker 1978). Eventually, the flow breaks up into rills. That overland flow occurs in such a widespread fashion has been questioned, particularly in well-vegetated areas where such flow occurs infrequently and covers only that 10 to 30 per cent of the area of a drainage basin closest to the stream sources (Kirkby 1969a). Under these conditions its occurrence is more closely related to the saturation of the soil and the fact that moisture storage capacity is exceeded rather than infiltration capacity. Although, as illustrated by the detailed studies of Dunne and Black (1970) in a small forested catchment in Vermont, the saturated area expands and contracts, being sensitive to heavy rain and snow melt, rarely can erosion by overland flow affect more than a small part of the hillslopes.

Since most of the observations testifying to the power of overland flow relate to semi-arid areas or to cultivated land with sparse plant cover, it would appear that vegetation is the critical factor. Some form of continuum exists, ranging from well-vegetated areas where overland flow occurs rarely and is mainly of the saturation type, to bare soil where it frequently occurs and is of the Hortonian type. Removal of the plant cover can therefore enhance erosion by overland flow. The change from one type of overland flow to another results from more rain reaching the ground surface, less being intercepted by the vegetation, and decreased infiltration as rainbeat and deposition of coarse material from the flow cause a surface crust to develop. Exceptions to this trend occur in areas where high rainfall intensities are recorded. Hortonian overland flow is widespread in the tropical rain forests near Babinda, northern Queensland, where 6-minute rainfall intensities of 60 to 100 mm/h are common, especially in the summer, and the saturated hydraulic conductivity of the soil at 200 mm depth is only 13 mm/h. As a result, a temporary perched water table develops in the soil soon after the onset of rain, subsurface flow commences and this quickly emerges on the soil surface (Bonell and Gilmour 1978).

Where runoff rates are relatively high over most of the hillside, overland flow or, more strictly, the combined action of overland flow and raindrop impact as interrill erosion, can be the dominant erosion process on the upper and middle slopes with deposition of material as colluvium on the footslopes. This appears to be true for many agricultural areas on non-cohesive soils. On loose, freshly-ploughed soils on colluvial deposits on 18–22° slopes in Calabria, Italy, van Asch (1983) found that overland flow accounted for 80–95 per cent of the sediment transport. On unvegetated sandy soils in Bedfordshire with an 11° slope, it accounts for 50–80 per cent (Morgan, Martin and Noble 1986). On a loam soil on a 14° slope in northern Belgium it accounts for 22–46 per cent of the total soil loss with rates ranging from 24 to 100 t/ha (Govers and Poesen 1988). Interrill processes can also be the main agent of erosion on well-vegetated slopes if the rainfall is very high.

2.4 Subsurface Flow

The lateral movement of water downslope through the soil is known as interflow. Where it takes place as concentrated flow in tunnels or subsurface pipes its erosive effects through tunnel collapse and gully formation are well known. Less is known about the eroding ability of water moving through the pore spaces in the soil, although it has been suggested that fine particles may be washed out by this process (Swan 1970). Pilgrim and Huff (1983) measured sediment concentrations as high as 1 g/l in subsurface flow through a silt-loam soil on a 17° slope under grass in California in storms of 10 mm/h intensity or less. The material, uniformly fine with particles ranging from 4 to 8 μm in diameter, was being detached by raindrop impact at the surface and then moved by the flow through the macropores in the soil. Macropores have also been observed to form the main pathways for interflow in tilled black cracking clays in the Darling Downs area of Queensland, Australia (Loch, Thomas and Donnollan 1987).

More important than the sediment concentrations, however, are the base mineral concentrations in the subsurface flow which can be twice those found in overland flow. Essential plant nutrients, particularly those added by fertilizers, can be removed, thereby impoverishing the soil and reducing its resistance to erosion. Even under the more natural conditions of a virgin tropical rain forest site on 8–14° slopes at Pasoh, Negeri Sembilan, Malaysia, where erosion rates are very low, the material removed as dissolved solids in the subsurface flow can amount to 15 to 23 per cent of the total sediment transport (Leigh 1982a).

2.5 Rill Erosion

As indicated earlier, it is widely accepted that rills are initiated at a critical distance downslope where overland flow becomes channelled. The break-up of shallow overland flow into small channels or micro-rills was examined by Moss, Green and Hutka (1982). They found that, in addition to the main flow path downslope, secondary flow paths developed with a lateral component. Where these converged, the increase in discharge intensified particle movement and small channels or trenches were cut by scouring. Studies of the hydraulic characteristics of the flow show that change from overland flow to rill flow passes through four stages: unconcentrated channel flow, overland flow with concentrated flow paths, micro-channels without headcuts and micro-channels with headcuts. The greatest differences exist between the first and second stages, suggesting that the flow concentrations within the overland flow should strictly be treated as part of an incipient rill system (Merritt 1984). In the second stage, small vortices appear in the flow which, in the third stage, develop into localized spots of turbulent flow characterized by roll waves (Rauws 1987) and eddies (Savat and De Ploey 1982). At the point of rill initiation, flow conditions change from subcritical to supercritical (Savat 1979). Although headcuts or knick points may form at this time and all rills are associated with headcut development, not all headcuts develop into rills (Slattery and Bryan 1992). The overall change in flow conditions through the four stages seems to take place smoothly as the Froude number increases from about 0.8 to 1.2 rather than occurring when a threshold value is reached (Torri, Sfalanga and Chisci 1987; Slattery and Bryan 1992). For this reason, attempts to explain the onset of rilling through the exceedance of a critical Froude number have not been successful and additional factors have had to be included when trying to define its value. Examples are the particle size of the material (Savat 1979) and the sediment concentration in the flow (Boon and Savat 1981).

Greater success has been achieved relating rill initiation to the exceedance of a critical shear velocity of runoff. Govers (1985) found that on smooth or plane surfaces, where all the shear velocity is exerted on the soil particles, the sediment concentration in the flow increased with shear velocity more rapidly once a critical value of about 3.0–3.5 cm/s was exceeded. At this point, the erosion becomes non-selective regarding particle size so that coarser grains can be as easily entrained in the flow and removed as finer grains. A value of about 3.5 cm/s for the critical shear velocity only applies to non-cohesive soils or soils which because they are highly sensitive to dispersal or to liquefaction resemble loose sediments. Rauws and Govers (1988) proposed that, for other soils except those with high clay contents, the critical shear velocity for rill initiation (u_{*cr}) is linearly related to the shear strength of the soil (τ_s; kPa) as measured at saturation with a torvane:

$$u_{\star cr} = 0.89 + 0.56\, \tau_s \qquad (2.25)$$

Using a shear vane, the equivalent equation is (Brunori, Penzo and Torri 1989):

$$u_{\star cr} = 0.9 + 0.3\, \tau_s \qquad (2.26)$$

An alternative but similarly-conceived approach relates rill initiation to a critical value of the ratio between the shear stress exerted by the flow (τ) and the shear strength of the soil (τ_s) measured with a shear vane. When $\tau/\tau_s > 0.0001{-}0.0005$, rills will form (Torri, Sfalanga and Chisci 1987). In all these relationships, it should be stressed that the shear velocity or shear stress is applied wholly to the soil particles and should be strictly known as grain shear velocity or grain shear stress.

All the above studies consider rills as a surface phenomenon formed from infiltration-excess overland flow. Observations in mid-Bedfordshire, England, suggest that this need not always be the case. Here rills develop from a sudden burst of water on to the surface near the bottom of the slope where a small cut is formed which rapidly extends headwards upslope as a channel (Morgan, Martin and Noble 1986). Such bursts may well be associated with saturation overland flow.

Once rills have been formed, their migration upslope occurs by the retreat of the headcuts on the steep banks at the top of the channel. The rate of retreat is controlled by the cohesiveness of the soil, the height and angle of the headwall slope, and the discharge and velocity of the flow (De Ploey 1989). Downslope extension of the rill is controlled by the shear stress exerted by the flow and the strength of the soil (Savat 1979). Shear stress also determines the rate of detachment of soil particles by flow within the rill which can be broadly described by an equation of the type (Foster 1982):

$$D_f = K_r(\tau - \tau_c) \qquad (2.27)$$

where K_r is a measure of the detachability of the soil and τ_c is the critical shear stress for the soil.

The transport capacity of rill flow can be approximately represented by Eqs. 2.22 or 2.23. Govers (1992) found experimentally that flow velocity in rills could be related to the discharge by the relationship:

$$v = 3.52\, Q^{0.294} \qquad (2.28)$$

and that this gave better predictions than the Manning equation (Eq. 2.13). This was because over a range from 2° to 8°, slope had no effect on flow velocity. Neither did the grain roughness nor the surface form of the soil. Govers (1992) therefore modified Eq. 2.22 by replacing the velocity term with Eq. 2.28 to read:

$$C_{max} = a(3.52\, Q^{0.294} - 0.0074)Q \qquad (2.29)$$

where a is dependent upon the grain size of the sediment and the value of 0.0074 is interpreted as a critical value of unit stream power. Although this equation expresses the maximum sediment concentration that can be carried by runoff in a rill, the actual sediment concentration and, therefore, the erosion may vary considerably from this. This is because the supply of sediment to the rill is not solely dependent upon the detachment of soil particles by the flow. Instead, the rill has to adjust continually for pulse influxes of sediment due to wash-in from interrill flow on the surrounding land, erosion and collapse of the head wall and collapse of the side walls. Mass failure of the side walls can contribute more than half of the sediment removed in rills, particularly when heavy rains follow a long dry period during which cracks have developed in the soil (Govers and Poesen 1988).

Since raindrop impact increases the transport capacity of the flow and, through the detachment of soil particles, causes higher sediment concentrations, Savat (1979) argued that the interaction of rainfall with the flow would enhance the probability of rilling. Quansah (1982) and Dunne and Aubry (1986), however, found that the particles detached by rain filled in the micro-channels as fast as they could form, so that rilling was inhibited. It appears that the two sets of processes compete (Moss, Green and Hutka 1982) so that either the micro-channels are short-lived because they drain away the overland flow, become laterally isolated and fill in, or the concentration of flow increases its erosive power and the channels deepen, widen and migrate both upslope and downslope.

Since rill flow is non-selective in the particle size that it can carry, large grains, even rock fragments up to 9 cm in diameter (Poesen 1987), can be moved. Meyer, Foster and Nikolov (1975) found that 15 per cent of the particles carried in rills on a 3.5° slope of tilled silt loam were larger than 1 mm in size and that 3 per cent were larger than 5 mm. On a 4.5° slope of bare untilled silt loam, 80 per cent of the sediment transported in rills was between 0.21 and 2.0 mm in size and most of the clay particles were removed as aggregates within this size range (Alberts, Moldenhauer and Foster 1980).

As expected from its considerable erosive power, rill erosion may account for the bulk of the sediment removed from a hillside, depending on the spacing of the rills and the extent of the area affected. Govers and Poesen (1988) found that the material transported in rills accounted for 54–78 per cent of the total erosion. In an intense hail storm, an event with a 50-year return period, near Prešov in the Czech Republic, on 14 April 1964, 91 ha of land with slopes of 9–18° and loamy soils devoted to oats and spring barley were damaged. About 58 per cent of this was affected by rilling (Midriak 1965). A cloud burst on 23 May 1958 near Banská Bystrica, also in the Czech Republic, nearly destroyed a field of potatoes on a 13–15° slope. About 60 per cent of the area was affected by rills which accounted for 70 per cent of the total sediment eroded (Zachar 1982). These figures contrast with the situation in mid-Bedfordshire, England, where rills accounted for only 20 to 50 per cent of the total erosion (Morgan, Martin and Noble 1986).

2.6 Gully Erosion

Gullies are relatively permanent steep-sided water courses which experience ephemeral flows during rainstorms. Compared with stable river channels which have a relatively smooth, concave-upwards long profile, gullies are characterized by a headcut and various steps or knick-points along their course. These rapid changes in slope alternate with sections of very gentle gradient, either straight or slightly convex in long profile. Gullies also have relatively greater depth and smaller width than stable channels, carry larger sediment loads and display very erratic behaviour so that relationships between sediment discharge and runoff are frequently poor (Heede 1975a). Gullies are almost always associated with accelerated erosion and therefore with landscape instability.

At one time it was thought that gullies developed as enlarged rills but studies in the gullies or *arroyos* of the southwest USA revealed that their initiation is a more complex process. In the first stage small depressions or knicks form on a hillside as a result of localized weakening of the vegetation cover by grazing or fire. Water concentrates in these depressions and enlarges them until several depressions coalesce and an incipient channel is formed. Erosion is concentrated at the heads of the depressions where near-vertical scarps develop over which supercritical flow occurs. Some soil particles are detached from the scarp itself but most erosion is associated with scouring at the base of the scarp which results in deepening of the channel and the undermining of the headwall, leading to collapse and retreat of the scarp upslope. Sediment is also produced further down the gully by bank erosion. This occurs partly by the scouring action of running water and the sediment it contains and partly by slumping of the banks following saturation during flow. Between flows sediment is made available for erosion by weathering and bank collapse. This sequence of gully formation, described by Leopold, Wolman and Miller (1964) in New Mexico, is shown in Fig. 2.4.

Not all gullies develop purely by surface erosion, however. Berry and Ruxton (1960), investigating gullies in Hong Kong which formed following clearance of natural forest cover, found that most water was removed from the hillsides by subsurface flow in natural pipes or tunnels and when heavy rain provided sufficient flow to flush out the soil in these, the ground surface subsided, exposing the pipe network as gullies. Numerous studies record the formation of gullies by pipe collapse in many different materials and climatic environments. The essential requirements are steep hydraulic gradients in a soil of high infiltration capacity through macropores but low intrinsic permeability so that water does not move readily into the matrix

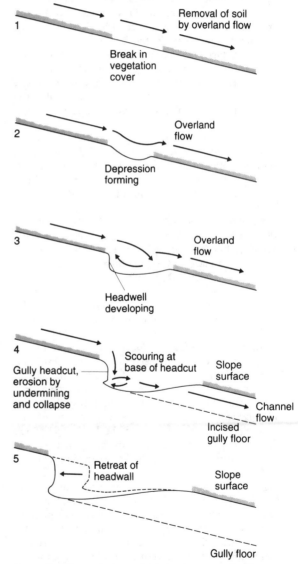

Figure 2.4 Stages in the surface development of gullies on a hillside (after Leopold, Wolman and Miller 1964).

(Crouch 1976; Bryan and Yair 1982). Suitable soils include those prone to cracking as a result of high sodium absorption, shrinkage on drying or release of pressure following unloading of overlying material.

Tunnel erosion has been widely reported in many hilly and rolling areas of Australia where it is associated with duplex soils. These are soils characterized by a sharp increase in clay content between the A and B horizons so that the upper layer, 0.03 to 0.6 m in depth, varies from a loamy sand to a clay loam and the lower layer ranges from a light to heavy clay. According to Downes (1946), overgrazing and removal of the vegetation cover cause crusting of the surface soil, resulting in greater runoff. This passes into the soil through small depressions, cracks and macropores but, on reaching the top of the B horizon, moves along it as subsurface flow. Localized dispersion of the clays in areas of moisture accumulation is followed by piping. Heavy summer rains cause the water in the pipes to break out on to the surface. Eventually, the roofs of the pipes collapse and gullying occurs (Floyd 1974).

A third way in which gullies are initiated is where linear landslides leave deep, steep-sided scars which may be occupied by running water in subsequent storms. This type of gully development has been described in Italy by Vittorini (1972) and in central Värmland, Sweden, by Fredén and Furuholm (1978). In the latter case, a 3–20-m deep, 20–40-m wide and 100-m long gully formed in glaciofluvial deposits following the 1977 spring snowmelt which caused the removal, as a mass flow, of 20,000 m^3 of saturated silt and sand in less than three days.

The main cause of gully formation is too much water, a condition which may be brought about by either climatic change or alterations in land use. In the first case, increased runoff may occur if rainfall increases or if less rainfall produces a reduction in the vegetation cover. In the second case, deforestation, burning of the vegetation and overgrazing can all result in greater runoff. If the velocity or tractive force of the runoff exceeds a critical or threshold value, gullying will occur. Since the exceedance of the threshold relates to changes in the landscape which are external to the processes operating in the gully itself, the threshold is described as extrinsic (Schumm 1979).

Attempts to relate gullying to changes in external factors have not proved entirely successful, however, because not all gullies in an area appear to respond in the same way. In order to explain the onset of instability in one gully whilst its neighbours remain stable, Schumm (1979) examined the role of intrinsic thresholds which are related to the internal working of the gully. From a review of studies in Wyoming, Colorado, New Mexico and Arizona, a discriminant function was established between stable and unstable conditions in terms of the size of the catchment area, which controls discharge, and the channel slope. When, for a given catchment area, channel slope exceeds a critical value, incision occurs creating a channel characterized by one or more head scarps. Subsequent scouring causes the gully to become very active: the channel widens, deepens and extends headwards. Over time, the channel slope is reduced promoting a consolidation phase as the gully stabilizes, the channel fills in, the sides and head wall become flatter and vegetation regrows. Deposition steepens the slope again and triggers a new phase of gullying. Thus gullies pass through successive cycles of erosion and deposition. It is not uncommon for the head of a gully to be extremely active whilst the lower section of the gully is stabilizing or for gullies to contain a sequence of alternating stable and unstable sections.

The interplay of extrinsic and intrinsic thresholds makes gully erosion an extremely complex process. It is not surprising that it remains poorly understood and that only a qualitative appreciation of the factors influencing gully behaviour has been obtained. Gullies may be classified according to their topographic position as valley floor and valley side gullies.

Valley floor gullies

Valley floor gullies can range from ephemeral gullies formed in topographic swales in the landscape during heavy rains to more permanent deeply-incised channels, 3–4 m deep. Ephemeral gullies (Poesen 1989) occur where runoff concentrates in the valley bottom, particularly where the surrounding hillslopes are convexo-concave, most of the land is under arable farming, the soils are either freshly-tilled and loose or crusted, and peak discharges reach several cumecs (m^3/s). They are essentially a surface phenomenon and can be regarded as rather large rills formed when the tractive force exerted by the flow exceeds the resistance of the soil. Although occasionally they can be several metres deep, they are generally limited in depth to no more than 25–30 cm by an underlying plough pan; in many cases they are wide and shallow. Once formed, they will generally be filled in by weathering processes and deep ploughing although they may re-form in the same place in subsequent storms.

Although the more permanent valley floor gullies may have been initiated by surface runoff, once they have attained a certain depth, tractive force plays only a minor role in their development. The main mechanism is the retreat of the headwall. In loess soils, this occurs mainly by slab failure as the wall is undermined by plunge-pool erosion and basal sapping, and vertical tension cracks develop on the slope above. Although head wall collapse may be the most important source of sediment in the gullies of western Iowa (Bradford and Piest 1980), measurements southwest of Sydney, Australia, show that erosion of the sidewalls subsequent to retreat of the headcut is much more important in contributing sediment to the gullies than the process of incision by the headcut itself (Blong, Graham and Veness 1982). In this case, however, the sediment appears to relate to erosion of the side walls by overland flow, interflow and rainsplash rather than their collapse through mass soil failure.

Even though flow along the floor of the gully may contribute very little sediment compared with head wall and side wall processes, it is still important in removing sediment from the gully, without which the gully would fill in and stabilize. Provided the material is flushed out, however, the gully can continue to retreat headwards until, as it moves back up the slope, the soil thins, the base of the gully intersects the bedrock and the height of the head wall reduces sufficiently to stabilize. Before this occurs, a valley floor gully will most likely have migrated into the surrounding hills where it will behave like a valley side gully.

Valley side gullies

Valley side gullies develop more or less at right-angles to the main valley line where local concentrations of surface runoff cut into the hillside, subsurface pipes collapse or local mass movements create a linear depression in the landscape. Valley side gullies may be both continuous, that is they discharge into the river at the bottom of the slope, or discontinuous, fading out into a depositional zone and not reaching the valley floor. Once formed, they can grow upslope by headward retreat and downslope by incision of the channel floor.

Among the most spectacular valley side gullies are the *lavakas* found in the deeply-weathered hills of the central part of the Malagasy Republic. They have developed on convex hillsides with basal slopes of 30–45° adjacent to wide flat-floored valleys. Under the average rainfall of 1,000 to 2,000 mm per year and maximum mean monthly temperatures of 18–26° C, the underlying metamorphic rocks have been chemically weathered to form a 5–25-m deep cover of saprolite. According to Wells and Andriamihaja (1991), gullies form where the vegetation is destroyed, probably by grazing, or along paths and tracks. Infiltration capacities in these areas are reduced to < 50 mm/h compared with > 300 mm/h on the surrounding land. With 10-minute rainfall intensities of 80 to 100 mm/h, sometimes 200 to 500 mm/h, runoff is quickly produced with sufficient force to develop rills within about 20 to 30 m of the crest, just where the slope steepens rapidly into the convexity. The rills enlarge over time, develop head scarps, and cut down into the underlying saprolite. Once this material is reached, the rate of downcutting, particularly in the head area, is very rapid until the channel reaches the groundwater zone above the bedrock. Thereupon, downcutting ceases but the gully continues to expand by sapping, undermining and collapse of the headwall and side walls (Wells, Andriamihaja and Rakotovolo-lona 1991). Since this occurs in the area of greatest relief, the gully develops its typical pear-shaped plan form, broadest towards the top of the slope and with a narrow outlet downslope. In this stage, upslope runoff contributes little, if anything, to the gully growth whereas localized concentrations of groundwater enhance headwater bifurcation and cause the perimeter of the gully to take on a feather-edged appearance when viewed in plan. Similar patterns of valley side gully development have been observed in the Appalachian Piedmont, southeast USA (Ireland, Sharpe and Eargle 1939), southeast Brazil (de Meis and de Moura 1984), Swaziland (WMS Associates 1988) and Zimbabwe (Whitlow and Bullock 1986).

In the examples just described, there is no evidence of piping. Yet piping is often associated with valley side gullies, even though the relationship can be complex. At Springdale in the Riverina area of New South Wales, some gullies clearly owe their origin to the collapse of tunnels, as evidenced by remnants of the soil forming bridges where the collapse has been incomplete, whilst others appear to have developed from tunnels initiated after the cutting of the main gully. Developed along permeable zones between layers of relatively impermeable material, they depend on the incision of the gully to expose the permeable seams and form the hydraulic gradient along which water can move (Crouch, McGarity and Storrier 1986).

Valley side gullies can often evolve into badlands, particularly where a fragile environment has been disturbed by unwise land use (López-Bermúdez and Romero-Díaz 1989). Once formed, badlands take a long time to heal and many of those found in the Mediterranean, although very spectacular, record rather low rates of erosion today and are relics of former periods of active gullying, for example those of the Guadix Basin in southeast Spain which were probably developed around 2000 BC (Wise, Thornes and Gilman 1982).

Although gullies can remove vast quantities of soil, according to Zachar (1982) gully densities are not usually greater than 10 km/km² and the surface area covered by gullies is rarely more than 15 per cent of the total area. This results in a considerable contrast between the erosion rate for an individual gully and its contribution to the overall soil loss of an area. Rates of headwall extension can be very rapid for relatively short periods of time. Whitlow and Bullock (1986) recorded retreats of 0.65 to 1.75 m between 7 February and 6 May 1986 in the Nyagui Gully system, Zimbabwe. Rates of 30 to 40 m per year are not uncommon in the surrounding area. Measurements on the Mbothoma Gully system, Swaziland, showed very rapid retreats of between 2.5 and 6.3 m per year from 1947 to 1961, followed by a slowing down to 0.13 to 0.52 m per year for the period 1961 to 1980 (WMS Associates 1988). Erosion from a gully developed on arable land near Cromer, Norfolk, England in 1975 was estimated at 195 t/ha (Evans and Nortcliff 1978) and soil loss from the side walls alone in the gullies near Bathurst, New South Wales over a 3-year period beginning April 1984, amounted to 1,100 t/ha (Crouch 1990a). These figures contrast with annual erosion rates for whole catchments for which typical figures include 3–5 t/ha for the gullied watersheds at Treynor, Iowa (Bradford and Piest 1980), 3–16 t/ha for the *lavakas* in the Malagasy Republic (Wells and Andriamihaja 1991) and 6.4 t/ha for a watershed near Gilgandra, New South Wales, of which 60 per cent came from the gully heads (Crouch 1990b). Wherever gullying occurs, however, it can completely destroy the landscape.

2.7 Mass Movements

Although mass movement has been widely studied by geologists, geomorphologists and engineers, it is generally neglected in the context of soil erosion. Yet Temple and Rapp (1972) have found that in the western Uluguru Mountains, Tanzania, landslides and mudflows are the dominant erosion processes. They occur in small numbers about once every 10 years. The quantity of sediment moved from the hillsides into rivers by mass movement is far in excess of that contributed by gullies, rills and overland flow. Further, less than one per cent of the slide scars are in areas of woodland, 47 per cent being on the cultivated plots and another 47 per cent on land lying fallow. The association of erosion with woodland clearance for agriculture is thus very clear. Further evidence for this is provided by Rogers and Selby (1980) in respect of shallow debris slides on clay and silty clay soils derived from greywackes in the Hapuakohe Range, south Auckland, New Zealand. Clearance of the forest for pasture causes a decline in the shear strength of the soil over the 5- to 10-year period needed for the tree roots to decay. As a result, landslides under pasture are triggered by a storm with a return period of 30

years whereas a storm with a 100-year return period is required to produce slides beneath the forest.

The stability of the soil mass on a hillslope in respect of mass movement can be assessed by a safety factor (*F*), defined as the ratio between the total shear strength (σ) along a given shear surface and the amount of stress (τ) developed along this surface. Thus

$$F = \frac{\sigma}{\tau} \qquad (2.30)$$

The slope is stable if $F > 1$ and failure occurs in $F \leqslant 1$. For the simple case of shallow or translational slides, *F* can be defined as:

$$F = \frac{c' + (\gamma z - \gamma_w h)\ \cos^2 \theta\ \tan \phi'}{\gamma z\ \sin \theta\ \cos \theta} \qquad (2.31)$$

where c' is the effective cohesion of the soil, γ is the unit weight of soil above the slide plane, z is the vertical depth of soil above the slide plane, γ_w is the unit weight of water, h is the height of the groundwater surface (piezometric surface) above the slide plane, θ is the slope angle and ϕ' is the effective angle of internal friction of the soil. Applications of this equation to slides in the Hapuakohe Range show that the value of *F* is particularly sensitive to changes in c' and z. Thus control measures should be directed at influencing these (Rogers and Selby 1980).

Mass movements, in the varied forms of creep, slides, rock falls and mudflows, are given detailed treatment in numerous books (Sharpe 1938; Zaruba and Mencl 1969; Brunsden and Prior 1984; Anderson and Richards 1987), but, for soil erosion studies, rather than stress the separate forms, it is more helpful to consider them as part of a continuum of flow phenomena, ranging from debris slides, in which the ratio of solid to liquid is high, through mudflows to running water which has a low solid–liquid ratio. The close relationship between mass movement and water erosion is illustrated by the studies of the so-called bottle slides in the Uluguru Mountains (Temple and Rapp 1972; Lundgren and Rapp 1974) which develop in areas of large subsurface pipes as a result of the flushing out of a muddy viscous mass of debris and subsequent ground collapse.

2.8 Wind Erosion

The main factor in wind erosion is the velocity of moving air. Because of the roughness imparted by soil, stones, vegetation and other obstacles, wind speeds are lowest near the ground surface. A plane of zero wind velocity can be defined at some height (z_0) above the mean aerodynamic surface. Above z_0, wind speed increases exponentially with height so that velocity values plot as a straight line on a graph against the logarithmic values of the height (Fig. 2.5). The change in velocity with height is expressed by the relationship (Bagnold 1941):

$$\bar{v}_z = \frac{2.3}{k} u_\star \, log \left(\frac{z}{z_0}\right) \qquad (2.32)$$

where $\bar{v}$ is the mean velocity at height z, k is the von Kármán universal constant for turbulent flow and is assumed to equal 0.4 for clear fluids, and $u_\star$ is the drag or shear velocity.

Although the movement of soil particles can be related to a critical wind velocity, many workers have attempted to define the conditions more precisely in terms of a critical value of the dimensionless shear stress. Using Eq. 2.10 and substituting ρ_a (the density of air) for ρ_w (the density of water), the critical value of Θ, the Shields parameter, for initiating soil particle movement approximates 0.1 (Bagnold 1941), which is much lower than that obtained for water. The difference may be due to the very great difference in the density of the fluids relative to the particle density. A sand grain in air is about 2,000 times more massive than the surrounding fluid whereas it is only about 2.6 times more massive than water (Bagnold 1979). As a result, much higher shear velocities are required to move particles by air and the initial particle motion is more violent. This violence is rapidly transmitted to neighbouring particles causing a general chain reaction of motion. At this point, grains are dislodged and entrained in the flow relatively easily and therefore at relatively low Shields numbers (Iversen 1985).

Although the Shields coefficient has been successfully applied to particle movement in air (Bagnold 1951; Iversen 1985), for most soil conservation work it is sufficient to relate the detachment of soil particles by wind to a critical value of the shear velocity, using it as a surrogate measure for the drag force exerted by the flow. Shear velocity is directly proportional to the rate of increase in wind velocity with the logarithm of height and is therefore the slope of the line in Fig. 2.5(d). Its value can be determined by measuring the wind speed at two heights but, in practice, by assuming $v = 0$ at height z_0, it can be obtained by measuring the speed at one height and applying the formula derived from Eq. 2.32:

$$u_\star = \frac{k}{2.3} \cdot \frac{\bar{v}_z}{log \, (z/z_0)} \qquad (2.33)$$

For use in estimates of sediment transport by wind, the velocities should be measured within 0.2 m of the ground surface (Rasmussen, Sørensen and Willetts 1985; Mulligan 1988).

Bagnold (1937) identifies two threshold velocities required to initiate grain movement. The static or fluid threshold applies to the direct action of the wind. The dynamic or impact threshold allows for the bombardment of the soil by grains already in motion. Impact thresholds are about 80 per cent of the fluid threshold velocities in value. In addition to detaching soil particles at a lower threshold velocity, the detachment potential of sediment-laden air is further enhanced by increases in shear velocity close to the ground surface by up to 14 per cent. This arises through the addition of the grain-borne shear stress to the air flow (Sørensen 1985). The critical shear velocities vary with the grain size of the material, being least

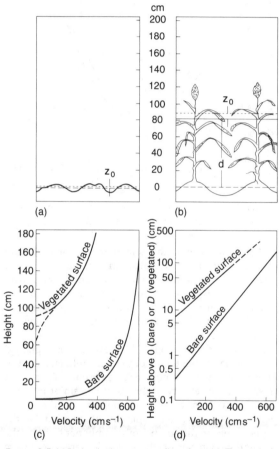

Figure 2.5 Wind velocity near a soil surface. (a) Zero wind velocity occurs at a height z_0 which lies above the height of the mean aerodynamic surface and below the high points. (b) A crop cover raises the height of the mean aerodynamic surface by a distance d and also increases the value of z_0 so that the plane of zero wind velocity ($d + z_0$) occurs at a height which is equal to about 70 per cent of the height of the plants. (c) Wind velocity profiles above a bare surface and a vegetated surface plotted with linear scales. (d) Wind velocity profiles plotted with a logarithmic scale for height (after Troeh, Hobbs and Donahue 1980).

for particles of 0.10 to 0.15 mm in diameter and increase with both increasing and decreasing grain size (Chepil 1945). The resistance of the larger particles results from their size and weight. That of the finer particles is due to their cohesiveness and the protection afforded by surrounding coarser grains (Fig. 2.6).

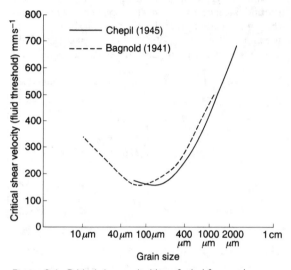

Figure 2.6 Critical shear velocities of wind for erosion as a function of particle size (after Savat 1982).

Once in motion, the transport of soil and sand particles by wind takes place in suspension, surface creep and saltation. Suspension describes the movement of fine particles, usually less than 0.2 mm in diameter, high in the air and over long distances. Surface creep is the rolling of coarse grains along the ground surface. Saltation is the process of grain movement in a series of jumps. The movement is explained by considering the theory of the Bernoulli effect. After a grain has been rolling along the ground for a short distance, the velocity of the air at any point near the grain is made up of two components, one due to the wind and the other to the spinning of the grain. On the upper side of the grain both components have the same direction but on the lower side are in opposite directions. As a result of greater velocity on the top surface of the grain, the pressure there is reduced whilst pressure at the lower surface increases. This difference in static pressure produces a lifting force and when this is sufficient to overcome the weight of the grain, the grain rises vertically. The existence of 'pressure lift' is short-lived and occurs only very close to the ground surface. The particle continues to rise, however, because of its initial vertical ejection velocity which is about twice the shear velocity of the air

(Willetts and Rice 1985). Drag with the surrounding air quickly reduces the vertical velocity which is also opposed by the settling velocity of the particle. Once in the air, however, the particle takes on a horizontal velocity, imparted by the wind, so that, whilst settling back to the ground, it is blown forwards. The result is an overall particle movement or jump comprising a vertical rise to a maximum height at which the settling velocity exceeds the vertical velocity followed by a falling path at an angle of 6° to 12° to the horizontal. Individual jump lengths for coarse grains vary from about 6 to 40 cm, increasing with the shear velocity (Sørensen 1985; Willetts and Rice 1988). On striking the ground, the impact energy of a saltating grain is distributed into a disruptive part which causes disintegration of the soil and a dispersive part which imparts a velocity to other soil particles and launches them into the air (Smalley 1970). In a soil blow, between 55 and 72 per cent of the moving particles are carried in saltation.

The rate at which soil particles are dislodged from a bare soil surface was found in wind tunnel experiments by Sørensen (1985) and Jensen and Sørensen (1986) to follow the relationship:

$$D_{fa} \propto u_\star^{2.8} \qquad (2.34)$$

where D_{fa} is the rate of detachment of soil particles in moving air. Although Willetts and Rice (1988) produce comparable data from their experiments, they caution that the overall database is too small to determine the value of the exponent in Eq. 2.34. It probably lies, however, between 2.0 and 3.0.

In contrast, general agreement exists among researchers that the transport capacity of moving air varies with the cube of the shear velocity. Based on this relationship and considering the transport of grains as representing a transfer of momentum from the air to the moving particles, Bagnold (1941) developed the following equation for determining the maximum sediment discharge per unit width (T_{fa}):

$$T_{fa} = C \, (d/D)^{\frac{1}{2}} \, (\rho_a/g) \, u_\star^3 \qquad (2.35)$$

where C is an experimentally-derived parameter relating to grain size, d is the average grain diameter of the material, D is a standard grain diameter of 0.25 mm, ρ_a is the density of the air and g is the gravitational constant. A similar equation was derived by Zingg (1953) from wind-tunnel experiments:

$$T_{fa} = C \, (d/D)^{\frac{3}{4}} \, \rho_a u_\star^3/g \qquad (2.36)$$

Taking data from studies in Florida, Texas, Alaska, Ecuador and Libya, Hsu (1973) successfully predicted sand transport from the relationship:

$$T_{fa} = C \left(\frac{u_\star}{(gd)^{\frac{1}{2}}} \right)^3 \qquad (2.37)$$

where $\ln C = 4.97\,d - 0.47$ when T_{fa} is in $\mathrm{g\,cm^{-1}s^{-1}}$ and $\ln C$ is the value of C in Napierian logarithms. Chepil (1945) avoided the need to calculate shear velocities and took account of erosion only occurring when windspeeds exceeded a threshold value. He developed a simple relationship:

$$T_{fa} = B(v - v_t)^3 \qquad (2.38)$$

where v is the wind velocity (cm/s) measured at a height of 1 m and v_t is the threshold velocity for windspeeds as measured at that height, usually taken as 400 cm/s. Chepil calculated $B = 52$ but Finkel (1959), in studies of barchan dunes in southern Peru, found that

$$B = \left(\frac{0.174}{log\ z/z_0}\right)^3 . C\left(\frac{d\rho_a^{1/2}}{Dg}\right) \qquad (2.39)$$

where $log\ z/z_0$ is as previously defined (Eq. 2.32).

It should be noted that whilst Eqs. 2.35 to 2.38 describe the transport capacity of moving air, they do not necessarily predict the actual rate of sediment transport which, particularly where the soil is very resistant, the soil surface has become armoured by rock fragments or a vegetation cover is present, may be lower. In these cases, the sediment transport will vary with the shear velocity of the wind raised to a power of less than 3.0.

Wind erosion impoverishes the soil and also buries the soil and crops on surrounding land. Although, as already seen, the most erodible particles are 0.10 to 0.15 mm in size, particles between 0.05 and 0.5 mm are generally selectively removed by the wind. Chepil (1946) found that areas of wind-blown deposits are enriched by particles in the 0.30 to 0.42 mm size range. Resistance to wind erosion increases rapidly when primary particles and aggregates larger than 1 mm predominate. If erosion results in armouring of the surface so that more than 60 per cent of the surface material is of this size, the soil is almost totally resistant to wind erosion.

CHAPTER 3

Factors Influencing Erosion

The factors controlling soil erosion are the erosivity of the eroding agent, the erodibility of the soil, the slope of the land and the nature of the plant cover.

3.1 Erosivity

3.1.1 Rainfall

Soil loss is closely related to rainfall partly through the detaching power of raindrops striking the soil surface and partly through the contribution of rain to runoff. This applies particularly to erosion by overland flow and rills for which intensity is generally considered to be the most important rainfall characteristic. The effect of rainfall intensity is illustrated by the data for 183 rain events which caused erosion at Zanesville, Ohio, between 1934 and 1942, which show that average soil loss per rain event increases with the intensity of the storm (Table 3.1; Fournier 1972).

Table 3.1 Relationship between rainfall intensity and soil loss

Maximum 5-min intensity (mm/h)	Number of falls of rain	Average erosion per rainfall (t/ha)
0–25.4	40	3.7
25.5–50.8	61	6.0
50.9–76.2	40	11.8
76.3–101.6	19	11.4
101.7–127.0	13	34.2
127.1–152.4	4	36.3
152.5–177.8	5	38.7
177.9–254.0	1	47.9

Data for Zanesville, Ohio, 1934–42 (after Fournier 1972).

The role of intensity is not always so obvious, however, as indicated by studies of erosion in mid-Bedfordshire. Taking data for the ten most erosive storms between May 1973 and October 1975 (Table 3.2), it can be seen that whilst intense storms, such as the one on 6 July 1973 of 34.9 mm, in which 17.7 mm fell at intensities greater than 10 mm/h, produce erosion, so do storms of long duration and low intensity,

Table 3.2 Rainfall and soil loss in Bedfordshire

Date	Rainfall (mm)	Intense rain (≥ 10 mm/h) (mm)	Rainfall duration (h)	Soil loss (g/cm-width)
6.7.73	34.9	17.7	8.00	70.70
5.5.73	6.7	—	4.16	61.72
21.5.73	7.1	3.9	2.00	55.98
27.8.73	16.8	4.6	6.50	50.90
6.5.73	2.2	1.0	0.33	40.00
8.8.74	17.0	8.8	13.50	36.78
16.7.74	7.6	6.4	13.00	28.66
13.7.74	16.5	9.0	10.50	24.71
19.6.73	39.6	—	23.83	20.38
27.6.73	18.6	2.3	13.83	18.99

Data for bare soil and 11° slope.

like the one of 19 June 1973 when 39.6 mm of rain fell in over 23 hours (Morgan, Martin and Noble 1986). It appears that erosion is related to two types of rain event, the short-lived intense storm where the infiltration capacity of the soil is exceeded, and the prolonged storm of low intensity which saturates the soil. In many instances it is difficult to separate the effects of these two types of events in accounting for soil loss.

The response of the soil to rainfall may also be determined by previous meteorological conditions. This can again be demonstrated by data for Zanes-

Table 3.3 Influence of antecedent rainfall conditions on soil loss

Date	Rainfall (mm)	Runoff (% of rainfall)	Erosion (g/m²)
9 June	19.3	25	1.5
10 June	13.7	66	4.0
11 June	23.8	69	8.9
15 June	14.0	65	4.2
17–18 June	13.0	50	4.6

Data for Zanesville, Ohio, June 1940, for five successive rainstorms on a plot of 20 m² (after Fournier 1972).

ville, Ohio (Table 3.3; Fournier 1972). Over the period 9 to 18 June 1940, the first rain fell on dry ground and, in spite of the quantity, little runoff resulted, most of the water soaking into the soil. In the second storm, however, 66 per cent of the rain ran off and soil loss almost trebled. The control in this case is the closeness of the soil to saturation which is dependent on how much rain has fallen in the previous few days. The pattern of low soil loss in the first and high loss in the second of a series of storms is reversed, however, where, between erosive storms, weathering and light rainfall loosen the soil surface. Most of the loose material is removed during the first runoff event leaving little for erosion in subsequent events. This sequence is illustrated by studies in the Alkali Creek watershed, Colorado (Heede 1975b), where a sediment discharge peak of 143 kg/s with a runoff from snowmelt of 2.21 m³/s was observed in one of the ephemeral gullies on 15 April 1964. This event followed one year without runoff. Next day peak runoff increased to 3.0 m³/s but the sediment discharge fell to 107 kg/s. Although this type of evidence clearly points to the importance of antecedent events in conditioning erosion, other observations deny its significance. No relationship was obtained between soil loss and antecedent precipitation in mid-Bedfordshire (Morgan, Martin and Noble 1986).

The question arises of how much rain is required to induce significant erosion. Hudson (1981) gives a figure, based on his studies in Zimbabwe, of 25 mm/h, a value which has also been found appropriate in Tanzania (Rapp et al. 1972a) and Malaysia (Morgan 1974). It is too high a value for western Europe, however, where it is only rarely exceeded. Arbitrary thresholds of 10, 6 and 1.0 mm/h have been used in England (Morgan 1980b), Germany (Richter and Negendank 1977) and Belgium (Bollinne 1977) respectively.

Threshold values vary with the erosion process. The figures quoted above are typical for erosion by overland flow, rills and mass movements which, as seen in Chapter 1, is characteristic of moderate events whereas higher magnitude events are required for the initiation of fresh gullies. The distinction between these types of event can be blurred, however, in those areas which, by world standards, regularly experience what may be described as extreme events. Starkel (1972) stresses the importance of regular gully erosion in the Assam Uplands where monthly rainfall may total 2,000 to 5,000 mm and in the Darjeeling Hills where over 50 mm of rain falls on an average of 12 days each year and rainfall intensities are often highest at the end of a rain event. In this very active landscape, overland flow and slope wash can start during rain storms of 50 mm with intensities greater than 30 mm/h; slides and slumps can occur after daily rains of 100 to 150 mm or a rainfall total of 200 mm in two or three

days; and debris flows and mudflows are generated when 500 to 1,000 mm of rain falls within two or three days (Froehlich and Starkel 1993). The effects of an extreme event may be long lasting and give rise to high soil losses for a number of years. The length of time required for an area to recover from a severe rain storm, flooding and gullying has not been fully investigated but, in a review of somewhat sparse evidence, Thornes (1976) quotes figures up to 50 years.

3.1.2 Rainfall erosivity indices

The most suitable expression of the erosivity of rainfall is an index based on the kinetic energy of the rain. Thus the erosivity of a rain storm is a function of its intensity and duration, and of the mass, diameter and velocity of the raindrops. To compute erosivity requires an analysis of the drop-size distributions of rain. Laws and Parsons (1943), based on studies of rain in the eastern USA, show that drop-size characteristics vary with the intensity of the rain with the median drop diameter (d_{50}) increasing with rainfall intensity. Studies of tropical rainfall (Hudson 1963) indicate that this relationship holds only for rainfall intensities up to 100 mm/h. At higher intensities, median drop size decreases with increasing intensity, presumably because greater turbulence makes larger drop sizes unstable. However, at intensities above 200 mm/h, coalescence of smaller drops takes place so that the median drop diameter begins to increase again (Carter et al. 1974). Considerable variability exists, however, because the relationship between median drop size and intensity is not constant; both median drop size and drop-size distribution vary for rains of the same intensity but different origins (Mason and Andrews 1960; Carter et al. 1974; Kinnell 1981; McIsaac 1990). The drop-size characteristics of convectional and frontal rain differ as do those of rain formed at the warm and cold fronts of a temperate depression.

In spite of the difficulties posed by these variations, it is possible to derive general relationships between kinetic energy and rainfall intensity. Based on the work of Laws and Parsons (1943), Wischmeier and Smith (1958) obtained the equation:

$$KE = 11.87 + 8.73 \, log_{10}I \qquad (3.1)$$

where I is the rainfall intensity (mm/h) and KE is the kinetic energy (J m^{-2} mm^{-1}). Many researchers (Mason and Ramandham 1953; Carte 1971; Houze et al. 1979; Styczen and Høgh-Schmidt 1988) consider the drop-size distribution of rainfall described by Marshall and Palmer (1948) as representative of a wide range of environments. The equivalent formula for calculating kinetic energy is:

$$KE = 8.95 + 8.44 \, log_{10}I \qquad (3.2)$$

For tropical rainfall, Hudson (1965) gives the equation:

$$KE = 29.8 - \frac{127.5}{I} \qquad (3.3)$$

based on measurements of rainfall properties in Zimbabwe. Zanchi and Torri (1980) carried out similar research in Italy and obtained:

$$KE = 9.81 + 11.25 \, log_{10}I \qquad (3.4)$$

and Onaga, Shirai and Yoshinaga (1988) give the following relationship for Okinawa, Japan:

$$KE = 9.81 + 10.6 \, log_{10}I \qquad (3.5)$$

Eqs. 3.1, 3.2 and 3.3 show that at intensities greater than 75 mm/h, the kinetic energy levels off at a value of about 29 $J m^{-2} mm^{-1}$ which seems to be representative for many locations (Kinnell 1987). However, Carter et al. (1974) found in the southern USA that kinetic energy increased to a maximum value at about 75 mm/h, decreased with further increases in intensity up to about 175 mm/h and then increased again at still higher intensities. Studies in Japan (Mihara 1951) and in the Marshall Islands (McIsaac 1990) give rainfall energies some 6–20 per cent lower than those calculated from Eq. 3.1 whilst the Italian and Okinawan research, represented by Eqs. 3.4 and 3.5 respectively, indicates energy values as high as 34 $J m^{-2} mm^{-1}$ when the intensity is 150 mm/h. Very high energy values have also been recorded from Nigeria. Kowal and Kassam (1976) report a 20-minute rain storm at Samaru with a peak intensity of 111 mm/h and energies ranging from 31.6 to 38.4 $J m^{-2} mm^{-1}$ whilst Osuji (1989) found that maximum energy values were generally about 35 $J m^{-2} mm^{-1}$ for intensities greater than 70 mm/h. Since rainfall energy varies with the density of the air raised to the 0.9 power, energy also increases with altitude. Tracy, Renard and Fogel (1984) found that the kinetic energy of rainfall at 900–1,800 m above sea level in Arizona was about 15 per cent higher than that predicted by Eq. 3.1.

To compute the kinetic energy of a storm, a trace of the rainfall from an automatically recording rain gauge is analysed and the storm divided into small time increments of uniform intensity. For each time period, knowing the intensity of the rain, the kinetic energy of rain at that intensity is estimated from one of the above equations and this, multiplied by the amount of rain received, gives the kinetic energy for that time period. The sum of the kinetic energy values for all the time periods gives the total kinetic energy of the storm (Table 3.4).

To be valid as an index of potential erosion, an

Table 3.4 Calculation of erosivity

Time from start (mm)	Rainfall (mm)	Intensity (mm/h)	Kinetic energy ($J m^{-2}$ mm^{-1})	Total kinetic energy (col 2 × col 4) (J/m^2)
0–14	1.52	6.08	8.83	13.42
15–29	14.22	56.88	27.56	391.90
30–44	26.16	104.64	28.58	747.65
45–59	31.50	126.00	28.79	906.89
60–74	8.38	33.52	26.00	217.88
75–89	0.25	1.00	—	—

Kinetic energy is calculated using Eq. 3.3.

Erosivity indices

Wischmeier index (EI30):

maximum 30-minute rainfall	= 26.16 + 31.50 mm
	= 57.66 mm
maximum 30-minute intensity	= 57.66 × 2
	= 115.32 mm/h
total kinetic energy	= total of column 5
	= 2,277.74 J/m^2
EI30	= 2,277.74 × 115.32
	= 262,668.98 $J mm \, m^{-2} h^{-1}$

Hudson index (KE > 25):

total kinetic energy for rainfall intensity ≥ 25 mm/h	= total of column 5, lines 2, 3, 4 and 5 only
	= 2,264.32 J/m^2

erosivity index must be significantly correlated with soil loss. Wischmeier and Smith (1958) found that soil loss by splash, overland flow and rill erosion is related to a compound index of kinetic energy (E) and the maximum 30-minute intensity (I_{30}). This index, known as EI_{30}, is open to criticism. First, being based on estimates of kinetic energy using Eq. 3.1, it is of suspect validity for tropical rains of high intensity as well as for high altitudes and areas like Japan and the Marshall Islands where rainfall energies are rather low. Second, it assumes that erosion occurs even with light intensity rain whereas Hudson (1965) showed that erosion is almost entirely caused by rain falling at intensities greater than 25 mm/h. The inclusion of I_{30} in the index is an attempt to correct for overestimating the importance of light intensity rain but it is not entirely successful because the ratio of intense erosive rain to non-erosive rain is not well correlated with I_{30} (Hudson, personal communication). In fact, there is no obvious reason why the maximum 30-minute intensity is the most appropriate parameter to choose. Stocking and Elwell (1973a) recommend its use only for bare soil conditions. With sparse and dense plant covers they obtain better correlations with soil loss using the maximum 15- and 5-minute intensities respectively. In order to overcome the likelihood of overestimating soil loss from high intensity rainfall, the recommended practice with the EI_{30} index is to use a maximum value of 28.3 J m^{-2}mm^{-1} for the E component for all rains above 76.2 mm/h and a maximum value of 63.5 mm/h for I_{30} term (Wischmeier and Smith 1978).

As an alternative erosivity index, Hudson (1965) uses $KE > 25$ which, to compute for a single storm, means summing the kinetic energy received in those time increments when the rainfall intensity equals or exceeds 25 mm/h (Table 3.4). When applied to data from Zimbabwe, a better correlation was obtained between this index and soil loss than between soil loss and EI_{30}. Stocking and Elwell (1973a) have reworked Hudson's data and, incorporating more recent information, have suggested that EI_{30} is the better index after all. Since they compute EI_{30} only for storms yielding 12.5 mm or more of rain and with a maximum 5-minute intensity greater than 25 mm/h, they have removed most of the objections to the original EI_{30} index, however, and produced an index which is philosophically very close to $KE > 25$. Hudson's index has the advantages of simplicity and less stringent data requirements. Although somewhat limiting for temperate latitudes, it can be modified by using a lower threshold value such as $KE > 10$ (Morgan 1980b).

By calculating erosivity values for individual storms over a period of 20 to 25 years, mean monthly and mean annual data can be obtained. Unfortunately, the EI_{30} and $KE > 25$ indices yield vastly different values because of the inclusion of I_{30} in the former. The two indices cannot be substituted for each other.

3.1.3 Wind erosivity

The kinetic energy (KE_a; J m^{-2}s^{-1}) of wind can be calculated from:

$$KE_a = \frac{\gamma_a u^2}{2g} \qquad (3.6)$$

where u is the wind velocity in m/s and γ_a is the specific weight of air defined in terms of temperature (T) in °C and barometric pressure (P) in kPa by the relationship (Zachar 1982):

$$\gamma_a = \frac{1.293}{1 + 0.00367\,T}\,\frac{P}{101.3} \qquad (3.7)$$

For $T = 15$°C and $P = 101.3$ kPa, kinetic energy $= 0.0625\,u^2$ J m^{-2}s^{-1} which converts to $227\,u^2$ J m^{-2}s^{-1}. Energy values for wind storms can be obtained by summing the energies for the different velocities weighted by their duration.

In practice, kinetic energy is rarely used as a basis for an index of wind erosivity and, instead, a simpler index based only on the velocity and duration of the wind (Skidmore and Woodruff 1968) has been devel-

oped. The erosivity of wind blowing in vector j is obtained from:

$$EW_j = \sum_{i=1}^{n} \nabla t^3_{ij} f_{ij} \qquad (3.8)$$

where EW_j is the wind erosivity value for vector j, ∇t is the mean velocity of wind in the ith speed group for vector j above a threshold velocity, taken as 19 km/h, and f_{ij} is the duration of the wind for vector j in the ith speed group. Expanding this equation for total wind erosivity (EW) over all vectors yields:

$$EW = \sum_{j=0}^{15} \sum_{i=1}^{n} \nabla t^3_{ij} f_{ij} \qquad (3.9)$$

where vectors $j = 0$ to 15 represent the 16 principal compass directions beginning with $j = 0 = $ E and working anticlockwise so that $j = 1 = $ ENE and so on.

3.2 Erodibility

Erodibility defines the resistance of the soil to both detachment and transport. Although a soil's resistance to erosion depends in part on topographic position, slope steepness and the amount of disturbance, for example during tillage, the properties of the soil are the most important determinants. Erodibility varies with soil texture, aggregate stability, shear strength, infiltration capacity and organic and chemical content.

The role of soil texture has been indicated in Chapter 2 where it was shown that large particles are resistant to transport because of the greater force required to entrain them and that fine particles are resistant to detachment because of their cohesiveness. The least resistant particles are silts and fine sands. Thus soils with a high silt content are highly erodible. Richter and Negendank (1977) show that soils with 40 to 60 per cent silt content are the most erodible. Evans (1980) prefers to examine erodibility in terms of clay content indicating that soils with a restricted clay fraction, between 9 and 30 per cent, are the most susceptible to erosion.

The use of the clay content as an indicator of erodibility is theoretically more satisfying because the clay particles combine with organic matter to form soil aggregates or clods and it is the stability of these which determines the resistance of the soil. Soils with a high content of base minerals are generally more stable as these contribute to the chemical bonding of the aggregates. Wetting of the soil weakens the aggregates because it lowers their cohesiveness, softens the cements and causes swelling as the water is adsorbed on the clay particles. Rapid wetting can also cause collapse of the aggregates through slaking. The wetting-up of initially dry soils results in greater aggregate breakdown than if the soil is already moist because, in the latter case, less air becomes trapped in the soil (Truman, Bradford and Ferris 1990). Aggregate stability also depends on the type of clay mineral present. Illite and smectite more readily form aggregates but the more open lattice structure of these minerals and the greater swelling and shrinkage which occurs on wetting and drying render the aggregates less stable than those formed from kaolinite.

In detail, however, the interactions between the moisture content of the soil and the chemical compositions of both the clay particles and the soil water are rather complex. This makes it difficult to predict how clays, particularly those susceptible to swelling, will behave. The identical treatment of different types of clay can have totally different effects (Thornes 1980). Although most clays lose strength when first wetted because the free water releases the bonds between the particles, some clays, under moist but unsaturated conditions, regain strength over time. This process, known as thixotropic behaviour, occurs because the hydration of clay minerals and the adsorption of free water promote hydrogen bonding (Grissinger and Asmussen 1963). Strength can also be regained if swelling brings about a reorientation of the soil particles from an alignment parallel to the eroding water to a more random orientation (Grissinger 1966). The strength of smectitic clays is largely dependent upon the sodium adsorption ratio. As this increases, i.e. the replacement of calcium and magnesium ions by sodium increases, so does water uptake and the likelihood of swelling and aggregate collapse. High salt concentrations in the soil water, however, can partly offset this effect so that aggregate stability is maintained at higher sodium adsorption ratios (Arulanandan, Loganathan and Krone 1975).

The shear strength of the soil is a measure of its cohesiveness and resistance to shearing forces exerted by gravity, moving fluids and mechanical loads. Its strength is derived from the frictional resistance met by its constituent particles when they are forced to slide over one another or to move out of interlocking positions, the extent to which stresses or forces are absorbed by solid-to-solid contact among the particles, cohesive forces related to chemical bonding of the clay minerals, and surface tension forces within the moisture films in unsaturated soils. These controls over shear strength are only understood qualitatively so that, for practical purposes, shear strength is expressed by an empirical equation:

$$\tau = c + \sigma \, tan \, \phi \qquad (3.10)$$

where τ is the shear stress required for failure to take place, c is a measure of cohesion, σ is the stress normal to the shear plane (all in units of force per unit area) and ϕ is the angle of internal friction. Both c and ϕ are best regarded as empirical parameters rather than as physical properties of the soil.

Increases in the moisture content of a soil decrease its shear strength and bring about changes in its behaviour. At low moisture contents the soil behaves as a solid and fractures under stress but with increasing moisture content it becomes plastic and yields by flow without fracture. The point of change in behaviour is termed the plastic limit. With further wetting, the soil will reach its liquid limit and start to flow under its own weight. The behaviour of a compressible soil when saturated depends on whether the water can drain. If drainage cannot take place and the soil is subjected to further loading, pressure will increase in the soil water, the compaction load will not be supported by the particles and the soil will deform, behaving as a plastic material. If drainage can occur, more of the load will be supported and the soil is more likely to remain below the plastic limit and retain a higher shear strength.

Although Chorley (1959) and Eyles (1968) demonstrated that shear strength affects the broad pattern of landforms with the stronger rocks and soils standing out as higher ground, it has, until recently, been little used as an indicator of soil erodibility. This is largely because valid data were difficult to obtain since they relate to a surface layer of soil only a few millimetres thick with resistance to shear at near-zero confining stresses and subjected to low velocity impacts, reaching a maximum of 6 to 9 m/s with raindrops. The mode of failure is usually static, being the result of overcoming the frictional resistance required to initiate motion rather than the dynamic force to maintain one particle sliding over another. As seen in Chapter 2, shear strength is being increasingly used, however, as a basis for understanding the detachability of soil particles by raindrop impact. Since the soils are usually saturated and the process is virtually instantaneous, there is no time for drainage and undrained failure occurs. Bradford, Truman and Huang (1992) found that soil strength measured with a drop-cone penetrometer after one hour of rainfall was a good indicator of a soil's resistance to splash erosion. The drop-cone apparatus simulates the same kind of failure mechanism, in terms of compression and shear, as the impact of a falling raindrop. In contrast, shear strength measured with a torvane has a poor correlation with splash erosion because the instrument does not adequately simulate the failure process.

The mechanism of soil particle detachment by surface flow involves different failure stresses on the soil surface compared with those generated by raindrop impact and Rauws and Govers (1988) show that these can be represented by measurements of the strength of the soil at saturation, made with a torvane. Eqs. 2.25 and 2.26 predict the critical shear velocity (u_{*c}; cm/s) for rill initiation on a smooth bare soil surface as a function of the strength or apparent cohesion of the soil measured at saturation by a torvane and a laboratory shear vane respectively.

Infiltration capacity, the maximum sustained rate at which soil can absorb water, is influenced by pore size, pore stability and the form of the soil profile. Soils with stable aggregates maintain their pore spaces better whilst soils with swelling clays or minerals that are unstable in water tend to have low infiltration capacities. Although estimates of the infiltration capacity can be obtained in the field using infiltrometers (Hills 1970), it was seen in Chapter 2 that actual capacities during storms are often much less than those indicated by field tests. Infiltration capacities of the Oligocene Tongrian Sands in Belgium are in excess of 200 mm/h according to field measurements but runoff can occur with rains of only 20 mm/h (De Ploey 1977). Similar discrepancies between measured infiltration capacities and rainfall intensities have been observed on soils of the Lower Greensand in Bedfordshire (Morgan, Martin and Noble 1986). Where soil properties vary with profile depth, it is the horizon with the lowest infiltration capacity which is critical. In the case of these sandy soils, the critical horizon is often the surface where, as described in Chapter 2, a crust of 2 mm thickness may be sufficient to decrease infiltration capacity enough to cause runoff, even though the underlying soil may be dry.

The organic and chemical constituents of the soil are important because of their influence on aggregate stability. Soils with less than 2 per cent organic carbon, equivalent to about 3.5 per cent organic content, can be considered erodible (Evans 1980). Most soils contain less than 15 per cent organic content and many of the sands and sandy loams have less than 2 per cent. Voroney, Van Veen and Paul (1981) suggest that soil erodibility decreases linearly with increasing organic content over the range of 0 to 10 per cent whereas Ekwue (1990) found that soil detachment by raindrop impact decreased exponentially with increasing organic content over a 0 to 12 per cent range. These relationships cannot be extrapolated, however, because some soils with very high organic contents, particularly peats, are highly erodible by wind and water whereas others with very low organic content can become very hard and therefore stronger under dry conditions. The role played by organic material depends on its origin. Whilst organic material from grass leys and farmyard manure con-

tributes to the stability of the soil aggregates, peat and undecomposed haulm merely protect the soil by acting like a mulch and do little to increase aggregate strength (Ekwue, Ohu and Wakawa 1993). Thus peat soils have very low aggregate stability.

Chemically, the most important control over erodibility is the proportion of easily dispersible clays in the soil. As seen above, a high proportion of exchangeable sodium can cause rapid deterioration in a soil's structure on wetting with consequent loss of strength, followed by the formation of a surface crust and decline in infiltration as the detached clay particles fill the pore spaces in the soil (Shainberg and Letey 1984). The addition of sodium-containing fertilizers to support crops such as tobacco can sometimes lead to quite small increases in exchangeable sodium yet result in very marked structural deterioration of a previously stable soil (Miller and Sumner 1988). Excess calcium carbonate within the clay and silt fractions of the soil also leads to high erodibility and appears to be the most important factor affecting the susceptibility of soils to erosion in southeast Spain (Barahona et al. 1990) and Morocco (Merzouk and Blake 1991).

Many attempts have been made to devise a simple index of erodibility based either on the properties of the soil as determined in the laboratory or the field, or on the response of the soil to rainfall and wind (Table 3.5). In a review of the indices related to water erosion, Bryan (1968) favours aggregate stability as the most efficient index. He uses the proportion of water-stable non-primary aggregates larger than 0.5 mm contained in the soil as an indicator of erodibility; the greater the proportion, the more resistant is the soil to erosion. A more commonly-used index is the K value which represents the soil loss per unit of EI_{30}, as measured in the field on a standard bare soil plot, 22 m long and at 5° slope. Estimates of the K value may be made if the grain-size distribution, organic content, structure and permeability of the soil are known (Wischmeier, Johnson and Cross 1971; Fig. 3.1).

Soil erodibility has been satisfactorily described by the K value for many agricultural soils in the USA (Wischmeier and Smith 1978) and the index has been found appropriate for some ferrallitic and ferruginous soils in West Africa (Roose 1977). Where K values have been determined from field measurements of erosion, they are valid. Difficulties arise, however, with attempts to predict the values from the nomograph (Fig. 3.1). Where it is applied to soils with similar characteristics to those in the USA, a close correlation exists between predicted and measured values, as found by Ambar and Wiersum (1980) on soils in west Java, Indonesia. Poorer predictions are

Table 3.5 Indices of soil erodibility for water erosion

Static Laboratory Tests

Dispersion ratio	$\dfrac{\%\ silt\ +\ \%\ clay\ in\ undispersed\ soil}{\%\ silt\ +\ \%\ clay\ after\ dispersal\ of\ the\ soil\ in\ water}$	Middleton (1930)
Clay ratio	$\dfrac{\%\ sand\ +\ \%\ silt}{\%\ clay}$	Bouyoucos (1935)
Surface aggregation ratio	$\dfrac{surface\ area\ of\ particles\ >0.05\,mm}{(\%\ silt\ +\ \%\ clay\ in\ dispersed\ soil)\ -\ (\%\ silt\ +\ \%\ clay\ in\ undispersed\ soil)}$	André and Anderson (1961)
Erosion ratio	$\dfrac{dispersion\ ratio}{colloid\ content/moisture\ equivalent\ ratio}$	Lugo-Lopez (1969)
Instability index (Is)	$\dfrac{\%\ silt\ +\ \%\ clay}{Ag_{air}\ +\ Ag_{alc}\ +\ Ag_{benz}}$ where Ag is the % aggregates >0.2 mm after wet sieving for no pretreatment and pretreatment of the soil by alcohol and benzene respectively	Hénin, Monnier and Combeau (1958)
Instability index (Is)	$\dfrac{\%\ silt\ +\ \%\ clay}{(\%\ aggregates\ >0.2\,mm\ after\ wet\ sieving)\ -0.9\ (\%\ coarse\ sand)}$	Combeau and Monnier (1961)
Pseudo-textural aggregation index (Ipta)	$\dfrac{MWDw\ -\ MWDt}{X\ -\ MWDt}\cdot 100$ where $MWDw$ is the mean weight diameter of the wet-sieving grain size distribution (mm), $MWDt$ is the mean weight diameter of the primary particle grain-size distribution (mm) and X is the maximum average grain-size diameter of the particles in the given grain-size distribution	Chisci, Bazzoffi and Mbagwu (1989)

Static Field Tests

Erodibility index	$\dfrac{1}{mean\ shearing\ resistance\ \times\ permeability}$	Chorley (1959)
Soil cohesion	direct measure of soil cohesion at saturation using a torvane	Rauws and Govers (1988)

Dynamic laboratory tests

Simulated rainfall test	comparison of erosion of different soils subject to a standard storm	Woodburn and Kozachyn (1956)
Water-stable aggregate (*WSA*) content	% *WSA* >0.5 mm after subjecting the soil to rainfall simulation	Bryan (1968)
Water drop test	% aggregates destroyed by a pre-selected number of impacts by a standard raindrop (e.g. 5.5 mm diameter, 0.1 g from a height of 1 m)	Bruce-Okine and Lal (1975)
Erosion index	$\dfrac{dh}{a}$ where d is an index of dispersion (ratio of % particles >0.05 mm without dispersion to % particles >0.05 mm after dispersion of the soil by sodium chloride); h is an index of water-retaining capacity (water retention of soil relative to that of 1 g of colloids); and a is an index of aggregation (% aggregates >0.25 mm after subjecting the soil to a water flow of 100 cm/min for 1 h)	Voznesensky and Artsruui (1940)

Dynamic Field Tests

Erodibility index (*K*)	mean annual soil loss per unit of EI_{30}	Wischmeier and Mannering (1969)

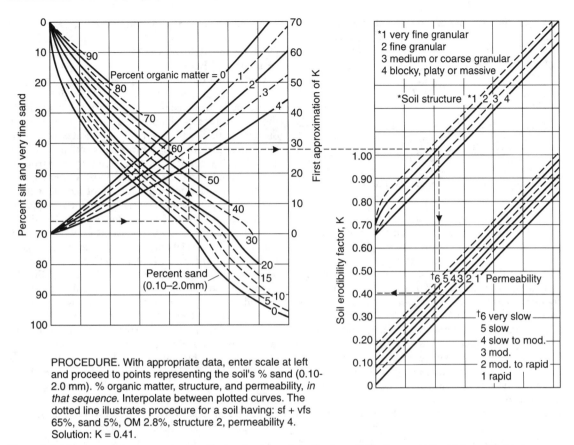

PROCEDURE. With appropriate data, enter scale at left and proceed to points representing the soil's % sand (0.10-2.0 mm). % organic matter, structure, and permeability, *in that sequence*. Interpolate between plotted curves. The dotted line illustrates procedure for a soil having: sf + vfs 65%, sand 5%, OM 2.8%, structure 2, permeability 4. Solution: K = 0.41.

Figure 3.1 Nomograph for computing the K value (metric units) of soil erodibility for use in the Universal Soil Loss Equation (after Wischmeier, Johnson and Cross 1971). Divide values by 1.3 to obtain K values in the original American units.

obtained where it is necessary to extrapolate the nomograph values. This applies to soils with organic contents above 4 per cent, swelling clays and those where resistance to erosion is a function of aggregate stability rather than primary particle size. Assessments of erodibility of many clay soils based on the destruction of aggregates, for example by the water-drop test, show little correlation with the K values estimated from the nomograph (De Meester and Jungerius 1978; Bergsma and Valenzuela 1981). Lindsay and Gumbs (1982) found the K index superior to the drop test, however, for assessing the erodibility of clay and clay loam soils in Trinidad.

The resistance of the soil to wind erosion depends upon dry rather than wet aggregate stability and on the moisture content, wet soil being less erodible than dry soil, but is otherwise related to much the same properties as affects its resistance to water erosion.

Chepil (1950), using wind tunnel experiments, has related wind erodibility of soils to various indices of dry aggregate structure but little of the work has been tested in field conditions. Wind tunnels have been used to determine the erodibility in $t\,ha^{-1}\,h^{-1}$ of soils in western Siberia and northern Kazakhstan for a range of windspeeds (Dolgilevich, Sofronova and Mayevskaya 1973). Similar studies by Chepil (1960) have been extrapolated to give an index in $t\,ha^{-1}\,y^{-1}$ based on climatic data for Garden City, Kansas. Values range from 84 to 126 for non-calcareous silty clay loams, silt loams and loams. They rise to 190 for sandy loams, clays and silty clays, 300 for loamy sand and 356–694 for sands. Both indexes are closely correlated with the percentage of dry stable aggregates larger than 0.84 mm (Table 3.6).

The indexes described above treat soil erodibility as constant over time. They thus ignore changes in

Table 3.6 Assessments of soil erodibility by wind

% dry stable aggregates >0.84 mm	>80	70–80	50–70	20–50	<20
erodibility (t ha^{-1} h^{-1})*	<0.5	0.5–1.5	1.5–5	5–15	>15
erodibility (t ha^{-1} y^{-1})†	<4	4–84	84–166	166–220	>220

* After Dolgilevich, Sofronova and Mayevskaya (1973) for windspeeds of 20–25 m/s.
† After Chepil (1960) for Garden City, Kansas.

organic content, aggregate structure, distribution of soluble and exchangeable ion content, moisture content and surface soil texture, particularly over short time periods. The most significant seasonal variations on agricultural land are associated with tillage operations which alter the bulk density and hydraulic conductivity of the soil. Martin and Morgan (1980) observed erodibility to be four times higher in summer than in winter on a bare uncultivated sandy soil in Bedfordshire, England, whilst Kwaad (1991) noticed erodibility in summer to be twice that in winter on silts and silt loam soils in Limbourg, The Netherlands. Although the seasonal variability reflects the different responses of the soils to intense convectional rainfall in summer producing runoff as Hortonian overland flow on the one hand and to low-intensity frontal rainfall in winter producing saturation overland flow on the other (Morgan, Martin and Noble 1986; Kwaad 1993), the true seasonal pattern in erodibility is likely to be more complex. Imeson and Kwaad (1990) show how the structural status of the Limbourg soils under maize declines rapidly following tillage in early May as a result of slaking and the development of a surface crust. Both the speed and extent of decline vary from one year to another depending upon the intensity and the amount of rain, the decline being greater when May and early June are wet. Offsetting this decline, however, is a change in aggregate stability of the soil which is lowest in May–June–July and highest in August–September, a change which may be related to bacterial activity in the soil.

Freezing and thawing also alter the erodibility of the soil through its effect on bulk density, hydraulic conductivity, shear strength and aggregate stability. Conditions of low bulk density and high soil moisture during periods of thaw produce a surface that is highly erodible. On agricultural soils in Ontario, Canada, the average value of the index, 1/shear strength, was 15 times higher in winter thaw conditions than in summer (Coote *et al.* 1988). Similar seasonally high erodi-

bility has been observed on loessial soils in Iowa (Spomer and Hjelmfelt 1983), in southwest Quebec (Kirby and Mehuys 1987) and southern Sweden (Alström and Bergman 1990).

Under natural conditions, the seasonal activity of burrowing animals is important, giving rise to considerable disturbance of the soil. Earthworms bring to the surface as casts as much as 2 to 5.8 t/ha of material on agricultural land (Evans 1948). Rates of 15 t/ha have been measured in temperate woodland in Luxembourg (Hazelhoff *et al.* 1981) and 50 t/ha in tropical forest in the Ivory Coast (Roose 1976). Other animals and their annual rates of production of sediment at the soil surface include ants, with 4 to 10 t/ha observed in Utah, USA (Thorp 1949); termites, with 1.2 t/ha in tropical forest in the Ivory Coast (Roose 1976); voles and moles, with 19 t/ha in temperate woodland in Luxembourg (Imeson 1976) and 6 t/ha from Pyrenean mountain voles in the Spanish Pyrenees (Borghi, Giannoni and Martinez-Rica 1990); and isopods and porcupines, with 0.03 to 0.7 t/ha on stony land in the Negev Desert, Israel (Yair and Rutin 1981). On coastal sand dunes in the western Netherlands, rabbits locally displaced between 0.09 and 5.1 t/ha of sediment from their burrows (Rutin 1992). In many cases, the material brought to the surface comprises loose sediment with low bulk density and cohesion which is rapidly broken down by splash erosion. The material contained in earthworm casts, however, consists of soil aggregates which are more stable under raindrop impact than the corresponding top soil, probably as a result of their higher organic content and secretions from the gut of the worm. Thus, earthworms have a positive effect on the stability of the soil aggregates and the hydraulic conductivity of the soil (Glasstetter and Prasuhn 1992).

3.3 Effect of Slope

Erosion would normally be expected to increase with increases in slope steepness and slope length as a result of respective increases in velocity and volume of surface runoff. Further, whilst on a flat surface raindrops splash soil particles randomly in all directions, on sloping ground more soil is splashed downslope than upslope, the proportion increasing as the slope steepens. The relationship between erosion and slope can be expressed by the equation:

$$E \propto tan^m \theta \, L^n \qquad (3.11)$$

where E is soil loss per unit area, θ is the slope angle and L is slope length. Zingg (1940), in a study of data from five experimental stations of the United States

Soil Conservation Service, found that the relationship had the form:

$$E \propto tan^{1.4}\theta\ L^{0.6} \qquad (3.12)$$

To express E proportional to distance downslope, the value of n must be increased by 1.0. Since the values for the exponents have been confirmed in respect of m by Musgrave (1947) and m and n by Kirkby (1969b), there is some evidence to suggest that Eq. 3.12 has general validity. Other studies, however, show that the values are sensitive to the interaction of other factors.

Exponents for slope steepness

Working with data from experimental stations in Zimbabwe, Hudson and Jackson (1959) found that m was close to 2.0 in value, indicating that the effect of slope is stronger under tropical conditions where rainfall is heavier. The effect of soil is illustrated by the laboratory experiments of Gabriels, Pauwels and De Boodt (1975) who show that the value of m increases with the grain size of the material, from 0.6 for particles of 0.05 mm to 1.7 for particles of 1.0 mm. The value of m may also be sensitive to slope itself, decreasing with increasing slope steepness. Horváth and Erődi (1962) found from their laboratory studies that $m = 1.6$ for slopes between 0° and 2.5°, 0.7 for slopes between 3° and 6.5°, and 0.4 for slopes over 6.5°. On still steeper slopes the value may be expected to decrease further as soil-covered slopes give way to rock surfaces and soil supply becomes a limiting factor. Heusch (1970) obtained a value of -3.8 for slopes of 6.5° to 33° in Morocco, attributing the negative exponent to decreasing surface runoff and increasing subsurface flow on the steeper slopes. In a detailed study of soil loss from 33 road cut-slopes on the Benin–Lagos Highway in Nigeria, Odemerho (1986) obtained values of $m = 1.09$ for slopes between 1.4° and 6°, 1.80 for slopes between 6° and 8.5°, -2.18 for slopes between 8.5° and 11°, and -1.39 for slopes between 11° and 26.5°. Combining the results of these studies suggests a curvilinear relationship between soil loss and slope steepness with erosion initially increasing rapidly as slope increases from gentle to moderate, reaching a maximum on slopes of about 8–10° and then decreasing with further increases in slope. Such a relationship would apply only to erosion by rainsplash and surface runoff; it would not apply to landslides, piping or to gully erosion by pipe collapse.

The exponents in Eq. 3.12 also vary in value with slope shape. Examining soil loss from 3 m long plots on slopes of average steepness between 2° and 8° under simulated rainfall, D'Souza and Morgan (1976) obtained values of m of 0.5 for convex slopes, 0.4 for straight slopes and 0.14 for concave slopes. No studies have been made of the influence of slope in plan, but Jackson (1984) found from erosion surveys and laboratory experiments that discharge varies with an index of contour curvature to the power of 5.5. If soil loss is assumed to vary with the square of the discharge, the value of m becomes 3.5. Contour curvature is here defined as the proportion of a circle centred on a point on a hillside that lies at a higher altitude than that point. The index ranges from 0 to 1 in value with values < 0.5 indicating diverging slopes, a value of 0.5 a straight slope and values > 0.5 converging slopes.

Few studies have examined the effect of varying plant cover and vegetation densities on the exponent values. Quinn, Morgan and Smith (1980) investigated the change in the value of m for soil loss from 1.2 m long plots with slopes of 5° to 30° under simulated rainfall in relation to decreasing grass cover brought about by human trampling. They found that $m = 0.7$ for fully grassed straight slopes, rose to 1.9 in the early stages of trampling before falling to 1.1 when only about 25 per cent of the grass cover remained. Lal (1976) obtained a value of $m = 1.1$ for both bare fallow and maize on erosion plots in Nigeria; use of a mulch, however, reduced m to 0.5.

Kirkby (1971) noted variations in the exponent values with the process of erosion. Values of m are 1.0 for soil creep, range between 1.0 and 2.0 for splash erosion and between 1.3 and 2.0 for erosion by overland flow, and may be as high as 3.0 for rivers. It was shown in Chapter 2 that m was about 0.3 to 1.0 in value for rainsplash and about 0.7 and 1.7–2.0 respectively for detachment and transport of soil particles by overland flow. Increases in slope steepness also cause an increase in the intensity of wind erosion on windward slopes and on the crests of knolls. Data presented by Chepil, Siddoway and Armbrust (1964) and Stredňanský (1977) show that $m = 0.4$ for slopes up to 2° and $m = 1.2$ for slopes from about 2° to 15°.

Exponents for slope length

The value of 0.6 for exponent n applies only to overland flow on slopes about 10–20 m long, with steepnesses greater than 3°. Wischmeier and Smith (1978) propose values of $n = 0.4$ for slopes of 3°, 0.3 for slopes of 2°, 0.2 for slopes of 1° and 0.1 for slopes of less than 1°. Kirkby (1971) suggests $n = 0$ for soil creep and splash erosion, ranges between 0.3 and 0.7 for overland flow and rises to between 1.0 and 2.0 if rilling occurs. This implies that the value of n is likely to vary with distance along a hillside as, for example, soil creep close to the summit gives way first to overland flow and then to rill flow. Without rills, n may become negative on slopes longer than about

10 m. The increasing depth of overland flow down-slope protects the soil from raindrop impact so that, even though the transporting capacity of the overland flow increases, erosion becomes limited by the rate of detachment which is decreasing with slope length (Gilley, Woolhiser and McWhorter 1985). Once rills form, soil loss will either increase with slope length (Meyer, Foster and Römkens 1975), particularly if the density of rills is very high, or decrease because, as the flow becomes concentrated, there is no longer suffi-cient flow on the interrill areas to remove all the material detached by rainsplash (Abrahams, Parsons and Luk 1991). Erosion may also decrease with increasing slope length if, as the slope steepens, the soil becomes less prone to crusting and infiltration rates remain higher than on the gentler-sloping land at the top of the slope (Poesen 1984). Similarly, if the slope declines in angle as length increases, soil loss may decrease as a result of deposition. Clearly, with such a great range of possible conditions, a single relationship between soil loss and slope length cannot exist.

3.4 Effect of Plant Cover

Vegetation acts as a protective layer or buffer between the atmosphere and the soil. The above-ground com-ponents, such as leaves and stems, absorb some of the energy of falling raindrops, running water and wind, so that less is directed at the soil, whilst the below-ground components, comprising the root system, con-tribute to the mechanical strength of the soil.

The importance of a plant cover in reducing erosion is demonstrated by the mosquito gauze experiment of Hudson and Jackson (1959) in which soil loss was compared from two identical bare plots on a clay loam soil. Over one plot was suspended a fine wire gauze which had the effect of breaking the force of the raindrops, absorbing their impact and allowing the water to fall to the ground from a low height as a fine spray. The mean annual soil loss over a 10-year period was 126.6 t/ha for the open plot and 0.9 t/ha for the plot covered by gauze. Similar results were obtained in a study carried out on a clay soil in Toscana, Italy (Zanchi 1983). The mean annual soil losses over a 6-year period were 43 t/ha for the bare plot and 3.8 t/ha for the covered plot.

Although numerous measurements have been made of erosion under different plant covers for comparison with that from bare ground, only a few researchers have examined the relationship between soil loss and changes in the extent of cover. Elwell (1981) favours an exponential decrease in soil loss with increasing percentage interception of rainfall energy and, there-fore, increasing percentage canopy cover. Such a relationship was suggested by Wischmeier (1975) as applicable to covers in direct contact with the soil surface and has been verified experimentally for crop residues (Laflen and Colvin 1981; Hussein and Laflen 1982) and grass covers (Lang and McCaffrey 1984; Morgan, McIntyre, Vickers, Quinton and Rickson 1994). Foster (1982) attributes the exponential form of the relationship for covers in proximity to the ground to the ponding of water behind the plant elements which reduces the effectiveness of raindrop impact. Since localized depressions exist on most soil surfaces and these are often filled or partially filled during rainstorms, especially where they lie beneath leaf drips, an exponential relationship could be expec-ted to apply to canopy effects too. Such relationships have been observed for *matorral* cover in Murcia, southeast Spain (Francis and Thornes 1990) and for tropical rain forest near Chieng-Mai, Thailand (Ruangpanit 1985).

Effect on rainfall

The effectiveness of a plant cover in reducing erosion by raindrop impact depends upon the height and continuity of the canopy, and the density of the ground cover. The height of the canopy is important because water drops falling from 7 m may attain over 90 per cent of their terminal velocity. Further, rain-drops intercepted by the canopy may coalesce on the leaves to form larger drops which are more erosive. Brandt (1989) shows that, for a wide range of plant types, leaf drips have a mean volume drop diameter between 4.5 and 4.9 mm, which is about twice that of natural raindrops. Raindrop-size distributions under a canopy are therefore characteristically bimodal with peaks at around 2 mm and 4.8 mm, corresponding to the direct throughfall and the leaf drainage respec-tively. The effects of these changes in drop-size dis-tribution have been studied mainly in relation to forest canopies. Chapman (1948) under pine forest in the USA, Wiersum, Budirijanto and Rhomdoni (1979) under *Acacia* forest in Indonesia, Mosley (1982) under beech forest in New Zealand, and Vis (1986) under tropical rain forest in Colombia all show that whilst interception by the canopy reduces the volume of rain reaching the ground surface, it does not significantly alter its kinetic energy which may even be increased compared with that in open ground. Mosley (1982) found that in a storm of 51 mm in 36 hours beginning on 8 July 1980, the amount of material detached by rainsplash under the beech canopy was 3.1 times that in open ground. Wiersum (1985) recorded 1.2 times as much detachment under the *Acacia* canopy as in open ground over 27 rain days with a total rainfall of 402 mm. Vis (1986) found that detachment under the rain forest was 2 to 16 per cent higher than that outside of the forest in a 15-month period covering three wet and two dry seasons. This evidence suggests that increasing tree canopies will bring about greater rates of detachment unless, as also observed by Wiersum (1985), the ground is protected by a litter layer.

Fewer investigations have been made to assess the effects of lower growing canopies. McGregor and Mutchler (1978) found that whilst cotton reduced the kinetic energy of the rainfall by 95 per cent under the canopy and 75 per cent overall, it was locally increased between the rows where the leaf drips were concentrated. Armstrong and Mitchell (1987) found that the detachment under soya bean was about 94 per cent of that in the open despite the very low canopy height. Some 40 per cent of the rainfall under the canopy was more erosive than that in open ground and 10 per cent was over three times as erosive. Somewhat lower detachment rates were measured in the field by Morgan (1985b) who found that, with a 90 per cent cover, detachment under soya bean was 0.2 times that in open ground for a 100 mm/h rainfall intensity and 0.6 times for a 50 mm/h intensity. Finney (1984) showed in a laboratory study that detachment rates from leaf drip were 1.7 and 1.3 times those in open ground for Brussels sprouts and sugar beet at 23 per cent and 16 per cent canopy cover respectively. In another laboratory study under Brussels sprouts, Noble and Morgan (1983) found that the average detachment rate under canopy covers between 10 and 25 per cent was the same as that in open ground. In a field study under maize Morgan (1985b) found that detachment with 88 per cent canopy cover at a height of 2 m was 14 times greater than that in open ground for a rainfall intensity of 100 mm/h and 2.4 times greater for an intensity of 50 mm/h.

In addition to modifying the drop-size distribution of the rainfall, a plant canopy changes its spatial distribution at the ground surface. Concentrations of water at leaf drip points can result in very high localized rainfall intensities which can considerably exceed infiltration capacities and play an important role in the generation of runoff. Under mature soya bean, Armstrong and Mitchell (1987) found that half of the rainfall reaching the ground in a storm of 25 mm/h did so at intensities greater than those in open ground and that 10 per cent of the rain had an intensity of 385 mm/h. Stemflow also concentrates rainfall at the ground surface. De Ploey (1982) found that the effective intensity of stemflow beneath a canopy of tussocky grass was 150 to 200 per cent greater than the rainfall intensity in open ground. Herwitz (1986) recorded local stemflows of between 830 and 18,878 mm/h in a rain storm of 118 mm/h in 6 minutes in a tropical rain forest in northern Queensland.

Effect on runoff

A plant cover dissipates the energy of running water by imparting roughness to the flow, thereby reducing its velocity. In most soil conservation work, the roughness is expressed as a value of Manning's n which represents the summation of roughness imparted by the soil particles, surface microtopography (form roughness) and vegetation, acting independently of each other. Typical values of Manning's n are given in Table 3.7 (Petryk and Bosmajian 1975; Temple 1982; Engman 1986). The level of roughness with different forms of vegetation depends upon the morphology and the density of the plants, as well as their height in relation to the depth of flow. When the flow depth is shallow, as with overland flow, the vegetation stands relatively rigid and imparts a high degree of roughness, represented for grasses by n values of 0.25 to 0.3. As flow depths increase, the grass stems begin to oscillate, disturbing the flow and causing n values to increase to around 0.4. With further increases in flow depth, the vegetation is submerged; the plants tend to lie down in the flow and offer little resistance, so that n values decrease rapidly (Ree 1949).

Greatest reductions in velocity occur with dense, spatially uniform, vegetation covers. Clumpy, tussocky vegetation is less effective and may even lead to concentrations in flow with localized high velocities between the clumps. When flow separates around a clump of vegetation, the pressure exerted by the flow is higher on the upstream face than it is downstream, and eddying and turbulence occur immediately downstream of the vegetation. Vortex erosion is induced both upstream and downstream (Babaji 1987). Detailed observations during laboratory experiments on overland flow (De Ploey 1981b) show that for slopes above about 8°, erosion under grass is higher than that from an identical plot without grass until the percentage grass cover reaches a critical value. Beyond this point, the grass has the expected protective effect.

Effect on air flow

Vegetation reduces the shear velocity of wind by imparting roughness to the air flow. Vegetation increases the roughness length, z_0 and raises the height of the mean aerodynamic surface by a distance, d, known as the zero plane displacement (Fig. 2.5). Estimates of d and z_0, however, can be obtained from the relationships:

$$d = H . F \qquad (3.13)$$

$$z_0 = 0.13 (H - d) \qquad (3.14)$$

where H is the average height of the roughness elements and F is the fraction of the total surface covered

Table 3.7 Guide values for Manning's n

Land use or cover	Manning's n
Bare soil	
roughness depth <25 mm	0.010–0.030
roughness depth 25–50 mm	0.014–0.033
roughness depth 50–100 mm	0.023–0.038
roughness depth >100 mm	0.045–0.049
Bermuda grass – sparse to good cover	
very short (>50 mm)	0.015–0.040
short (50–100 mm)	0.030–0.060
medium (150–200 mm)	0.030–0.085
long (250–600 mm)	0.040–0.150
very long (>600 mm)	0.060–0.200
Bermuda grass – dense cover	0.300–0.480
Other dense sod-forming grasses	0.390–0.630
Dense bunch grasses	0.150
Kudzu	0.070–0.230
Lespedeza	0.100
Natural rangeland	0.100–0.320
Clipped rangeland	0.020–0.240
Wheat straw mulch	
2.5 t/ha	0.050–0.060
5.0 t/ha	0.075–0.150
7.5 t/ha	0.100–0.200
10.0 t/ha	0.130–0.250
Chopped maize stalks	
2.5 t/ha	0.012–0.050
5.0 t/ha	0.020–0.075
10.0 t/ha	0.023–0.130
Cotton	0.070–0.090
Wheat	0.100–0.300
Sorghum	0.040–0.110
Concrete or asphalt	0.010–0.013
Gravelled surface	0.012–0.030
Chisel-ploughed soil	
<0.6 t/ha residue	0.006–0.170
0.6–2.5 t/ha residue	0.070–0.340
2.5–7.5 t/ha residue	0.190–0.470
Disc-harrowed soil	
<0.6 t/ha residue	0.008–0.410
0.6–2.5 t/ha residue	0.100–0.250
2.5–7.5 t/ha residue	0.140–0.530
No tillage	
<0.6 t/ha residue	0.030–0.070
0.6–2.5 t/ha residue	0.010–0.130
2.5–7.5 t/ha residue	0.160–0.470
Bare mouldboard-ploughed soil	0.020–0.100
Bare soil tilled with coulter	0.050–0.130

After Petryk and Bosmajian (1975), Temple (1982) and Engman (1986).

by those elements (Abtew, Gregory and Borelli 1989).

The effect of the vegetation can be described by a frictional drag coefficient (Cd) exerted by the plant layer in bulk and computed from:

$$Cd = \frac{2u^{\star 2}}{u^2} \qquad (3.15)$$

where u is the mean velocity measured at a height z which equals 1.6 times the average height of the roughness elements. The coefficient generally decreases in value from about 0.1 in light winds to 0.01 in strong winds for a wide range of crops (Uchijima 1976; Ripley and Redman 1976; Morgan and Finney 1987) but both Randall (1969) in apple orchards and Bache (1986) with cotton canopies found that Cd could also increase with windspeed. When the bulk drag coefficient exceeds 0.0104 in value, no regional-scale wind erosion will occur (Lyles, Schrandt and Schmeidler 1974).

Instead of considering these bulk drag coefficients, more insight can be gained by examining conditions close to the ground surface. Contrary to expectations, wind velocity is rarely zero below the zero plane displacement but is at a low level and generally constant with height until very close to the ground surface when, because of the sparser vegetation cover at the bottom of the plant layer, it may increase slightly before finally reducing to zero (Landsberg and James 1971). Drag coefficients within the plant layer (Cd') can be calculated from:

$$Cd' = \frac{2 u^{\star 2}}{\int_0^h u^2 A(z) \, dz} \qquad (3.16)$$

where h is the height of the vegetation, $A(z)$ is the leaf area per unit volume for the vegetation at height z and dz is the difference in height between z and the ground surface. For a wide range of crops, values of Cd' within the lowest 0.5 m of the plant layer decrease from about 0.1 in low windspeeds to about 0.001 in high windspeeds. However, when the wind is moderate to strong and consistent over time and the crops are at an early stage of growth, the drag coefficient is found to increase with windspeed (Morgan and Finney 1987). This is probably due to waving of the leaves disturbing the surrounding air and creating a wall effect which acts as a barrier to the air flow. The result is that the windspeed is reduced close to the ground surface but remains the same or even increases at the canopy level, thereby increasing the drag or shear velocity and enhancing the risk of erosion. The effect is particularly marked in crops of young sugar beet and onions. Similar increases in the drag coefficient with windspeed within a crop have been reported for maize by Wright and Brown (1967).

Effect on slope stability

It was shown in Chapter 2 that forest covers generally help to protect the land against mass movements partly through the cohesive effect of the tree roots. The fine roots, 1–20 mm in diameter, interact with the soil to form a composite material in which root fibres of relatively high tensile strength reinforce a matrix of lower tensile strength. In addition, soil strength is increased by the adhesion of soil particles to the roots. Roots can make significant contributions to the cohesion of a soil, even at low root densities and in materials of low shear strength. Increases in cohesion in forest soils due to roots can range from 1.0 to 17.5 kPa (Greenway 1987) although the local variability in this may be as high as 30 per cent (Wu 1995). Grasses, legumes and small shrubs can reinforce a soil down to depths of 0.75 to 1.0 m and trees can enhance soil strength to depths of 3 m or more. The magnitude of the effect depends upon the angle at which the tree roots cross the potential slip plane, being greatest for those at right-angles, and whether the strain exerted on the slope is sufficient to mobilize fully the tensile strength of the roots. The effect is limited where roots fail by pull-out because of insufficient bonding with the soil, as can occur in stony materials, or where the soil is forced into compression instead of tension, as can occur at the bottom of a hillslope, and the roots fail by buckling.

A vegetation cover should theoretically contribute to slope stability as a result of evapotranspiration producing a drier soil environment so that a higher intensity and longer duration rainfall are required to induce slope failure compared with an unvegetated slope (Greenway 1987). Also, since soil moisture depletion can affect depths well below those reached by the roots, increases in slope stability should extend some 4–6 m below ground level. In practice, however, as found by Terwilliger (1990) in southern California, soil moisture levels reach similar levels after a few storms, regardless of the presence or absence of vegetation, so that under the conditions when the risk of mass movement is highest, the drying effect of vegetation is unlikely to play a role.

Following observations on the forested slopes of the Serra do Mar, east of Santos, Brazil, De Ploey (1981b) proposed that trees could sometimes have a triggering effect on landslides through an increase in loading (surcharge) brought about by their weight and an increase in infiltration which allows more water to penetrate the soil, lowering its shear strength. Bishop and Stevens (1964) show that large trees can increase the normal stress on a slope by up to 5 kPa but that less than half of this contributes to an increase in shear stress and the remainder has the beneficial effect of increasing the frictional resistance of the soil. Whilst generally, surcharge enhances slope stability, under certain circumstances it can be detrimental. Trees planted only at the top of a slope can reduce stability, as can trees planted on steep slopes with shallow soils characterized by low angles of internal friction. In the case of the Serra do Mar, the landslides occurred in a soil with an angle of internal friction of less than 20°, on slopes greater than 20°, and after two days on which respectively 260 mm and 420 mm of rain fell.

Very few studies have attempted to quantify the effect of vegetation on slope stability. Theoretically, increases in the factor of safety of a slope of 55–60 per cent have been computed (Bache and MacAskill 1984; Styczen and Morgan 1995) but, in reality, the increase is more likely to be in the 20–30 per cent range (Greenway 1987; Wu 1995).

Summary review

Plant covers can play an important role in reducing erosion provided that they extend over a sufficient proportion of the soil surface. Overall, forests are the most effective but a dense growth of grass may be almost as efficient and quicker to obtain. Agricultural crops vary in their effectiveness depending on their stage of growth and the amount of bare ground exposed to erosion. For adequate protection at least 70 per cent of the ground surface must be covered (Elwell and Stocking 1976), but reasonable protection can be achieved with 40 per cent cover. The effects of vegetation are clearly not straightforward, however, and under certain conditions a plant cover can exacerbate erosion. When using plant covers as a basis for erosion control, it is vital that these conditions be understood and specified.

CHAPTER 4

Erosion Hazard Assessment

The assessment of erosion hazard is a specialized form of land resource evaluation, the objective of which is to identify those areas of land where the maximum sustained productivity from a given land use is threatened by excessive soil loss. The assessment aims at dividing a land area into regions, similar in their degree and kind of erosion hazard, as a basis for planning soil conservation work.

4.1 Generalized Assessments

4.1.1 Erosion intensity

Two indexes of erosion intensity were used by Morgan (1974) to assess erosion risk in Peninsular Malaysia. They were drainage density, defined as the length of streams per unit area, and drainage texture, defined as the number of first-order streams per unit area which, being equivalent to the density of source points, is analogous to gully density. The two indexes are virtually uncorrelated with each other and probably relate to different controlling mechanisms (Morgan 1976). High values of drainage density are associated with the transport of runoff from regular, moderate rainfalls whereas high values of texture are a response to a more seasonal rainfall regime with rains of greater intensity. Drainage density is perhaps better regarded as an index of runoff and drainage texture as an index of erosion.

A similar technique is to map the density of gullies directly. Józefaciuk and Józefaciuk (1993) used this method, plotting the gullies on to 1:25,000 scale maps and then generalizing the information to produce a map for Poland. They chose a density > 0.5 km/km^2 as representing areas where erosion protection measures were needed. Where sufficient data are available on sediment yields in rivers they may also be used to map regional variations in erosion although, as indicated in Chapter 1, they may be unreliable as statements of the rate of erosion on hillslopes. Maps have been prepared from sediment yield information for the countries of former Yugoslavia (Jovanović and Vukčević 1958), Romania (Diaconu 1969) and South Africa (Rooseboom and Annandale 1981).

4.1.2 Using erosivity indexes

Erosivity data can be used as an indicator of regional variations in erosion potential. Stocking and Elwell (1976) present a generalized picture of erosion risk in Zimbabwe, based on mean annual erosivity values, showing that high-risk areas are in the Eastern Districts, the region east of Masvingo, and the High and Middle Veld north and east of Harare (Fig. 4.1). The area south and west of Bulawayo, by contrast, has a much lower risk. Temporal variations in erosion risk are revealed by the mean monthly erosivity values. Hudson (1981; Fig. 4.2) contrasts the erosivity patterns of Bulawayo and Harare. At Bulawayo, erosivity is low at the beginning of the wet season and increases as the season progresses. By the time the maximum values are experienced, the plant cover has had a chance to become established, giving protection against erosion and lowering the risk. At Harare, however, erosivity is highest at the time of minimal vegetation cover. The seasonal pattern of erosivity was also analysed in Kenya by Rowntree (1983) taking data for the Katumani Research Station near Machakos. The mean annual erosivity of 164,352 J mm m^{-2}h^{-1} is much higher than the values recorded in Zimbabwe. The highest mean monthly value, 69,688 J mm m^{-2}h^{-1}, occurs in April, the beginning of the period when the potential soil moisture deficit is severe and the vegetation cover poorly established. Cultivation of the land at this time creates a great risk of erosion. May has a much lower erosivity, only 20,961 J mm m^{-2}h^{-1}, and is also a month with a good vegetation cover. Erosivity is very low during the summer with mean monthly values of less than 1,000 but rises again in the short rainy season, reaching 15,306 in November before falling to mean monthly values of about 10,000 during the winter.

Maps of erosivity using the rainfall erosion index, R, have been produced for the USA (Wischmeier and Smith 1978). Originally R was calculated as mean annual $EI_{30}/100$ with E being the rainfall energy in foot-tons per acre and I_{30} being the maximum 30-minute rainfall intensity in inches per hour; this can be described as an R value in American units.

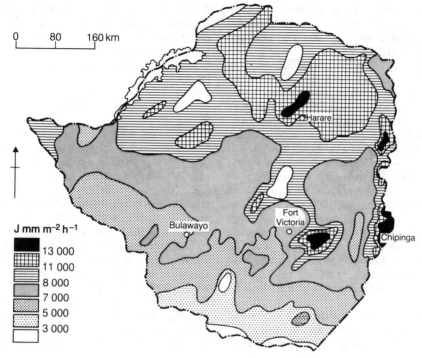

Figure 4.1 Mean annual erosivity in Zimbabwe (after Stocking and Elwell 1976).

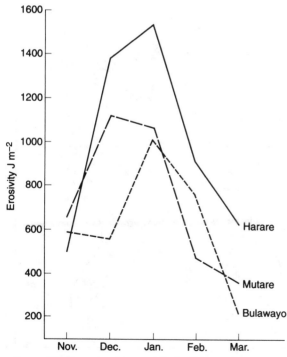

Figure 4.2 Mean monthly erosivity for 3 towns in Zimbabwe (after Hudson 1965).

With E calculated in J/m^2 and I_{30} in mm/h, an R value can be derived in metric units as mean annual $EI_{30}/1,000$. The conversion is 1 unit of metric R = 1.73 units of American R. Since maps for many parts of the world exist in both units, care should be taken to ensure which one is being used. The metric version of the map for the USA is shown in Fig. 4.3. Values range from about 50 in the arid west to over 1,000 along the Gulf Coast. These are rather low compared with those occurring in more tropical climates. Values in India range from about 250 in western Rajasthan to over 1,250 along the coasts of Maharashtra and Karnataka (Singh, Babu and Chandra 1981; Fig. 4.4). Those in Zambia range from 500 in the south to about 800–900 in the Copper Belt and over 1,000 in the western part of Northern Province (Lenvain, Sakala and Pauwelyn 1988). Very low values are characteristic of western Europe. Those in France range from 50 in the north and west to 340 in the southern Massif Central and Languedoc (Pihan 1979). The values in northern France are matched by those in Belgium and The Netherlands where mean annual R values range between 60 and 75, rising to 100–150 in the Ardennes (Laurant and Bollinne 1978) and in eastern Poland where the values range from 43 on the northern plains to nearly 100 in the mountainous south (Banasik and Górski 1993). In South Africa (Smithen and Schulze

1982) values are lowest in the southwest of Cape Province where they are about 50 to 100 and rise eastwards reaching over 400 in eastern Transvaal and the midlands of Natal (Fig. 4.5).

In many countries insufficient rainfall records from autographic gauges are available to calculate erosivity nationwide. In such cases, an attempt is made, for the recording stations where erosivity can be determined, to find a more widely available rainfall parameter which significantly correlates with erosivity and from which erosivity values might be predicted using a best-fit regression equation. Considerable caution should be taken when estimating erosivity in this way because the results are only valid if the rainfall parameters used are themselves significantly correlated with soil loss; otherwise the exercise is pointless (Hudson 1981). Roose (1975), examining data in the Ivory Coast and Burkina Faso, found that mean annual R values (in American units) could be approximated by the mean annual rainfall totals (mm) multiplied by 0.5. This relationship was used to produce a map of R values for west Africa.

A similar approach was adopted to compile and map data on mean annual erosivity for Peninsular Malaysia (Morgan 1974) using, for simplicity, the KE > 25 index. Rainfall records for ten stations of the Malaysian Meteorological Service with autographic rain gauges were examined for 1965, 1966 and 1969 and the following relationship established between mean annual erosivity (EVa) and mean annual rainfall (P; mm):

$$EVa = 9.28 \ P - 8838.15 \qquad r = 0.81 \quad (4.1)$$

This equation was used to produce a map of erosivity from mean annual rainfall data for the whole country (Fig. 4.6). It should be emphasized that whilst Eq. 4.1 appears to work well for Peninsular Malaysia, applying it to other countries is less satisfactory. With rainfall totals below 900 mm, the equation yields estimates of erosivity which are obviously nonsense.

Equations have also been developed relating mean annual precipitation to mean annual erosivity using the R index. The work in different countries is not always comparable, however, because of different assumptions made when calculating the R value. In Belgium, the equation relates mean annual R for all storms greater than 1.27 mm to mean annual precipitation (Bollinne, Laurant and Boon 1979). In Bavaria, Germany, the R values are calculated for

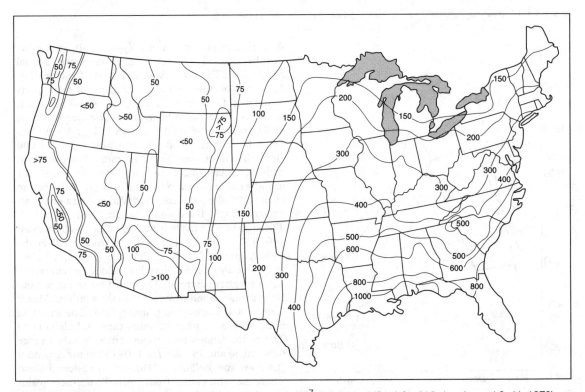

Figure 4.3 Mean annual values of the rainfall erosion index R (10^7 J/ha) in the USA (after Wischmeier and Smith 1978). Divide values by 1.7 to obtain R values in original American units (10^2 foot-tons/acre).

storms with total rainfalls greater than 10 mm and I_{30} values greater than 10 mm/h (Rogler and Schwertmann 1981).

Erosion risk in Great Britain was assessed by calculating and mapping values of the *KE > 10* index for each 10×10 km grid square of the grid referencing system used by the Ordnance Survey. The annual values are rather low (Morgan 1980b; Fig. 4.7), rising above $1,400\,\text{J/m}^2$ only in parts of the Pennines, the Welsh mountains, Exmoor and Dartmoor. They are less than $900\,\text{J/m}^2$ along most of the west coast and fall below 700 in the Outer Hebrides, Orkneys, Shetlands and on the north coast of Scotland. Values over much

of eastern and southern England are around 1,100 to 1,300. Since these are the main areas devoted to arable farming, it is here that the greatest risk of agricultural soil erosion occurs.

4.1.3 Rainfall aggressiveness

The most commonly used index, shown to be significantly correlated with sediment yields in rivers (Fournier 1960), is the ratio p^2/P, where p is the highest mean monthly precipitation and P is the mean annual precipitation. It is strictly an index of the concentration of precipitation into a single month and

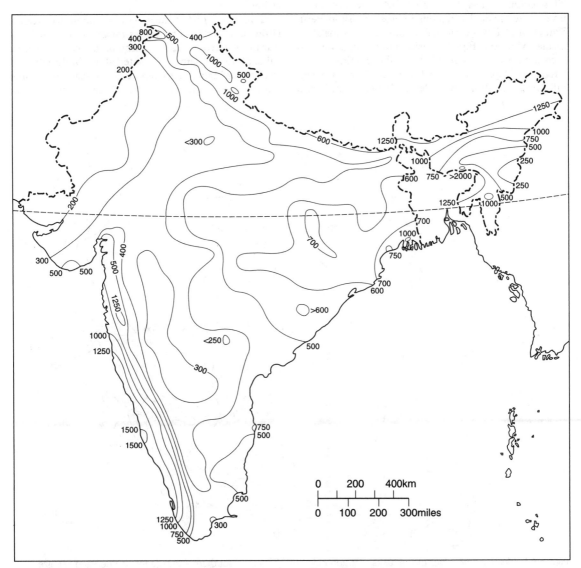

Figure 4.4 Mean annual values of the rainfall erosion index R (10^7 J/ha) in India (after Singh, Babu and Chandra 1981).

thereby gives a crude measure of the intensity of the rainfall and, in so far as a high value denotes a strongly seasonal climatic regime with a dry season during which the plant cover decays, of erosion protection by vegetation. Using data from 78 drainage basins, Fournier (1960) derived the following empirical relationship between mean annual sediment yield (Qs; g/m^2), mean altitude (H; m) and mean slope of the basin (S; degrees):

$$log\ Qs = 2.65\ log\ p^2/P + 0.46\ (log\ H)\ (tan\ S) - 1.56 \tag{4.2}$$

This equation has been used by Low (1967) to investigate regional variations in erosion risk in Peru. Values of p^2/P were calculated and mapped for Peninsular Malaysia (Fig. 4.8) using rainfall data for 680 recording stations compiled by the Drainage and Irrigation Department (1970). Since there is a significant correlation in Malaysia between p^2/P and

drainage texture (r = 0.38; n = 39; Morgan 1976) and, as seen in Section 4.1.1, drainage texture is analogous to gully density, p^2/P may be regarded as an indicator of the risk of gully erosion. In contrast, mean annual erosivity values reflect the risk of erosion by rainsplash, overland flow and rills. By superimposing the maps of p^2/P and erosivity, a composite picture of erosion risk over the whole country is obtained (Fig. 4.9). As there will always be at least a moderate risk of erosion in any tropical environment, the three categories of erosion risk recognized in this reconnaissance survey are designated severe, high and moderate, rather than introduce any category which is described as low.

A map of p^2/P values was prepared for Kenya (Rowntree 1983). The values range from 13 at Lodwar in the north to 123 at Lamu on the coast. Using a value of 50, the same as adopted above in the study of Peninsular Malaysia, to denote high erosion risk, three sensitive areas are identified. These are the coastal

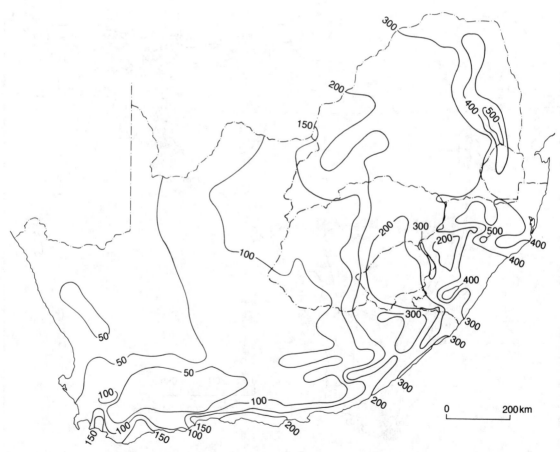

Figure 4.5 Mean annual values of the rainfall erosion index R (10^7 J/ha) in southern Africa (after Smithen and Schulze 1982).

belt, the Kitui-Yatta region and the area of the Aberdare Mountains and Mount Kenya.

4.1.4 Factorial scoring

A simple scoring system for rating erosion risk was devised by Stocking and Elwell (1973b) for Zimbabwe. Taking a 1:1,000,000 base map, the country is divided on a grid system into units of $184\,km^2$. Each unit is rated on a scale from 1 to 5 in respect of erosivity, erodibility, slope, ground cover and human occupation, the latter taking account of the density and type of settlement. The scoring is arranged so that 1 is associated with a low risk of erosion and 5 with a high risk. The five factor scores are summed to give a total score which is compared with an arbitrarily chosen classification system to categorize areas of low, moderate and high erosion risk. The scores are

mapped and areas of similar risk delineated (Fig. 4.10).

Several problems are associated with this technique. First, the classification may be sensitive to different scoring systems. For example, the use of different slope groups may yield different assessments of the degree of erosion risk. Second, each factor is treated independently whereas there is often interaction between the factors. For example, slope steepness may be much more important in areas of high than in areas of low erosivity. Third, the factors are combined by addition. There is no reason why this should be a more appropriate method of combining them than multiplication, although multiplication often results in the score for one factor dominating the total score and, for that reason, is difficult to use with zero values in the scoring system. Fourth, each factor is given equal weight. Despite these difficulties, the technique

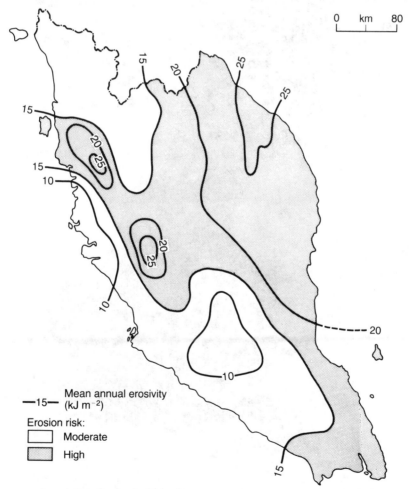

Figure 4.6 Mean annual erosivity in Peninsular Malaysia.

is easy to use and has the advantage that factors which cannot be easily quantified in any other way can be readily included.

The technique was adopted for an assessment of erosion risk in the Mediterranean countries of the European Union within the context of the CORINE programme (Briggs and Giordano 1992). Soil erodibility is evaluated in four classes, depending upon the texture, depth and stoniness of the soil; erosivity in three classes; and slope angle in four classes. The scores for these three factors are multiplied to give a combined score for evaluating the potential risk of soil erosion. By including an assessment of land cover, a further score is obtained which expresses the actual risk of erosion. Data are obtained for 1 km × 1 km grid squares and the information processed using the ARC/INFO Geographical Information System to provide maps of potential and actual erosion risk at a scale of 1:1,000,000. Although field checks in sample areas supposedly confirm that the broad patterns of erosion risk are realistic, no information is provided to support this. When sections of the map are examined in detail

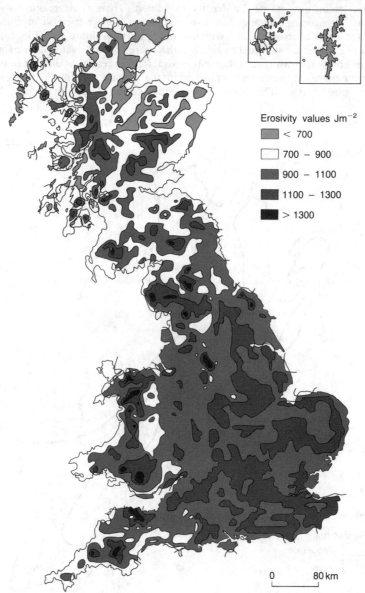

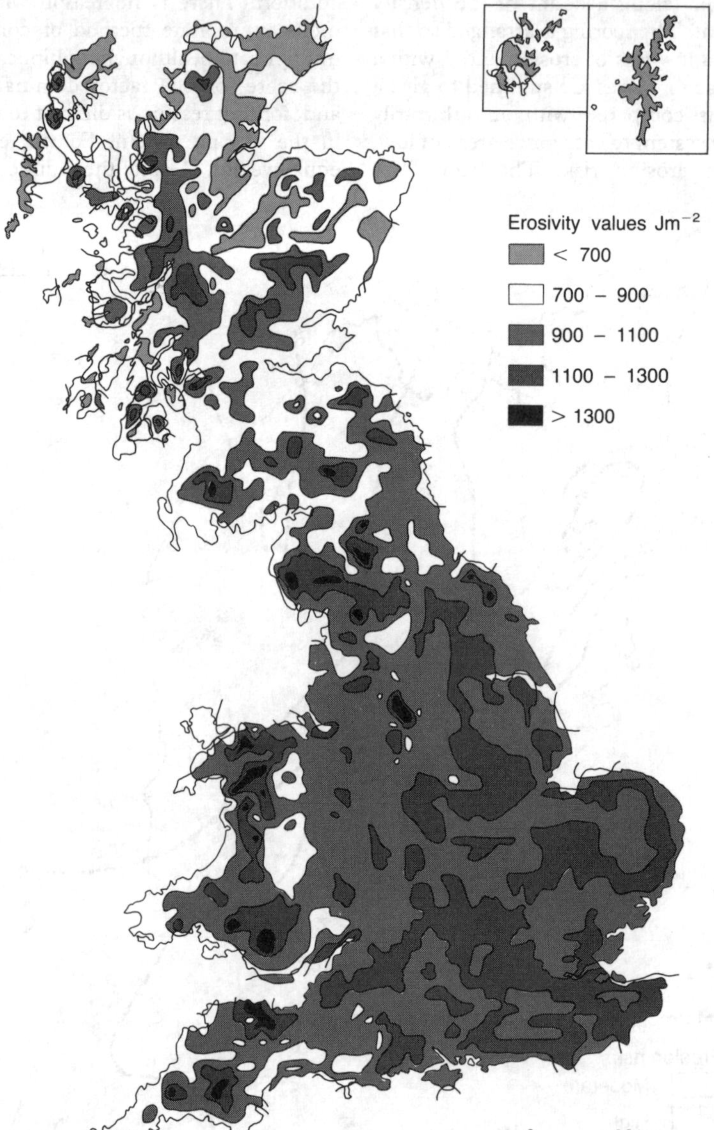
Erosivity values Jm^{-2}
< 700
700 – 900
900 – 1100
1100 – 1300
> 1300
0
80 km

Figure 4.7 Mean annual erosivity (*KE* > *10*) in Great Britain (after Morgan 1980b).

and compared with regional evaluations using a similar methodology, there is often little similarity between the two (Morgan 1993). Nevertheless, the result is the most comprehensive attempt to evaluate the situation in southern Europe. If used carefully, factorial scoring can provide a valuable general appreciation of erosion risk and indicate vulnerable areas where more detailed assessments should be made.

4.2 Semi-detailed Assessment

4.2.1 Land capability classification

Land capability classification was developed by the United States Soil Conservation Service as a method of assessing the extent to which limitations such as erosion risk, soil depth, wetness and climate hinder the agricultural use that can be made of the land. The United States classification (Klingebiel and Montgomery 1966) has been adapted for use in many other countries (Hudson 1981).

The objective of the classification is to regionalize an area of land into units with similar kinds and degrees of limitation. The basic unit is the capability unit. This consists of a group of soil types of sufficiently similar conditions of profile form, slope and degree of erosion as to make them suitable for similar crops and warrant the use of similar conservation measures. The capability units are combined into subclasses according to the nature of the limiting factor and these, in turn, are grouped into classes based on the degree of limitation. The United States system

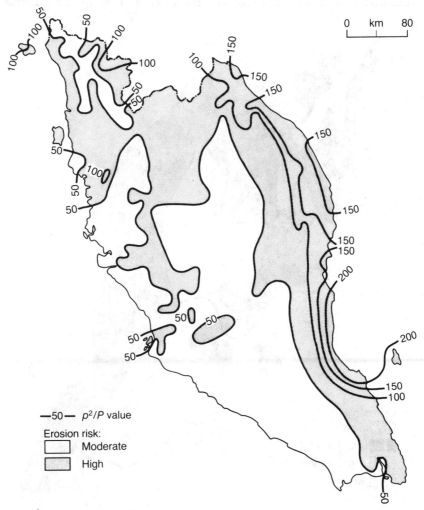

Figure 4.8 Values of p^2/P in Peninsular Malaysia.

recognizes eight classes arranged from Class I, characterized by no or very slight risk of damage to the land when used for cultivation, to Class VIII, very rough land which can be safely used only for wildlife, limited recreation and watershed conservation. The first four classes are designated as suitable for arable farming (Table 4.1). Assigning a tract of land to its appropriate class is aided by the use of a flow diagram (Fig. 4.11).

Although the inclusion of many soil properties in the classification may seem to render it useful for land use planning generally, it must be appreciated that, as befits a classification evolved in the wake of the erosion scare in the United States in the 1930s, its bias is towards soil conservation. This bias is illustrated by the dominance of slope as a factor determining capability class and by the emphasis given to soil conservation in the recommendations on how each class of

land should be treated. Attempts to use the classification in a wider sphere have only drawn attention to its limitations. The classification does not specify the suitability of land for particular crops; a separate land suitability classification has been devised to do this (FAO 1976). The assigning of a capability class is not an indicator of land value which may reflect the scarcity of a certain type of land, nor is it a measure of whether the farmer can make a profit, which is much influenced by the market prices of the crops grown and the farmer's skill. The stress laid by the classification on arable farming can also be a disadvantage. Insufficient attention is given to the recreational use of land. The capability classification implies that land is set aside for recreation only when it is too marginal for arable or pastoral farming but such land is often marginal for recreational use too. This fact highlights the difficulty of incorporating agricultural and non-

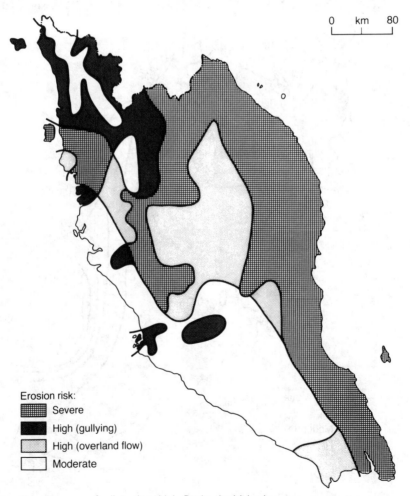

0   km   80

Erosion risk:

▦ Severe

■ High (gullying)

▨ High (overland flow)

☐ Moderate

Figure 4.9 Reconnaissance survey of soil erosion risk in Peninsular Malaysia.

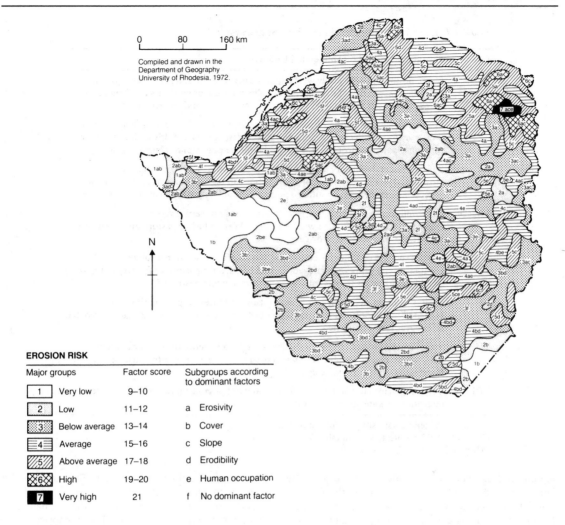

EROSION RISK

Major groups		Factor score	Subgroups according to dominant factors
1	Very low	9–10	
2	Low	11–12	a Erosivity
3	Below average	13–14	b Cover
4	Average	15–16	c Slope
5	Above average	17–18	d Erodibility
6	High	19–20	e Human occupation
7	Very high	21	f No dominant factor

0   80   160 km

Compiled and drawn in the
Department of Geography
University of Rhodesia. 1972.

N

Categories		Erosivity (J mm m^{-2} h^{-1})	Cover (mm of rainfall) and basal cover est. (%)	Slope (degrees)	Erodibility	Human ooccupation*
Low	I	below 5000	above 1000 7–10	0–2	orthoferralitic regosols	Extensive large scale commercial ranching National Parks or Unreserved
Below average	II	5000–7000	800–1000 5–8	2–4	paraferralitic	Large scale commercial farms
Average	III	7000–9000	600–800 3–6	4–6	fersiallitic	Low density CLs (below 5 p.p.km^2) and SCCF
Above average	IV	9000–11 000	400–600 1–4	6–8	siallitic vertisols lithosols	Moderately settled CLs (5–30 p.p.km^2)
High	V	above 11 000	below 400 0–2	above 8	non-calcic hydromorphic sodic	Densely settled CLs (above 30 p.p.km^2)

(*Notes:* Cover, Erodibility and Human occupation are only tentative and cannot as yet be expressed on a firm quantitative basis)
*p.p.km^2 = persons per square kilometre   CL = Communal Lands   SCCF = Small Scale Commercial Farms

Figure 4.10 Erosion survey of Zimbabwe (after Stocking and Elwell 1973b).

Table 4.1 Land capability classes (United States system)

Class	Characteristics and recommended land use
I	Deep, productive soils easily worked, on nearly level land; not subject to overland flow; no or slight risk of damage when cultivated; use of fertilizers and lime, cover crops, crop rotations required to maintain soil fertility and soil structure.
II	Productive soils on gentle slopes; moderate depth; subject to occasional overland flow; may require drainage; moderate risk of damage when cultivated; use of crop rotations, water-control systems or special tillage practices to control erosion.
III	Soils of moderate fertility on moderately steep slopes, subject to more severe erosion; subject to severe risk of damage but can be used for crops provided plant cover is maintained; hay or other sod crops should be grown instead of row crops.
IV	Good soils on steep slopes, subject to severe erosion; very severe risk of damage but may be cultivated if handled with great care; keep in hay or pasture but a grain crop may be grown once in 5 or 6 years.
V	Land is too wet or stony for cultivation but of nearly level slope; subject to only slight erosion if properly managed; should be used for pasture or forestry but grazing should be regulated to prevent plant cover from being destroyed.
VI	Shallow soils on steep slopes; use for grazing and forestry; grazing should be regulated to preserve plant cover; if plant cover is destroyed, use should be restricted until cover is re-established.
VII	Steep, rough, eroded land with shallow soils; also includes droughty or swampy land; severe risk of damage even when used for pasture or forestry; strict grazing or forest management must be applied
VIII	Very rough land; not suitable even for woodland or grazing; reserve for wildlife, recreation or watershed conservation.

Classes I–IV denote soils suitable for cultivation.
Classes V–VIII denote soils unsuitable for cultivation.
(Modified from Stallings 1957).

agricultural activities in a single classification. One approach to this problem is provided by the Canada Land Inventory (McCormack 1971) which employs four separate classifications covering agriculture, forestry, recreation and wildlife. In the UK, separate classifications exist for evaluating the suitability of land for picnic sites and camping sites (George and Jarvis 1979).

As a result of these criticisms, the land capability classification has tended, in recent years, to be discredited, often unfairly, as many of the criticisms arise from attempts to use the classification for purposes for which it was not intended. The value of land capability assessment lies in identifying the risks attached to cultivating the land and in indicating the soil conservation measures which are required. Improvements to the classification rest on making the conservation recommendations more specific, as is the case with the treatment-oriented scheme developed in Taiwan and tested on hilly land in Jamaica (Sheng 1972a; Table 4.2; Fig. 4.12).

Information contained in land capability surveys can be combined with that on erosivity to give a more detailed assessment of erosion risk. The susceptibility of soils to water and wind erosion is one factor considered in the description of the management characteristics of the capability classes mapped by the Soil Survey of England and Wales (1979). It is also considered in the descriptions of the Soil Associations contained in the legend to the 1:250,000 Soil Map of England and Wales (Soil Survey of England and Wales 1983). The information can be combined with that on rainfall erosivity (Fig. 4.7) and wind velocity to produce a map showing areas vulnerable to soil erosion (Morgan 1985a; Fig. 4.13). A threshold value of $1,100 \, J/m^2$ for mean annual $KE > 10$ is selected to denote a high risk of water erosion. Values of $900-1,100 \, J/m^2$ and less than $900 \, J/m^2$ indicate moderate and low risk respectively. It is assumed that the whole country is likely to experience wind speeds greater than the threshold level for erosion at least once a year. The inclusion of land capability in the assessment procedure means that the map indicates areas with a risk of erosion when the land is used in accordance with its capability rating. Erosion in these areas can be attributed to mismanagement of the land

Land Capability Class	I	II		III		IV	
Permissible slope	0°–1°	0°–1°	1°–2.5°	0°–2.5°	2.5°–4.5°	0°–4.5°	4.5°–7°
Minimum effective depth (Texture here refers to average textures)	1m of Cl or heavier	50cm of Sal or heavier	50cm of Sacl or heavier	50cm of S or LS 25cm of Sal or heavier	(a) 50cm of Sacl (b) 25cm of Cl or heavier	25cm of any texture	25cm of Sacl or heavier
Texture of surface soil	Cl or heavier	Sal or heavier S, or Ls if upper subsoil is Sal or heavier	Sal or heavier	No direct limitations	(a) Sal or heavier (b) Cl or heavier	No direct limitations	Sal or heavier
Permeability 5 or 4 to at least–	1m	50cm	50cm	No direct limitations	No direct limitations	No direct limitations	
Not worse than 3 to–		1m	1m	1m or 50cm if average texture is Cl is heavier	1m		
Physical characteristics of the surface soil– Permissible symbols	Not permitted	t1	t1	t1 and t2	t1 and t2	t1 and t2	t1 and t2
Erosion– Permissible symbols	1	1 and 2	1 and 2	1,2 and 3	1,2 and 3	1,2 and 3	1,2 and 3
Wetness criteria– Permissible symbols	Not permitted	w1	w1	w1	w1	w1 and w2	w1 and w2

S = Sand
Sal = Sandy loam
Ls = Loamy sand
Cl = Clay loam
Sacl = Sandy clay loam

Figure 4.11 Criteria and flow chart for determining land capability class according to the Classification of the Department of Conservation and Extension, Zimbabwe (after Hudson 1981).

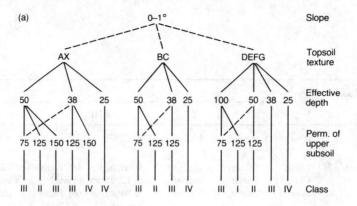

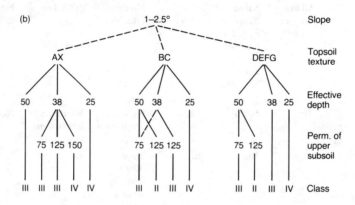

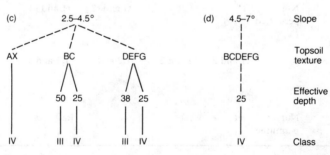

Figure 4.11 Continued

Effective depth (m)

1	Deep	More than 1.5m
2	Moderately deep	1m to 1.5m
3	Moderately shallow	50cm to 1m
4	Shallow	25cm to 50cm
5	Very shallow	Less than 25cm

Texture of surface soil

A	Sand	More than 85% sand
X	Loamy sand	80–85% sand
B	Sandy loam	Less than 20% clay; 50–80% sand
C	Sandy clay loam	20–30% clay; 50–80% sand
D	Clay loam	20–30% clay; less than 50% sand
E	Sandy clay	More than 30% clay; 50–70% sand
F	Clay	30–50% clay; less than 50% sand
G	Heavy clay	More than 50% clay

Permeability	Description	Rate of flow*
	Very slow	Less than 1.25
	Slow	1.25 to 5
	Moderately slow	5 to 20
	Moderate	20 to 65
	Moderately rapid	65 to 125
	Rapid	125 to 250
	Very rapid	Over 250

*The rate of flow in mm per hour through saturated undisturbed cores under a head of 12.5mm of water.

Physical characteristics of surface soil

t1	Slightly unfavourable physical conditions. The soil has a tendency to compact and seal at the surface and a good tilth is not easily obtained.
t2	Unfavourable physical conditions. Compaction and sealing of the surface soil are more severe. A hard crust forms when the bare soil is exposed to rain and sun and poor emergence of seedlings can severely reduce the crop. On ploughing, large clods are turned up which are not easily broken.

Erosion

1	No apparent, or slight, erosion.
2	Moderate erosion: moderate loss of topsoil generally and/or some dissection by run-off channels or gullies.
3	Severe erosion, severe loss of topsoil generally and/or marked dissection by run-off channels or gullies.
4	Very severe erosion: complete truncation of the soil profile and exposure of the subsoil (B horizon) and/or deep and intricate dissection by run-off channels or gullies.

Wetness

w1	Wet for relatively short and infrequent periods.
w2	Frequently wet for considerable periods.
w3	Very wet for most of the season.

Figure 4.11 Continued

Additional Requirements

*Factors affecting cultivation

g		Gravelly or stony
b	Downgrade Class I to II	Very gravelly or stony
o		Bouldery
s		Very bouldery
v	Class VI	Outcrops
r		Extensive outcrops

*Permeability

75 to 500–
otherwise Class IV
Not applicable to
basalts or norites

*Erosion

Class: I : 1
       II : 1; 2
       III : 1; 2; 3

*'t' Factors

Class: II : t1
       III : t1; t2
       IV : t1; t2

*Wetness

Class: II : $w1$
       III : $w1$
       IV : $w2$
       V : $w3$

$w2$ downgrades Class II and III to IVw unless the land is already Class IV on code, in which case it remains as Class IV.

*Note

Any land not meeting the minimum requirements shown on this sheet is Class VI.

Figure 4.11 Continued

whereas that which takes place on land being used for an activity not in accordance with the land capability classification can be attributed to misuse of the land.

A system of land capability classification has been devised by the Soil Conservation Service of New South Wales, Australia, for the planning of urban land use with particular reference to erosion control (Hannam and Hicks 1980). An example of its application to a small area at Guerilla Bay on the south coast of New South Wales is presented in Fig. 4.14. A map of landform regions is produced by combining information on slope, divided into seven classes, and topographic position, here called terrain, divided into six classes. The potential hazards related to urban land use are tabulated for each region. This information is combined with data on soils, paying attention to erosion hazard and limitations to urban development, to produce a map showing urban capability using five classes. A suitable form of land use is then determined for each area consistent with its capability class, physical limitations and the degree of disturbance it can sustain without causing excessive erosion and sedimentation.

4.2.2 Land systems classification

Land systems analysis is used to compile information on the physical environment for the purpose of resource evaluation. The land is classified into areal units, termed land systems, which are made up of smaller units, land facets, arranged in a clearly recurring pattern. Since land facets are defined by their uniformity of landform, especially slope, soils and plant community, land systems comprise an assemblage of landform, soil and vegetation types. Land systems classification has been described and discussed in many texts (Cooke and Doornkamp 1974; Young 1976) as a method of integrated resources survey, a role which lies beyond the scope of this book. What is important here, considering that many of the factors examined in land systems analysis are relevant to the soil erosion system, is its value for erosion risk evaluation. The extent to which land systems conform to discrete units of kind and degree of erosion hazard was examined by Higginson (1973) in the Hunter Valley, New South Wales, Australia. He carried out an erosion survey of the area from aerial photographs, using a classification recognizing seven grades of erosion severity, and analysed the distribution of erosion classes falling within each land system. Far from each land system being distinct, he found that the land systems could be combined into four groups. It would seem from this, therefore, that the relationship between erosion risk and land systems is poor and that the land systems classification provides only a generalized assessment. In contrast, in a similar study carried out in the Middle Veld of Swaziland (Morgan, Rickson, McIntyre and Brewer 1994), it was found that the ten land systems examined were generally distinct from each other in terms of erosion severity, both for 1972 and 1990, and in their rates of change between the two dates. It was concluded that land

Table 4.2 Treatment-oriented land capability classification

Group	Class	Characteristics and recommended treatments
Suitable for tillage	C1	Up to 7° slope; soil depth normally over 10 cm; contour cultivation; strip cropping; broad-base terraces.
	C2	Slopes 7–15°; soil depth over 20 cm; bench terracing (construction by bulldozers); use of four-wheel tractors.
	C3	Slopes 15–20°; soil depth over 20 cm; bench terracing on deep soil (construction by small machines); silt-pits on shallower soils; use of small tractors or walking tractors.
	C4	Slopes 20–25°; soil depth over 50 cm; bench terracing and farming operations by hand labour.
	P	Slopes 0–25°; soil depth too shallow for cultivation; use for improved pasture or rotational grazing system; zero grazing where land is wet.
	FT	Slopes 25–30°; soil depth over 50 cm; use for tree crops with bench terracing; inter-terraced areas in permanent grass; use contour planting; diversion ditches; mulching.
	F	Slopes over 30° or over 25° where soil is too shallow for tree crops; maintain as forest land.
Wetland, liable to flood; also stony land	P	Slopes 0–25°; use as pasture.
	F	Slopes over 25°; use as forest.
Gullied land	F	Maintain as forest land.

After Sheng (1972a).
The scheme is most suitable for hilly lands in the tropics.

Soil Depth \ Slope	1. Gentle sloping < 7°	2. Moderate sloping 7°–15°	3. Strongly sloping 15°–20°	4. Very strongly sloping 20°–25°	5. Steep 25°–30°	6. Very steep > 30°
Deep (D) > 36 in. (> 90 cm)	C1	C2	C3	C4	FT	F
Moderately deep (MD) 20–36 in. (50–90 cm.)	C1	C2	C3	C4 / P	FT / F	F
Shallow (S) 8–20 in. (20–50 cm.)	C1	C2 / P	C3 / P	P	F	F
Very shallow (VS) < 8 in. (< 20 cm.)	C1 / P	P	P	P	F	F

Figure 4.12 Chart for determining land capability class according to the treatment-oriented scheme of Sheng (1972a; Table 4.2).

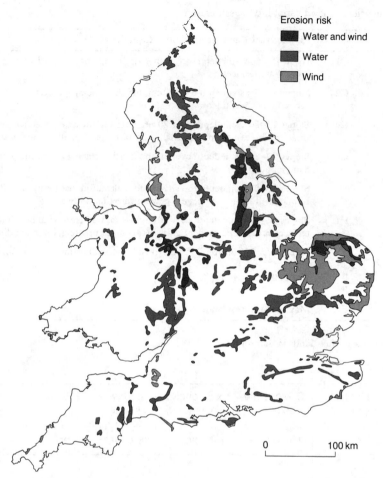

Figure 4.13 Areas susceptible to soil erosion in England and Wales (after Morgan 1985a).

systems could be interpreted as basic dynamic units reflecting differences in both the extent of erosion at any one time and its evolution over time.

4.2.3 Soil erosion survey

The first types of erosion survey were essentially static in concept and consisted of mapping, often from aerial photographs, the sheet wash, rills and gullies occurring in an area (Jones and Keech 1966). Erosion hazard was estimated by calculating simple indices such as gully density. No attempt was made to map the factors that influence erosion and which could therefore provide the basis for predicting change. The approach to analysing change was the sequential survey in which mapping was carried out at regular intervals, using photography of different dates. In this way changes in gully density could be examined in

relation to changing agricultural practices and increasing population pressure (Keech 1969). A similar study was carried out for 1,800 km² of the Middle Veld in Swaziland, centred on Manzini (Morgan, Rickson, McIntyre and Brewer 1994), in which areas of sheet wash, individual rills and gullies were mapped from 1:30,000 scale black-and-white panchromatic aerial photographs for 1972 and 1990. The resultant maps were then reduced to fit the 1:50,000 scale topographical maps. A 0.5×0.5 km grid was placed over the maps and each square was then assigned to one of four erosion severity classes. The information was placed into a SPANS Geographical Information System and maps produced of erosion severity for the two dates.

A more direct approach to gaining an understanding of the dynamics of erosion is to map both the erosion features and the factors influencing them, and to seek relationships between the two. A geomorpho-

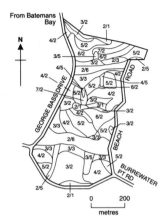

LANDFORM

Slope		Terrain	
0–1%	1	Hillcrest	1
1–5%	2	Sideslope	2
5–10%	3	Footslope	3
10–15%	4	Drainage plan	5
15–20%	5	Swamp	6
20–25%	6		
25–30%	7		

LANDFORM

Physical Criteria and Urban Landuse

Slope class	Terrain component	Potential hazards related to topographic location and slope and which will affect urban landuse	Suitable urban landuse
0–5%	Drainage plain	Flooding, seasonally high water-tables, high shrink-swell soils, high erosion hazard	Drainage reserves/stormwater disposal
	Floodplain	Flooding, seasonally high water-tables, high shrink-swell soils, saline soils, gravelly soils	Open space areas, playing fields
	Hillcrests Sideslopes	Shallow soils, stony/gravelly soils Overland flow, poor surface drainage and profile damage	Residential: all types of recreation; large-scale industrial, commercial and institutional development
	Footslopes	Impedence in lower terrain positions, deep soils Others–swelling soils, erodible soils, dispersible soils	
5–10%	Hillcrests Sideslopes Footslopes	Shallow soils Overland flow Deep soils, poor drainage Others–swelling soils, erodible soils, dispersible soils	Residential subdivisions, detached housing, medium-density housing/ unit complexes, modular industrial, active recreational pursuits
10–15%	Sideslopes	Overland flow Geological constraints–possibility of mass movement Swelling soils Erodible soils	Residential subdivisions, detached housing, medium-density housing/ unit complexes, modular industrial, passive recreational
15–20%	Sideslopes	Overland flow Geological constraints–possibility of mass movement Swelling soils Erodible soils	Residential subdivisions, detached housing, medium-density housing/ unit complexes, modular industrial, passive recreational
20–25%	Sideslopes	Geological constraints Mass movement High to very high erosion hazard	Residential subdivision, passive recreational
25–30%	Sideslopes	Geological constraints Possible mass movement High to very high erosion hazard	Upper limit for selective residential use, low-density housing on lots greater than 1 ha, passive recreation
>30%	Sideslopes	Geological constraints Mass movement Severe erosion hazard	Recommend against any disturbance for urban development

SOILS

Summary of Properties of Soils

Map unit	Dominant soils	Lithology and physiography	Erosion hazard	Limitations
A	Shallow gravelly soils	Metasediments; cherts, phyllites, etc. with quartz veination Ridges, sideslopes and some footslopes	High	Impeded soil drainage, shallow soil depth, high stone and gravel contents
B	Swamp alluvial soils	Metasediments; cherts, phyllites, etc. with quartz veination Alluvial parent materials, swamp and drainage plains	Very high to extreme	Seasonally high water-tables, poor to impeded soil drainage
C	Yellow duplex soils	Metasediments; cherts, phyllites, etc. Crests, sideslopes and some footslopes	High to very high	Low to moderate shrink-swell potential, poor soil drainage
D	Drainage plain alluvial soils	Metasediments; cherts, phyllites, etc. Alluvial/colluvial parent materials and surface materials	Very high	Seasonally high water-tables, poor to impeded soil drainage

Figure 4.14 Urban capability classification for soil erosion control (after Hannam and Hicks 1980).

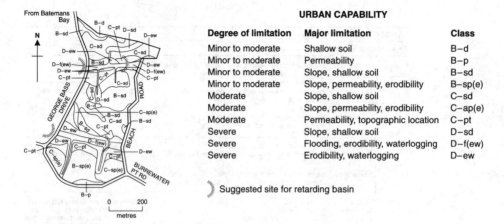

URBAN CAPABILITY

Degree of limitation	Major limitation	Class
Minor to moderate	Shallow soil	B–d
Minor to moderate	Permeability	B–p
Minor to moderate	Slope, shallow soil	B–sd
Minor to moderate	Slope, permeability, erodibility	B–sp(e)
Moderate	Slope, shallow soil	C–sd
Moderate	Slope, permeability, erodibility	C–ap(e)
Moderate	Permeability, topographic location	C–pt
Severe	Slope, shallow soil	D–sd
Severe	Flooding, erodibility, waterlogging	D–f(ew)
Severe	Erodibility, waterlogging	D–ew

〰 Suggested site for retarding basin

Urban Capability
Definitions of the classes used in urban capability assessment are:

Class A – Areas with little or no physical limitations to urban development.

Class B – Areas with minor to moderate physical limitations to urban development. These limitations may influence design and impose certain management requirements on development to ensure a stable land surface is maintained both during and after development.

Class C – Areas with moderate physical limitations to urban development. These limitations can be overcome by careful design and by adoption of site management techniques to ensure the maintenance of a stable land surface.

Class D – Areas with severe physical limitations to urban development which wil be difficult to overcome, requiring detailed site investigation and engineering design.

Class E – Areas where no form of urban development is recommended because of very severe physical limitations, to such development, that are very difficult to overcome.

Figure 4.14 Continued

logical mapping system for this purpose has been developed by Williams and Morgan (1976) to portray information on the distribution and type of erosion, erosivity, runoff, slope length, slope steepness, slope curvature in profile and plan, relief, soil type and land use. As much detail as possible is shown on a single map but, to avoid clutter, overlays can be used and are recommended for erosivity, soils and slope steepness. The legend (Fig. 4.15) is designed for use with colour: blue for water features, including river terraces; brown for relief features such as contours and crest lines; red for accelerated erosion by water or wind; yellow for aeolian features; green for plant covers; and black for features of the cultural environment, including badly-located and ill-designed conservation structures. An example, taken from an erosion survey of central Pahang, Malaysia, is shown in Figure 4.16.

Although much information is obtained from aerial photographs, this needs to be checked and supplemented by additional data collected in the field where the severity of erosion can be rated by a simple scoring system taking account of the exposure of tree roots, crusting of the soil surface, formation of splash pedestals, the size of rills and gullies, and the type and structure of the plant cover (Table 4.3). Observations are made using quadrat sampling over areas of $1\,m^2$ for ground cover, crusting and depth of ground lowering, $10\,m^2$ for shrub cover, and $100\,m^2$ for tree cover and the density of rills and gullies. A pro-forma for recording data in the field is shown in Figure 4.17. In interpreting the results of field survey, it is important to place the data in its time perspective, particularly with respect to likely seasonal variations in the ecology.

4.2.4 Summary

The maps produced in soil erosion surveys can be used in several ways. First, they depict the location and nature of erosion and distinguish between natural and accelerated erosion. Spatial variations in erosion intensity can be examined relative to the topographical, ecological and cultural information shown on the maps. Second, they not only indicate the location of an erosion problem but also enable the significance of that location in an erosion sequence to be appreciated. Often, complete sequences from erosion to deposition can be discerned in downslope, downstream or down-

SYMBOL	FEATURE	COLOUR
~~	Perennial water course	blue
~-----	Seasonal water course	blue
— — —	Crest line	brown
~~	Contour line	brown
vvvvvvv	Major escarpment	brown
ᵕᵕᵕᵕᵕ	Convex slope break	brown
ⱽⱽⱽⱽ	Concave slope break	brown
―╫―╫―	Waterfall	blue
―――╫―	Rapids	blue
— — —	Edge of flood plain	blue
▼▼▼▼▼	Edge of river terrace	blue
—··—··—	Back of river terrace	blue
≛ ≛ ≛	Swamp or marsh	blue
⊞⊞⊞⊅	Active gully	red
⊞⊞⊞⊅	Stable gully	blue
ᐟᐟᐟᐟ	Active rills	red
≡ ≡ ≡	Sheetwash/rainsplash (inter-rill erosion)	red
~~↓~~	River bank erosion	red
⌒⌒⌒⌒	Landslide or slump scar	red
⊂ᶜᶜ	Landslide or slump tongue	red
ᶜᶜᶜᶜᶜ	Small slides, slips	red
⊞⊞⊞	Colluvial or alluvial fans	brown
▨▨▨	Sedimentation	brown
~―·~―	Landuse boundary (landuse denoted by letter e.g. R – rubber; F – forest; P – grazing land; L – arable land.)	green
―――――	Roads and tracks	black
―+――+―	Railway	black
⊥ ⊥ ⊥	Cutting	black
▼▼▼▼	Embankment	black
▪ ▪ ▪ ▪	Buildings	black
⌒――	Terrace	black
═══	Waterway	black

SLOPES

☐	0°–1°
▦	2°–3°
▨	4°–8°
▩	9°–14°
■	15°–19°
▦	over 19°

Figure 4.15 Legend for mapping soil erosion.

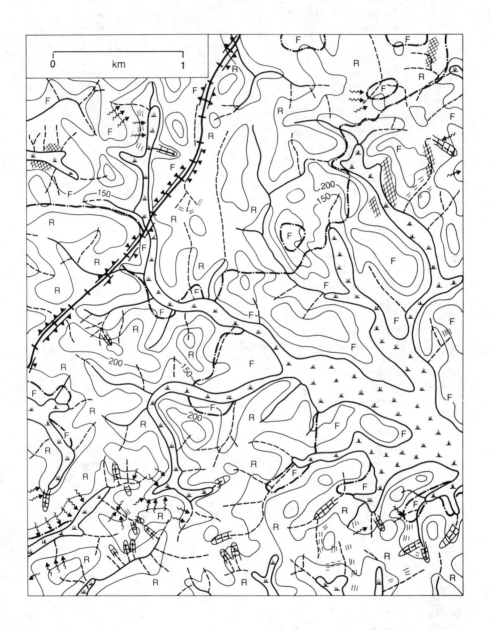

Figure 4.16 Extract of the soil erosion survey map, central Pahang. Contour interval 50 ft.

Recorder :	Area :	FACET NO.
Date :	Air photo no :	
Altitude :	Grid Reference :	

Present landuse															

Climate	Month	J	F	M	A	M	J	J	A	S	O	N	D	Erosivity
	Rainfall (mm)													
	Mean temp (C)													
	Maximum intensity													

Vegetation	Type	% Ground cover	% Tree and shrub cover

Slope	Position	Degree	Distance from crest	Shape

Soil	Depth	Surface texture		Erodibility
		Permeability	Clay fraction	

Erosion	

REMARKS	

EROSION CODE	0	½	1	2	3	4	5

Figure 4.17 Proforma for recording soil erosion in the field (devised by Baker personal communication).

Table 4.3 Coding system for soil erosion appraisal in the field

Code	Indicators
0	No exposure of tree roots; no surface crusting; no splash pedestals; over 70% plant cover (ground and canopy).
½	Slight exposure of tree roots; slight crusting of the surface; no splash pedestals; soil level slightly higher on upslope or windward sides of plants and boulders; 30–70% plant cover.
1	Exposure of tree roots, formation of splash pedestals, soil mounds protected by vegetation, all to depths of 1–10 mm; slight surface crusting; 30–70% plant cover.
2	Tree root exposure, splash pedestals and soil mounds to depths of 1–5 cm; crusting of the surface; 30–70% plant cover.
3	Tree root exposure, splash pedestals and soil mounds to depths of 5–10 cm; 2–5 mm thickness of surface crust; grass muddied by wash and turned downslope; splays of coarse material due to wash and wind; less than 30% plant cover.
4	Tree root exposure, splash pedestals and soil mounds to depths of 5–10 cm; splays of coarse material; rills up to 8 cm deep; bare soil.
5	Gullies; rills over 8 cm deep; blow-outs and dunes; bare soil.

wind directions. An understanding of such sequences is essential for predicting the broader, regional effects of changes in land use or the installation of conservation works. Third, length, area and height measurements can be made from the maps. These, together with the knowledge of relief which the maps provide, are essential information for designing such conservation measures as terraces, grass waterways and contour bunds. Fourth, the maps contain much of the material required for assessing land capability. Thus, through erosion survey, investigations of the severity of erosion, land capability and conservation treatment can be better linked, aiding the soil conservationist in the delineation of areas suitable for arable farming, grazing and non-agricultural activities. In addition to these four uses, the information shown on the maps can be incorporated in models of the soil erosion system.

CHAPTER 5

Modelling Soil Erosion

The techniques described in the last chapter enable the risk of soil erosion to be assessed. Before planning conservation work it is helpful if this assessment can be transformed into a statement of how fast soil is being eroded. Estimates of the rate of soil loss may then be compared with what is considered acceptable and the effects of different conservation strategies can be determined. What is required, therefore, is a method of predicting soil loss under a wide range of conditions.

Most of the models used in soil erosion studies are of the empirical grey-box type (Table 5.1). They are based on defining the most important factors and, through the use of observation, measurement, experiment and statistical techniques, relating them to soil loss. In recent years significant advances have been made in our knowledge of the mechanics of erosion processes and, as a result, greater emphasis is now being placed on developing white-box and physically-based models. Along with this goes a switch from using statistical techniques to employing mathematical ones frequently requiring the solution of partial-differential equations.

5.1 Defining Objectives

Models are of necessity simplifications of reality. Decisions need to be made on the suitable level of complexity or simplicity depending on the objective. The starting point for all modelling must therefore be a clear statement of the objective which may be prediction or explanation. Potential users will have different objectives. Managers, planners and policy-makers require relatively simple predictive tools to aid decision-making, albeit about rather complex systems. Researchers seek models that describe how the system functions, that might enlighten understanding of the system and emphasize degrees of uncertainty in speculations on how the system responds to change. This chapter is concerned only with predictive models since these are the most useful for practical applications but, increasingly, many of these models are developed by researchers and have a sound research base reflecting our knowledge of how the erosion system works.

When predicting erosion, decisions need to be made on whether the prediction should be for a year, a day, a storm or for short time periods within a storm; and whether it should be for a field, a hillslope or a drainage basin. These differences in temporal and spatial perspectives will influence the processes which

Table 5.1 Types of models

Type	Description
Physical	Scaled-down hardware models usually built in the laboratory; need to assume dynamic similitude between model and real world.
Analogue	Use of mechanical or electrical systems analogous to system under investigation, e.g. flow of electricity used to simulate flow of water.
Digital	Based on use of digital computers to process vast quantities of data.
(a) Physically-based	Based on mathematical equations to describe the processes involved in the model, taking account of the laws of conservation of mass and energy.
(b) Stochastic	Based on generating synthetic sequences of data from the statistical characteristics of existing sample data; useful for generating input sequences to physically-based and empirical models where data only available for short period of observation.
(c) Empirical	Based on identifying statistically significant relationships between assumed important variables where a reasonable database exists. Three types of analysis are recognized: *black-box*: where only main inputs and outputs are studied; *grey-box*: where some detail of how the system works is known; *white-box*: where all details of how the system operates are known.

After Gregory and Walling (1973).

need to be included in the model, the way they are described and the type of data required for model validation and operation. Good scientific practice should force all modellers to specify the design requirements of their models before they are developed. A user should be able to expect that a model has been tested for the conditions for which it was designed and, from this information, should then be able to select the most appropriate model from those available to meet a specific set of objectives. Any attempt to use a model for conditions outside those specified should be viewed as bad practice and, at best, speculative.

The objective has implications for how well a model must perform. *Screening models* are simple in concept and designed to identify problem areas. Usually it is sufficient if predictions are of the right order of magnitude. *Assessment models* need to predict with greater accuracy because they are used for evaluating the severity of erosion under different management systems and may be intended as design tools for the selection of conservation practices. For examining the on-site consequences of erosion, such as loss of soil depth and declines in productivity, predictions of erosion are required for periods of 20 to 30 years either on an annual basis or as an annual average over the time span. More detailed, event models are needed, however, for assessing the off-site effects, for example, pesticide use which may create a problem for only a few days following application, and comparing the pollutant loads in rivers from different agricultural management systems on the adjacent hillsides.

Our understanding of erosion processes is greatest over very short time periods of only a few minutes. Whilst it may be feasible to apply this understanding to slightly longer periods, continuous extrapolation is not possible. A single event or a storm is probably the upper limit to which relationships established for instantaneous conditions can be applied. Thus longer-term modelling can only be achieved by summing the predictions for individual storms. The alternative is to develop models empirically using data collected on an annual or mean annual basis.

The scale of operation must also be considered. The detailed requirements for modelling erosion over a large drainage basin differ from those demanded by models of soil loss from a short length of hillslope or at the point of impact of a single raindrop. Scale influences the number of factors which need to be incorporated in the model, those that can be held constant and those that can be designated primary factors around which the model must be constructed.

In addition to defining the area over which the model should operate, the behaviour of the model at the boundary of that area needs to be considered. The simplest assumption is that a model applies to a single area of land with a well-defined boundary across which there is no transfer of water or sediment. Such an assumption is clearly unrealistic for most hillslope applications since water and sediment pass downslope from one segment to another and, where the slope is variable in plan, may also move across the slope to concentrate in hollows or areas of flow convergence. Most erosion models allow for transfer of material across the lower boundary but few deal with more complicated boundaries such as the junction between two systems or process domains, for example between the hillslope and the river channel. Also most soil erosion models consider the boundaries as fixed in their location over time, ignoring the possible movement of the lower hillslope boundary upslope as a result of bank erosion and the gradual lowering of the summit by erosion.

The model should be formulated conceptually, representing it by a flow chart such as that shown in Fig. 5.1. Viewing the model in this way enables the struc-

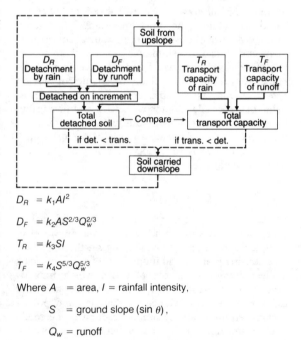

$D_R = k_1 A I^2$

$D_F = k_2 A S^{2/3} Q_w^{2/3}$

$T_R = k_3 S I$

$T_F = k_4 S^{5/3} Q_w^{5/3}$

Where A = area, I = rainfall intensity,

$\quad\quad\quad S$ = ground slope $(\sin \theta)$,

$\quad\quad\quad Q_w$ = runoff

Figure 5.1 Flow chart for the model of the processes of soil erosion by water (after Meyer and Wischmeier 1969).

ture of the system, the logical order of transfers of energy and matter through the system, the system variables and the interactions between the variables to be defined. It is also a good test of the level of scientific understanding of the system and the degree to which the model might need to be simplified because of insufficient knowledge. A model can be simplified by concentrating on those processes which have the greatest influence over the output and ignoring those which have very little effect.

Once the processes operating within the model have been identified, they need to be described mathematically. The methods used can range from simple equations, often expressing a statistical relationship, to complex equations related to the fundamental physics or mechanics of the process. The former are more common in empirical models whereas the latter provide the foundation for physically-based models.

5.2 Empirical Models

The simplest model is a black-box type relating sediment loss to either rainfall or runoff. A typical relationship is:

$$Qs = aQ^b \qquad (5.1)$$

where Qs is the sediment discharge and Q is the water discharge. Jovanović and Vukčević (1958), using data for 16 gauging stations in the former Yugoslavia, established that $b = 2.25$ whilst, according to Leopold, Wolman and Miller (1964), b ranges in value from 2.0 to 3.0. The value of a is an index of erosion severity. Thus, in the Yugoslav example, $a > 0.0007$ denotes excessive soil loss and $a < 0.0003$ indicates a low erosion rate. It is not always possible to establish the values with confidence. The relationship between sediment and water discharge may vary with the volume of runoff and therefore change seasonally. During a single storm the value of b often differs on the rise of the flood wave from that on the recession (Gregory and Walling 1973). The main disadvantage of this type of model is that it gives no indication of why erosion takes place.

Greater understanding of the causes of erosion is achieved by grey-box models. These frequently culminate in expressing the relationship between sediment loss and a large number of variables with a regression equation. One example is Eq. 4.2 of Fournier (1960) relating mean annual sediment yield to rainfall, altitude and slope.

A problem with these empirical models is that the equations cannot be extrapolated beyond their data range with confidence, either to more extreme events or to other geographical areas. Although Eq. 4.2 is often assumed to have general validity, its limitations were often recognized by Fournier (1960) who produced four other regression equations relating sediment yield to the p^2/P index for specific conditions of relief and climate (Fig. 5.2)

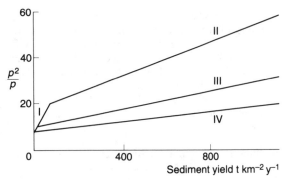

I   Low relief (mean valley floor gradient < 1:100), $p^2/P \leq 20$

    Sediment yield $= 6.14 \dfrac{p^2}{P} - 49.78$

II Low relief (mean valley floor gradient < 1:100), $p^2/P > 20$

    Sediment yield $= 27.12 \dfrac{p^2}{P} - 475.40$

III High relief (mean valley floor gradient $\geq$ 1:100), $P > 600$ mm

    Sediment yield $= 52.49 \dfrac{p^2}{P} - 513.20$

IV High relief (mean valley floor gradient $\geq$ 1:100), $200 \leq P \leq 600$ mm

    Sediment yield $= 91.78 \dfrac{p^2}{P} - 737.62$

Method is inapplicable to areas where $P < 200$ mm

Figure 5.2 Relationship between mean annual suspended sediment yield of rivers and rainfall aggressiveness (after Fournier 1960).

5.2.1 Universal Soil Loss Equation

All the predictive equations discussed so far refer to drainage basins and do not provide a suitable technique for assessing soil loss from smaller areas such as hillslopes and fields. The first attempt to develop a soil loss equation for hillslopes was that of Zingg (1940) who related erosion to slope steepness and slope length (Eq. 3.11). Further developments led to the addition of a climatic factor based on the maximum 30-minute rainfall total with a 2-year return period (Musgrave 1947), a crop factor, to take account of the protection-effectiveness of different crops (Smith 1958), a conservation factor and a soil erodibility factor. Changing the climatic factor to the rainfall erosivity index (R) ultimately yielded the Universal Soil Loss Equation (USLE) which, following considerable experience with its use, was later modified and up-dated (Wischmeier and Smith 1978). The equation is:

$$E = R . K . L . S . C . P \qquad (5.2)$$

where E is the mean annual soil loss, R is the rainfall erosivity factor, K is the soil erodibility factor, L is the slope length factor, S is the slope steepness factor, C is the crop management factor and P is the erosion-control practice factor. Since K represents mean annual soil loss per unit of R, E has the same units as

K. Thus, if K is in t/ha for one unit of metric R, multiplication by the metric R value will give the value of E in t/ha. If K is in US tons/acre per unit of American R, multiplication by the American R value yields a value of E in US tons/acre. In both cases, the E value is then adjusted by multiplying by the values of L, S, C and P which are dimensionless coefficients. The individual factors in the equation are derived as follows.

R. This is the rainfall erosivity index, described in Section 4.1.2, based on mean annual EI_{30}. If E is in foot-tons/acre and I_{30} is in in/h,

$$R = EI_{30}/100 \qquad (5.3)$$

and R is in American units. If E is in J/m^2 and I_{30} is in mm/h,

$$R = EI_{30}/1,000 \qquad (5.4)$$

and R is in metric units.

K. This is the soil erodibility index (Section 3.2) defined as mean annual soil loss per unit of R for a standard condition of bare soil, recently tilled up-and-down slope with no conservation practice and on a slope of 5° and 22 m length. K can be defined for a unit of both American and metric R. Wherever possible, K should be based on measured values. If the value is obtained from the nomograph (Fig. 3.1), it should be remembered that this is an estimated value and will be subject to error.

LS. The factors of slope length (L) and slope steepness (S) are combined in a single index which expresses the ratio of soil loss under a given slope steepness and slope length to the soil loss from the standard condition of a 5° slope, 22 m long, for which $LS = 1.0$. The appropriate value can be obtained from nomographs (Wischmeier and Smith 1978) or from the equation:

$$LS = \left(\frac{x}{22.13}\right)^n (0.065 + 0.045s + 0.0065s^2) \qquad (5.5)$$

where x is the slope length (m) and s is the slope gradient in per cent. As seen in Section 3.3, the value of n should be varied according to the slope steepness.

C. The crop management factor represents the ratio of soil loss under a given crop to that from bare soil. Since soil loss varies with erosivity and the morphology of the plant cover (Section 3.4), it is necessary to take account of changes in these during the year in arriving at an annual value. The year is divided into periods corresponding to different stages of crop growth. For annuals, these are: (i) fallow – inversion

ploughing to seed-bed establishment; (ii) seed-bed – secondary tillage for formation of the seed-bed to 10 per cent crop cover; (iii) establishment – 10–50 per cent crop cover; (iv) development – 50–75 per cent crop cover; (v) maturity – 75 per cent crop cover to harvest; and (vi) residue or stubble – harvest to ploughing or new seeding. For the last period, three options are considered: leaving the residue in the field without seeding; leaving the residue and seeding an off-season crop; and removing the residue and leaving the ground bare. For a given crop, separate ratio values are obtained for each period from tables (Wischmeier and Smith 1978) summarizing data collected over many years by the United States Soil Conservation Service at their experimental stations. The values vary not only with the crop but also, for a single crop, with yield, plant density and the nature of the previous crop. The individual values for each period are weighted according to the percentage of the mean annual R value falling in that period and summed to give the annual C value.

The method just described allows C factor values to be determined for the crop rotations and management practices found in the USA. For other countries, the detailed information for computing the C factor in this way does not exist and it is more appropriate to use average annual values. Table 5.2 gives typical ranges of values for different crops and management systems.

P. Values for the erosion-control practice factor are obtained from tables of the ratio of soil loss where the practice is applied to the soil loss where it is not. With no erosion-control practice, $P = 1.0$. Values cover contouring and contour strip-cropping and vary with the slope steepness (Table 5.3). Where diversion terracing is adopted, the value for contouring is used for the P factor and the LS factor is adjusted for the slope length which represents the horizontal spacing between the terraces. This method is not appropriate for bench terracing for which values for different types have been established from research in Taiwan (Chan 1981a).

As the example in Table 5.4 shows, the equation is normally used to predict mean annual soil loss. Since it was derived from and tested on data from experimental stations in the USA which, when combined, represent over 10,000 years of records, it is widely accepted as reliable. It has become the standard technique of soil conservation workers. The equation can be rearranged so that, if an acceptable value of E is chosen, the slope length (L) required to reduce soil loss to that value can be calculated. In this way, appropriate terrace spacings can be determined. Alternatively, the C value may be predicted and the tables searched to find the most suitable cropping practice which will give that value.

Although the equation is described as universal, its database, though extensive, is restricted to the USA east of the Rocky Mountains. The base is further restricted to slopes where cultivation is permissible, normally 0 to 7°, and to soils with a low content of montmorillonite; it is also deficient in information on the erodibility of sandy soils. Several attempts have

Table 5.2 C-factor values for the Universal Soil Loss Equation

Practice	Average annual C-factor
Bare soil	1.00
Forest or dense shrub, high mulch crops	0.001
Savanna or prairie grass in good condition	0.01
Overgrazed savanna or prairie grass	0.10
Maize, sorghum or millet: high productivity conventional tillage	0.20–0.55
Maize, sorghum or millet: high productivity conventional tillage	0.50–0.90
Maize, sorghum or millet: low productivity no or minimum tillage	0.02–0.10
Maize, sorghum or millet: high productivity chisel ploughing into residue	0.12–0.20
Maize, sorghum or millet: low productivity chisel ploughing into residue	0.30–0.45
Cotton	0.40–0.70
Meadow grass	0.01–0.025
Soya beans	0.20–0.50
Wheat	0.10–0.40
Rice	0.10–0.20
Groundnuts	0.30–0.80
Palm trees, coffee, cocoa with crop cover	0.10–0.30
Pineapple on contour: residue removed	0.10–0.40
Pineapple on contour: with surface residue	0.01
Potatoes: rows downslope	0.20–0.50
Potatoes: rows across-slope	0.10–0.40
Cowpeas	0.30–0.40
Strawberries: with weed cover	0.27
Pomegranate: with weed cover	0.08
Pomegranate: clean-weeded	0.56
Ethiopian tef	0.25
Sugar cane	0.13–0.40
Yams	0.40–0.50
Pigeon peas	0.60–0.70
Mungbean	0.04
Chilli	0.33
Coffee: after first harvest	0.05
Plantains: after establishment	0.05–0.10
Papaya	0.21

After Wischmeier and Smith (1978); Roose (1977); Singh, Babu and Chandra (1981); El-Swaify, Dangler and Armstrong (1982); Hurni (1987); Hashim and Wong (1988).

Table 5.3 P-factor values for the Universal Soil Loss Equation

Erosion-control practice	P-factor value
Contouring: 0–1° slope	0.60★
Contouring: 2–5° slope	0.50★
Contouring: 6–7° slope	0.60★
Contouring: 8–9° slope	0.70★
Contouring: 10–11° slope	0.80★
Contouring: 12–14° slope	0.90★
Level bench terrace	0.14
Reverse-slope bench terrace	0.05
Outward-sloping bench terrace	0.35
Level retention bench terrace	0.01
Tied-ridging	0.10–0.20

★ Use 50% of the value for contour bunds or if contour strip cropping is practised.
After Wischmeier and Smith (1978); Roose (1977); Chan (1981a).

been made to apply the equation more widely. Extensive data have been collected on R, K, C and P factor values so that it can be used in West Africa (Roose 1975) and in India (Singh, Babu and Chandra 1981). Modifications have been made to the R value and data collected on C factor values so that it can be applied to Bavaria (Schwertmann, Vogl and Kainz 1987) where it has been used in association with erosion surveys to quantify the risk of erosion (Auerswald and Schmidt 1986).

In addition to the limitations of its database, there are theoretical problems with the equation. Soil erosion cannot be adequately described merely by multiplying together six factor values. There is considerable interdependence between the variables. Some of these are considered. For instance, rainfall influences the R and C factors and terracing the L and P factors. Other interactions however, such as the greater significance of slope steepness in areas of intense rainfall (Section 3.3), are ignored. The rainfall erosion index is based on studies of drop-size distributions of rains which, as seen in Section 3.1.2, may have limited applicability. One important factor to which soil loss is closely related, namely runoff, is not dealt with explicitly but is incorporated within the R factor.

During the 1970s, the Universal Soil Loss Equation was widely used for estimating sheet and rill erosion in national assessments of soil erosion in the USA, based on estimates at more than one million sample points. The information provided the base for formulating soil conservation policies. By necessity, the equation was applied to conditions beyond its database, for example to rangeland and forest land in the western USA. As a result of the experience, a number of changes have been made which are now incorporated

Table 5.4 Prediction of soil loss using the Universal Soil Loss Equation

Problem
Calculation of mean annual soil loss on a 100-m long slope of 7° on soils of the Rengam Series under maize cultivation with contour bunds spaced at 20 m intervals, near Kuala Lumpur.

Equation
Mean annual soil loss = $R \times K \times LS \times C \times P$

Estimating *R* (Rainfall erosion index)
Method 1:

Mean annual precipitation (P)	=	2,695 mm
From Roose (1975), mean annual rainfall erosion index (R) in US units	=	0.5 P
	=	0.5 × 2,695
	=	1,347.5
Conversion to metric units	=	1,347.5 × 1.73
	=	<u>2,331.18</u>

Method 2:

From Morgan (1974), mean annual erosivity ($KE > 25$)	=	9.28 P − 8,838
	=	(9.28 × 2,695) − 8,838
	=	16,171.6 J/m^2
Multiply by I_{30} (use 75 mm/h; maximum value recommended by Wischmeier and Smith 1978)	=	16,171.6 × 75
	=	1,212,870
Divide by 1,000 to give R value in metric units	=	<u>1,212.87</u>

Method 3:

From Foster *et al.* (1981), mean annual EI_{30} (kg.m.mm)/(m^2.h)	=	0.276 $P \times I_{30}$
	=	0.276 × 2,695 × 75
	=	55,786.5
With these units, divide by 100 to give R value in metric units	=	<u>557.9</u>

<u>Best estimate</u>: discard result from Method 3 which is rather low. Take average value of Methods 1 and 2:

	R =	<u>1,772</u>

Estimating *K* (Soil erodibility index)
From Whitmore and Burnham (1969), the soils have a 43% clay, 8% silt, 9% fine sand and 40% coarse sand; organic content about 3%.

Using the nomograph (Fig. 3.2), gives a first approximation K value	=	<u>0.05</u>

Estimating *LS* (Slope factor)
For slope length (l) and slope steepness (s) in metres and per cent respectively,
$LS = (l/22)^{0.5} (0.065 + 0.045s + 0.0065s^2)$
With contour bunds at 20 m spacing, l = 20 m and s = 12% (approximation of 7°)
$LS = (20/22)^{0.5} (0.065 + (0.045 \times 12) + (0.0065 \times 12^2))$
$LS = 0.95 \times 1.54$
$LS = \underline{1.46}$

Estimating *C* (Crop management factor)
According to Table 5.2, the C value for maize ranges between 0.2 and 0.9, depending on the productivity. For many tropical farming conditions, C for maize lies between 0.4 and 0.9 (Roose 1975), depending on the cover.

During the 3-month period from seeding to harvest, the cover is likely to vary from 9 to 45 per cent in the first month, to 55 to 93 per cent in the second month, and 45 to 57 per cent in the third month. Therefore, we might assume C values of 0.9, 0.4 and 0.7 for the three respective months.

Maize can be planted at any time of year in Malaysia but assume planting after the April rains, allowing growth, ripening and harvesting in June and July which are the driest months. Land is under dense secondary growth prior to planting (assume C = 0.001) and allowed to revert to the same after harvest (assume C = 0.1).

Table 5.4 Continued

Of the mean annual precipitation, 32 per cent falls between January and April inclusive, 10 per cent in May, 6 per cent in June, 7 per cent in July, and 45 per cent between August and December. Assuming that erosivity is directly related to precipitation amount, these values can be used to describe the distribution of the R factor throughout the year.

From these data, the following table is constructed.

Months	C value	Adjustment factor (% R value)	Weighted C value (col 2 $\times$ col 3)
January–April	0.001	0.32	0.00032
May	0.9	0.10	0.09
June	0.4	0.06	0.024
July	0.7	0.07	0.049
August–December	0.1	0.45	0.045
		TOTAL	0.20832

C factor value for the year   =   <u>0.028</u>

Estimating P (Erosion-control practice factor)
From Table 5.3, P value for contour bunds   =   <u>0.3</u>

Soil loss estimation
Mean annual soil loss   = $1{,}772 \times 0.05 \times 1.46 \times 0.208 \times 0.3$
=   <u>8.07 t/ha</u>

in the Revised Universal Soil Loss Equation (RUSLE) (Renard *et al.* 1991). These include revisions to the R factor values in the USA; the development of a seasonally variable K factor, obtained by weighting instantaneous estimates of K by the proportion of the annual R for successive 15-day periods; modifications to the LS factor to take account of the susceptibility of the soils to rill erosion; and a new procedure for computing the C factor value through the multiplication of various sub-factor values. The sub-factors included in the C factor take account of prior land use, crop canopy, surface cover (mulches and ground vegetation) and surface roughness. The documented model is available as an interactive program designed to run on a personal computer with a DOS or UNIX operating system.

The Universal Soil Loss Equation was developed as a design tool for conservation planning but, because of its simplicity, attempts have been made to use it as a research technique. Applying the equation to purposes for which it was not intended, however, cannot be recommended (Wischmeier 1978). Since it was designed for interrill and rill erosion, it should not be used to estimate sediment yield from drainage basins or to predict gully or stream-bank erosion. Care should be taken in using it to estimate the contributions of hillslope erosion to basin sediment yield because it does not estimate deposition of material or

incorporate a sediment delivery ratio. Since the equation was developed to estimate long-term mean annual soil loss, it cannot be used to predict erosion from an individual storm. If applied in this way, it provides an estimate of the average soil loss expected from a number of such storms and this may be quite different from the actual soil loss in any single storm. The equation should not be applied to conditions for which factor values have not been determined and therefore need to be estimated by extrapolation. These qualifications apply equally to the Revised Universal Soil Loss Equation.

5.2.2 SLEMSA

The Soil Loss Estimator for Southern Africa (SLEMSA) was developed largely from data from the Zimbabwe highveld to evaluate the erosion resulting from different farming systems so that appropriate conservation measures could be recommended. The technique has since been adopted throughout the countries of Southern Africa. The equation is (Elwell 1978a):

$$Z = K.X.C \qquad (5.6)$$

where Z is mean annual soil loss (t/ha), K is mean annual soil loss (t/ha) from a standard field plot, 30 m long, 10 m wide, at 2.5° slope for a soil of known erodibility (F) under a weed-free bare fallow, X is a dimensionless combined slope length and steepness factor and C is a dimensionless crop management factor.

K. The value of K is determined by relating mean annual soil loss to mean annual rainfall energy (E) using the exponential relationship:

$$ln\,K = b\,lnE + a \qquad (5.7)$$

where E is in J/m^2, and the values of a and b are functions of the soil erodibility factor (F):

$$a = 2.884 - 8.2109\,F \qquad (5.8)$$

$$b = 0.4681 + 0.7663\,F \qquad (5.9)$$

X. The topographic factor (X) adjusts the value of soil loss calculated for the standard condition to that for the actual conditions of slope steepness and slope length. The value of X is obtained from:

$$X = (L)^{\frac{1}{2}}(0.76 + 0.53s + 0.076s^2)/\,25.65 \qquad (5.10)$$

where L is the slope length (m) and s is the percentage slope.

The procedure is modified if contour ridges are used as a conservation measure. In this case, instead of defining L and s in the downslope direction, L becomes the maximum length of the ridges and s is the across-slope grade (in per cent) of the ridge. If the across-slope grade is less than 1 per cent, the value of X is first determined for a 1 per cent grade and then multiplied by the term, Y, defined by:

$$Y = s\,/\,(0.572s + 0.428) \qquad (5.11)$$

C. The crop management factor (C) adjusts the value of soil loss for the standard bare soil condition to that from a cropped field. The value is dependent upon the percentage of the rainfall energy intercepted by the crop (i). For crops and natural grassland with $i < 50$ per cent and for dense pastures and mulches when $i \geqslant 50$ per cent, the following relationship is used:

$$C = e^{(-0.06i)} \qquad (5.12)$$

For crops and natural grasslands when $i < 50$ per cent, the relationship is:

$$C = (2.3 - 0.01i)\,/\,30 \qquad (5.13)$$

The value of i is obtained by weighting the percentage crop cover in each ten-day period by the percentage of the mean annual energy (E) occurring in that period and summing the values.

Guide values for soil erodibility (F) and average percentage crop covers for use with SLEMSA are

given in Table 5.5. An example of its use is shown in Table 5.6.

5.2.3 The Morgan, Morgan and Finney method

Morgan, Morgan and Finney (1984) developed a model to predict annual soil loss from field-sized areas on hillslopes which, whilst endeavouring to retain the simplicity of the Universal Soil Loss Equation, encompasses some of the recent advances in understanding of erosion processes. The model was compiled by bringing together the results of research by geomorphologists and agricultural engineers.

The model separates the soil erosion process into a water phase and a sediment phase (Fig. 5.3). The sediment phase is a simplification of the scheme described by Meyer and Wischmeier (1969; Fig. 5.1). It considers soil erosion to result from the detachment of soil particles by raindrop impact and the transport of those particles by overland flow. The processes of splash transport and detachment by runoff are ignored. Thus, the sediment phase comprises two predictive equations, one for the rate of splash detachment and one for the transport capacity of overland flow. The inputs to these equations of rainfall energy and runoff volume respectively are obtained from the water phase. The model uses six operating functions (Table 5.7) for which 15 input parameters are required (Table 5.8). Typical values for soil parameters are given in Table 5.9 and for the E_t/E_0 ratio, interception and the crop management factors in Table 5.10.

The effects of soil conservation practices can be allowed for within the separate phases of the model. For example, the introduction of agronomic measures to control erosion will bring about changes in evapotranspiration, interception and crop management which will affect respectively the volume of runoff, the rate of detachment and the transport capacity.

The model compares the predictions of detachment by rainsplash and the transport capacity of the runoff and assigns the lower of the two values as the annual rate of soil loss, thereby denoting whether detachment or transport is the limiting factor. The predictions obtained by the model are most sensitive to changes in annual rainfall and soil parameters when erosion is transport-limited and to changes in rainfall interception and annual rainfall when erosion is detachment-limited. Thus, good information on rainfall and soils is required for successful prediction (Morgan, Morgan and Finney 1984). An example of the use of the model is presented in Table 5.11. Like the Universal Soil Loss Equation, the model cannot be used for predicting sediment yield from drainage basins or soil loss from individual storms. The model has been used

Table 5.5 Input values for soil erodibility and crop cover for use in SLEMSA

Soil erodibility (*F* factor)

Soil texture	Soil type	*F* Value
light	sands	4
	loamy sands	
	sandy loams	
medium	sandy clay loam	5
	clay loam	
	sandy clay	
heavy	clay	6
	heavy clay	

Subtract the following from the *F* value:

1   for light-textured soils consisting mainly of sands and silts

1   for restricted vertical permeability within one metre of the surface or for severe soil crusting

1   for ridging up-and-down the slope

1   for deterioration in soil structure due to excessive soil loss in the previous year (>20 t/ha) or for poor management

0.5   for slight to moderate surface crusting or for soil losses of 10–20 t/ha in the previous year.

Add the following to the *F* value:

2   for deep (>2 m) well-drained, light-textured soils

1   for tillage techniques which encourage maximum retention of water on the surface, e.g. ridging on the contour

1   for tillage techniques which encourage high surface infiltration and maximum water storage in the profile, e.g. ripping, wheel-track planting

1   for first season of no tillage

2   for subsequent seasons of no tillage

Crop cover ratings (*C*)

Crop	Average percentage cover
cotton	40–65
cowpeas	40–55
tobacco	11–54
sorghum	50–70
sunflower	20–59
groundnuts	55–65
velvet beans	46–70
coffee	60–80
maize	42–80
rotational grass	80–98
soya beans	40–65
rice	70–78
weed fallow	100

After Elwell and Wendelaar (1977); Elwell (1978b).

Table 5.6 Prediction of soil loss using SLEMSA

Problem

Calculation of mean annual soil loss on a 100-m long slope of 7° on soils of the Rengam Series under maize cultivation with contour bunds at 20 m spacing, near Kuala Lumpur (compare with Table 5.4).

Basic equation

Soil loss $(Z) = K \cdot X \cdot C$

Estimating *K*

Mean annual precipitation (P) = 2,695 mm
Mean annual rainfall energy (E) = $9.28\,P - 8,838$
= $(9.28 \times 2,695) - 8,838$
= 16,172 J/m^2

Soil erodibility (F)
From Table 5.5, for a sandy clay, F = 5

Calculation of K
$$ln\,K = b\,ln\,E + a$$
where $a = 2.884 - 8.1209\,F$
$b = 0.4681 + 0.7663\,F$

For the example here, a = $2.884 - (8.1209 \times 5)$
= -37.7205
and b = $0.4681 + (0.7663 \times 5)$
= 4.2996
$ln\,K$ = $(4.2996 \times 9.6910) - 37.7205$
= 3.9469
which gives K = <u>51.78</u>

Estimating *X*

For unridged land
$X = \sqrt{l}\,(0.76 + 0.53\,s + 0.0765\,s^2)/25.65$
where l is slope length (m) and s is slope steepness (%)
for contour bunds every 20 m, $l = 20$ m and $s = 7°$ or 12 per cent
$X = \sqrt{20}\,(0.76 + (0.53 \times 0.12) + (0.0765 \times 0.12^2)/25.65$
= $4.47 \times 18.064/25.65$
= <u>3.15</u>

Estimating *C*

Calculation of proportion of rainfall intercepted by the plant cover (i) by weighting the percentage crop cover for each cropping season by the proportion of the mean annual rainfall occurring in that season.

For the conditions described in Table 5.4 and taking values from Table 5.5

Months	% Rainfall	% Cover	Value of i (col 2 × col 3)
January–April	0.32	100	32.00
May–July	0.23	45	10.35
August–December	0.45	90	40.50
		Total	82.85

C = $(2.3 - 0.01i)/30$
= $(2.3 - \{0.01 \times 82.85\})/30$
= <u>0.049</u>

Soil loss estimation

Mean annual soil loss = $51.78 \times 3.15 \times 0.049$
= <u>7.99 t/ha</u>

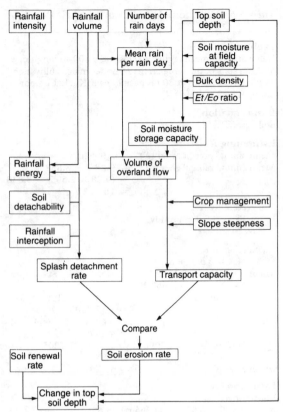

Figure 5.3 Flow chart for the Morgan, Morgan and Finney method of predicting soil loss.

Table 5.7 Operating functions for the Morgan, Morgan and Finney method of predicting soil loss

Water phase
$$E = R (11.9 + 8.7 \log_{10}I)$$
$$Q = R \, exp \, (-R_c/R_o)$$
where
$$R_c = 1000 \, MS.BD.RD \, (E_t/E_o)^{0.5}$$
$$R_o = R/R_n$$
Sediment phase
$$F = K \, (E \, e^{-aA})^b \, . \, 10^{-3}$$
$$G = C \, Q^d \sin S \, . \, 10^{-3}$$

E = kinetic energy of rainfall (J/m^2)
Q = volume of overland flow (mm)
F = rate of soil detachment by raindrop impact (kg/m^2)
G = transport capacity of overland flow (kg/m^2)
Values of exponents: $a = 0.05$; $b = 1.0$; $d = 2.0$

to predict soil loss in Indonesia (Besler 1987) and incorporated with remote sensing and a GIS for quantifying erosion surveys (De Jong and Riezebos 1992).

Table 5.8 Input parameters to the Morgan, Morgan and Finney method of predicting soil loss

MS	Soil moisture content at field capacity or 1/3 bar tension (% w/w).
BD	Bulk density of the top soil layer (Mg/m^3).
RD	Top soil rooting depth (m) defined as the depth of soil from the surface to an impermeable or stony layer; to the base of the A horizon; to the dominant root base; or to 1.0 m, whichever is the shallowest.
	Reasonable values are 0.05 for grass and cereal crops and 0.1 for trees and tree crops.
SD	Total soil depth (m) defined as the depth of soil from the surface to bedrock.
K	Soil detachability index (g/J) defined as the weight of soil detached from the soil mass per unit of rainfall energy.
W	Rate of increase in soil depth by weathering at the rock–soil interface (mm/y).
V	Rate of increase in the top soil rooting layer (mm/y) as a result of crop management practices and the natural breakdown of vegetative matter into humus.
S	Steepness of the ground slope expressed as the slope angle.
R	Annual rainfall (mm).
R_n	Number of rain days in the year.
I	Typical value for intensity of erosive rain (mm/h). Use 11 for temperate climates, 25 for tropical climates and 30 for strongly seasonal climates such as the Mediterranean.
A	Percentage rainfall contributing to permanent interception and stemflow.
E_t/E_o	Ratio of actual (E_t) to potential (E_o) evaporation.
C	Crop cover management factor.
	Combines the C and P factors of the Universal Soil Loss Equation to give ratio of soil loss under a given management to that from bare ground with downslope tillage, other conditions being equal.
N	Number of years for which the model is to operate.

5.2.4 Wind Erosion Prediction Equation

A similar technique to the Universal Soil Loss Equation has been developed for predicting wind erosion (Woodruff and Siddoway 1965; Skidmore and Williams 1991) taking account of soil erodibility (I), wind energy, expressed by a climatic factor (C), surface roughness (K), length of open wind blow (L) and the vegetation cover (V). The equation is:

$$WE = f \, (I, \, C, \, K, \, L, \, V) \qquad (5.14)$$

where WE is the mean annual wind erosion (t/ha), I is a mean annual value expressed in t/ha, L is in m, V is

Table 5.9 Typical values for soil parameters used in the Morgan, Morgan and Finney method of predicting soil loss

Soil	MS	BD	K
Clay	0.45	1.1	0.02
Clay loam	0.40	1.3	0.4
Silty clay	0.30	–	–
Sandy loam	0.28	1.2	0.3
Silt loam	0.25	1.3	–
Loam	0.20	1.3	–
Fine sand	0.15	1.4	0.2
Sand	0.08	1.5	0.7

Sources: summarized in Morgan, Morgan and Finney (1982).

the quantity of vegetative cover expressed as small grain equivalent in kg/ha and C and K are dimensionless. The equation allows for interactions between the factors and so it cannot be solved by multiplying the values of the various factors together. Factor relationships are rather complex and predictions can only be obtained by using complicated nomographs or specially developed equations in a set sequence (Table 5.12).

5.3 Physically-based Models

With greater concern over the last decade about the off-site consequences of erosion and identification of non-point source pollution, efforts have been made to develop models which will predict the spatial distribution of runoff and sediment over the land surface

Table 5.10 Typical values for plant parameters used in the Morgan, Morgan and Finney method of predicting soil loss

	A	E_t/E_o	C
Wet rice		1.35	0.1–0.2
Wheat	43	0.59–0.61	0.1–0.2 (winter sown)
			0.2–0.4 (spring sown)
Maize	25	0.67–0.70	0.2
Barley	30	0.56–0.60	0.1–0.2
Millet/sorghum		0.62	0.4–0.9
Cassava/yam			0.2–0.8
Potato	12	0.70–0.80	0.2–0.3
Beans	20–25	0.62–0.69	0.2–0.4
Groundnut	25	0.50–0.87	0.2–0.8
Cabbage/Brussels sprouts	17	0.45–0.70	
Banana		0.70–0.77	
Tea		0.85–1.00	0.1–0.3
Coffee		0.50–1.00	0.1–0.3
Cocoa		1.00	0.1–0.3
Sugar cane		0.68–0.80	
Sugar beet	12–22	0.73–0.75	0.2–0.3
Rubber	20–30	0.90	0.2
Oil palm	30	1.20	0.1–0.3
Cotton		0.63–0.69	0.3–0.7
Cultivated grass		0.85–0.87	0.004–0.01
Prairie/savanna grass	25–40	0.80–0.95	0.01–0.10
Forest/woodland	25–35 (coniferous and tropical)	0.90–1.00	0.001–0.002 (with undergrowth)
	15–25 (temperate broad-leaved)		0.001–0.004 (no undergrowth)
Bare soil	0	0.05	1.00

Sources: summarized in Morgan, Morgan and Finney (1982).
Note: C values should be adjusted by the following P ratios if mechanical soil conservation measures are practised:
   contouring: multiply by 0.6
   contour strip-cropping: multiply by 0.35
   terracing: multiply by 0.15

during individual storms in addition to total runoff and soil loss. In meeting these objectives, empirical models possess severe limitations. They cannot be universally applied, they are not able to simulate the movement of water and sediment over the land and they cannot be used on scales ranging from individual fields to small catchments. A more physically-based approach to modelling is thus required. In reality,

Table 5.11 Prediction of soil loss using the Morgan, Morgan and Finney method

Problem

Calculation of mean annual soil loss on a 100 m long slope of 7° on soils of the Rengam Series under maize cultivation with contour bunds at 20 m spacing, near Kuala Lumpur (compare with Tables 5.4 and 5.6)

Estimating input parameter values

MS	Estimate for a sandy clay; assume similar to a sandy loam. From Table 5.9	$= 0.28$
BD	Estimate for a sandy clay; assume similar to a clay loam. From Table 5.9	$= 1.3$
RD	Guide value from Table 5.8	$= 0.05$
K	Estimate for a sandy clay; since clay content will resist detachment but sand content will not, assume similar to a sandy loam. From Table 5.9	$= 0.3$
S	Slope angle of 7° gives sine value	$= 0.1219$
R	Annual rainfall	$= 2,695$
R_n	Number of rain days	$= 185$
I	Typical intensity of erosive rain; guide value from Table 5.8	$= 25$
A	Percentage rainfall contributing to interception is calculated for land use conditions given in Table 5.4, taking data from Table 5.10	

Months	Land use	A value	Weighted A value (col 1 × col 3)
January–April (4)	forest	25	100
May–July (3)	maize	25	75
August–December (5)	forest	25	125
		TOTAL	300

Average weighted A value = 300/12   = 25

E_t/E_o Calculation of time-weighted average value, taking data from Table 5.10

Months	Land use	E_t/E_o value	Weighted E_t/E_o value
January–April (4)	forest	0.90	3.6
May–July (3)	maize	0.70	2.1
August–December (5)	forest	0.90	4.5
		TOTAL	10.2

Average weighted E_t/E_o value = 10.2/12   = 0.85

C Calculation of time-weighted average value, taking data from Table 5.10

Months	Land use	C Value	Weighted C value
January–April (4)	forest	0.001	0.004
May–July (3)	maize	0.2	0.6
August–December (5)	forest regrowth	0.1	0.5
		TOTAL	1.104

Average weighted C value = 1.104/12   = 0.09

Estimating rainfall energy (E)

From Table 5.6

$$E = 2,695 \, (11.9 + 8.7 \log_{10} 25)$$
$$= 64,847 \text{ J/m}^2$$

Estimating volume of overland flow (Q)

From Table 5.6

$$R_c = 1,000 \times 0.28 \times 1.3 \times 0.05 \times 0.85^{0.5}$$
$$= 16.78 \text{ mm}$$
$$R_n = 2,695/185$$
$$= 14.57$$
$$Q = 2,695 \cdot e^{-16.78/14.57}$$
$$= 851.9 \text{ mm}$$

Estimating rate of soil detachment (F)

From Table 5.6

$$F = 0.3 \times (64,847 \times e^{-0.05 \times 25})^{1.0} \times 10^{-3}$$
$$= 5.57 \text{ kg/m}^2$$

Estimating transport of capacity of overland flow (G)

From Table 5.6

$$G = 0.09 \times 851.9^2 \times 0.1219 \times 10^{-3}$$
$$= 7.96 \text{ kg/m}^2$$

Soil loss estimation

Mean annual soil loss equals the lower of the values of R and G

$$= 5.57 \text{ kg/m}^2$$

Convert to t/ha = 10 × 5.57
$$= \underline{55.7 \text{ t/ha}}$$

physically-based models still rely on empirical equations to describe erosion processes and so should strictly be termed process-based models.

Physically-based models incorporate the laws of conservation of mass and energy. Most of them use a particular differential equation known as the continuity equation which is a statement of the conservation of matter as it moves through space over time. The equation can be applied to soil erosion on a small segment of a hillslope as follows. There is an input of material to the segment from the segment upslope and an output of material to the segment downslope. Any

Table 5.12 Prediction of wind erosion

Basic equation

$$WE = f(I, C, K, L, V)$$

Estimating *I* (soil erodibility index)
Values represent the potential annual soil loss (t/ha) for a level bare dry field near Garden City, Kansas. They have been determined from wind tunnel studies (Section 3.2; Table 3.6) and may be estimated by knowing the percentage of dry stable aggregates larger than 0.84 mm in the soil. The values are adjusted by a term *Is* to take account of the steepness of the windward slope (Woodruff and Siddoway 1965).

Estimating *C* (climatic index)
The index takes account of wind erosion as a function of wind velocity and soil moisture content. The former is expressed by the mean annual wind velocity (*v*; m/s) measured at a height of 9 m and the latter by the Thornthwaite precipitation effectiveness (*P − E*) index. The index is then expressed as a percentage of its value of 2.9 for Garden City, Kansas. Thus:

$$C = \frac{100}{2.9} \cdot \frac{v^3}{(P-E)^2}$$

where

$$P - E = 115 \sum_{i=1}^{12} \left(\frac{P}{T-10}\right)^{10/9}$$

in which *P* is the mean precipitation for month *i* (in) and *T* is the mean temperature for month *i* (°F). In application, a minimum value of 18.4 is adopted for mean monthly *T − 10* and a maximum value of 13 mm is used for mean monthly precipitation.

Estimating *K* (ridge roughness index)
The roughness of ridges produced by tillage and planting equipment is expressed by a roughness factor (*R*) calculated from

$$R = 4H^2/I$$

where *H* is the ridge height (mm) and *I* is the distance between the ridges. Values of *K* are expressed by:

$K = 1$	$R < 2.27$
$K = 1.125 - 0.153 \ln R$	$2.27 \leq R < 89$
$K = 0.336 \exp (0.00324\, R)$	$R \geq 89$

Estimating *L* (length of open wind blow)
The value of *L* is calculated as a function of equivalent field length (*D*) and the distance sheltered by any trees, shelterbelts, field hedges or windbreaks. The equivalent field length is calculated from measurements of actual field length (*l*; m), field width (*w*; m), field orientation expressed as the clockwise angle between field length and north (φ; rad), and wind direction clockwise from north (θ; rad):

$$D = \frac{l\,w}{l \left| \cos\left(\frac{\pi}{2} + \theta - \phi\right)\right| + w \left| \sin\left(\frac{\pi}{2} + \theta - \phi\right)\right|}$$

The value of L is then determined from:

$$L = D - 10H$$

where *H* is the height of the shelterbelt.

Estimating *V* (vegetation cover)
The vegetation cover index depends on standing live biomass, standing dead residue and flattened crop residue. The original work on the effects of vegetation was carried out for flattened wheat straw and so all weights of living or dead vegetative matter need to be converted into flattened wheat straw equivalents defined as 254 mm tall stalks lying flat on the soil in rows perpendicular to the wind direction with 254 mm row spacing and stalks oriented perpendicular to the wind direction. Values of *V* for various crops are given in Woodruff and Siddoway (1965) and for perennial rangeland grasses in Lyles and Allison (1980). For a wide range of crops, an estimate of the flattened wheat straw or small grain equivalent weight of the residue (*SGe*; kg/ha) can be obtained from:

$$SGe = 0.162\, Rw/d + 8.708\, (Rw/d\gamma)^{1/2} - 271$$

where *Rw* is the weight of the standing residue of the crop (kg/ha), *d* is the average stalk diameter (cm) and γ is the average specific weight of the stalks (Mg/m³) (Lyles and Allison 1981).

Applying the equation
Step 1: $E1 = I \cdot Is$
Step 2: $E2 = E1 \cdot K$
Step 3: $E3 = E2 \cdot C$
Step 4: $E4 = (F^{0.3484} + E3^{0.3484} - E2^{0.3484})^{2.87}$
  where $F = E2\,\{1 - 0.1218\,(L/Lo)^{-0.3829}\,\exp(-3.33\,L/Lo)\}$
  $Lo = 1.56 \times 10^6\,(E2)^{-1.26}\,\exp(-0.00156\,E2)$
Step 5: $E5 = g\,E4^h$
  where $g = 952.5^{(1-h)}$

$$h = \frac{1}{0.0537 + 0.9436 \exp(-0.000112\,V)}$$

Step 6: $WE = E5$

difference between the input and output relates to either erosion or deposition on the segment. Using the symbol ∂ to denote a change in the value, the continuity equation can be expressed as a mass balance (Bennett 1974; Kirkby 1980b):

$$\frac{\partial(AC)}{\partial t} + \frac{\partial(QC)}{\partial x} - e(x,t) = q_s(x,t) \qquad (5.15)$$

where A is the cross-sectional area of the flow, C is the sediment concentration in the flow, t is time, x is horizontal distance downslope, e is the net pick-up rate or erosion of sediment on the slope segment and q_s is the rate of input or extraction of sediment per unit length of flow from land external to the segment, for example from the sides on a convergent slope surface. On a plane slope segment, $q_s = 0$.

Almost all physically-based erosion models owe their origin to the relatively simple scheme developed by Meyer and Wischmeier (1969; Fig. 5.1) to test whether a mathematical approach to simulating erosion was feasible. This model uses four equations to describe the separate processes of soil particle detachment by rainfall, soil detachment by runoff, transport capacity of rainfall and transport capacity of runoff. By applying the model to consecutive downslope segments in turn, sediment is routed over the land surface and the pattern of erosion evaluated along a complete slope profile.

Because of its simplicity, the model has significant limitations. Steady-state conditions are assumed for rainfall intensity, infiltration rate and runoff rate. No other erosion processes and no weathering are allowed for. The soil is assumed bare of plant cover and no account is taken of tillage practices. Surface depression storage is ignored. There is no removal of material from the base of the slope and the altitudes of the highest and lowest points of the slope remain fixed through time. All these limitations are, of course, recognized by the authors.

The Meyer and Wischmeier scheme (Fig. 5.1) was extended by Beasley, Huggins and Monke (1980) to form ANSWERS (Areal Nonpoint Source Watershed Environment Response Simulation), designed to simulate the hydrological and erosional behaviour of small agricultural catchments during and immediately after a rainfall event. ANSWERS has the advantage of being easy to link to raster-based Geographical Information Systems (De Roo 1993). When it was first developed, it was considered by Foster (1982) as one of the most complete erosion models. Since then, a large number of process-based models have become available, many still in various stages of testing and development. Some examples are presented here.

5.3.1 CREAMS

CREAMS (Chemicals, Runoff and Erosion from Agricultural Management Systems) is a field-scale model, developed in the USA, to assess non-point source pollution and to investigate quantitatively the environmental consequences of different agricultural practices (Knisel 1980). The model consists of three components: hydrology, erosion and chemistry. The hydrology component is used to estimate runoff volume and peak runoff rate, infiltration, evapotranspiration, soil water content and percolation on a daily basis. The erosion component provides an estimate of erosion and sediment yield including the particle-size distribution of the eroded material at the edge of a field, also on a daily basis. The chemistry component includes elements for plant nutrients and pesticides and gives estimates of storm loads and average concentrations of adsorbed and dissolved chemicals in the runoff, sediment and percolated water.

The erosion component applies the continuity equation for sediment transport downslope in the form (Foster and Meyer 1972):

$$dQs / dx = Di + Dr \qquad (5.16)$$

where Qs is the sediment load per unit width per unit time, x is the distance downslope, Di is the delivery rate of particles detached by interrill erosion to rill flow and Dr is the rate of detachment or deposition by rill flow. This relationship does not consider interaction between detachment and sediment load; this is allowed for by adopting the equation (Foster and Meyer 1972):

$$(Dr / Dc) + (Qs / Tc) = 1 \qquad (5.17)$$

where Dc and Tc are respectively the detachment and transport capacities of the rill flow. This implies that if a flow is carrying less material than it has the capacity to transport, it will detach more particles to fill this capacity at a rate which is dependent upon the size of the deficit (see Section 2.2; Eq. 2.18). Alternatively, if the sediment load is greater than the transport capacity, deposition will occur. The rate of deposition (Dp) is determined by (Foster and Meyer 1975):

$$Dp = \alpha (Tc - Qs) \qquad (5.18)$$

where α is a first-order reaction coefficient. Its value is defined by:

$$\alpha = \xi Vs/Qw \qquad (5.19)$$

in which Vs is the particle fall velocity, Qw is the rate of runoff per unit width and $\xi = 0.5$ for overland flow and 1.0 for channel flow. The deposition process is modelled separately for three particle-size and eight aggregate-size groups. The initial particle input to the model is the distribution of sediment-size classes in the original soil. Since no selectivity is assumed in the detachment process, this is also the particle-size distribution of the detached material. A new particle-size distribution of the sediment load is calculated when deposition occurs.

The erosion component of CREAMS (Foster et al. 1981) has elements for overland flow, defined as including interrill and rill erosion; channel flow, as in a terrace channel or grass waterway; and deposition of sediment where water is impounded, as at a fence line, road culvert or tile-outlet terrace. The overland flow element operates by estimating detachment by interrill and rill flow and comparing the quantity of detached particles with the sediment transport capacity of the rill to determine the rates of erosion and deposition. This implies that all the material detached on the interrill area is delivered to the rills. Interrill detachment rate (Di; g m^{-2}s^{-1}) is estimated from:

$$Di = 4.57 \ (EI) \ (sin \ \theta + 0.014) \ KCP \ (Qp/Qw) \quad (5.20)$$

where EI is the storm EI_{30} value (mJ mm m^{-2}h^{-1}), θ is the slope angle, Qp is peak runoff rate (m/s), Qw is runoff volume (m^3/m^2) and K, C and P are factors from the Universal Soil Loss Equation. The K value has units of g/EI_{30} and can be obtained by multiplying K values in American units by 131.7. The P factor is used only for contouring because strip-cropping and deposition in terrace channels are taken account of directly in the model. The rill detachment rate (Df; g m^{-2}s^{-1}) is estimated from:

$$Df = 6.86 \times 10^6 \ m \ Qw \ Qp^{1/2} \ (x/22.1)^{m-1} \ sin^2 \ \theta \ KCP$$
$$(Qp/Qw) \quad (5.21)$$

where m is the slope length exponent from the Universal Soil Loss Equation and x is slope length (m).

The transport capacity of the flow is calculated from the Yalin (1963) equation, modified to cover different particle size classes. The equation requires an estimate of the shear stress exerted on the soil by the flow (τ_{soil}). For different cover and management conditions, this is estimated from:

$$\tau_{soil} = \gamma \ h \ S \ (n_b/n_c)^{0.9} \quad (5.22)$$

where γ is the weight density of water, h is the depth of flow, S is slope, n_b is the value of Manning's n for bare soil and n_c is the value of Manning's n for rough or vegetated surfaces. The depth of flow (h) is obtained from the Manning equation:

$$h = (Qw \ n_b \ / \ S^{0.5})^{0.6} \quad (5.23)$$

where Qw is the discharge rate per unit width.

The CREAMS model can be operated without the need to calibrate it against observed data (Knisel 1980) to provide a reasonable representation of the hydrological and erosion system and give a qualitative appraisal of the effects of different management strategies. It is not then a predictive model in absolute quantity (Knisel and Svetlosanov 1982). For use as a predictive tool, calibration with locally available data is desirable. CREAMS is a daily simulation model and estimates of longer-term erosion rates are obtained by summing the daily predictions. The erosion component can be operated in two ways: on its own using observed data for rainfall and runoff; or in relation to the hydrology component, using the latter to predict values for EI_{30}, runoff volume and runoff peak from observed rainfall. Since the hydrology component uses a runoff predictor which estimates the median rather than the actual response to a rainfall event, predictions of actual storm erosion cannot be obtained when the erosion component is operated in this way. When applied in this form to a hillslope in Bedfordshire, England, the model predicted erosion on ten occasions between May 1973 and July 1979 compared with 113 recorded events. Since five of the predicted events were for days when no erosion occurred, it was concluded that the model could not be relied upon to predict erosion on the right days and certainly not to the correct magnitude (Morgan, Morgan and Finney 1987). Mean annual predictions of soil loss for several sites in Bedfordshire, however, were extremely good (Table 5.13). When the erosion component was used with observed values for rainfall and runoff, reasonable predictions of soil loss were obtained for 31 storms with a significant correlation ($r = 0.87$; $p < 0.001$) between the predicted and observed values (Morgan 1980c).

CREAMS is one of the most widely tested process-based models. In addition to the above studies, trials with the erosion component have been carried out in the USA (Foster and Lane 1981), Czech Republic (Holý et al. 1982) and Lithuania (Kairiukstis and Golubev 1982).

5.3.2 WEPP

WEPP (Water Erosion Prediction Project) is a process-oriented model, based on modern hydrological and erosion science, designed to replace the Universal Soil Loss Equation for the routine assessment of soil erosion by organizations involved in soil and water conservation and environmental planning and assessment (Nearing et al. 1989). The overall package contains three computer models: a profile version, a watershed version and a grid version. The profile version estimates soil detachment and deposition along a hillslope profile and the net total soil loss from the end of the slope. It can be applied to areas up to about 260 ha in size. The watershed and grid versions allow estimations of net soil loss and deposition over small catchments and can deal with ephemeral gullies

Table 5.13 Comparison of observed mean annual runoff and soil loss in Bedfordshire with values predicted by CREAMS model

Site	Runoff (l/m-width)		Soil loss (kg/m²)	
	observed	predicted	observed	predicted
Silsoe, bare sandy soil, 11° slope, 2 plots	64.6 70.4	152.7	3.28 4.44	4.76
Silsoe, grass, sandy soil, 7° and 11° slope, 2 plots	18.6 14.4	43.6	0.15 0.31	0.04
Maulden, woodland, sandy loam soil, 16° slope	4.8	1.1	0.001	0.002
Woburn, oats–wheat–beans rotation, sandy loam soil, 7° slope	10.9	9.8	0.06	0.09
Ashwell, wheat, chalky soil, 10° slope	4.7	5.1	0.07	0.09
Pulloxhill, barley, clay soil, 10° slope	1.1	0.6	0.07	0.10
Meppershall, wheat–barley rotation, clay soil, 10° slope	5.6	4.2	0.05	0.27

formed along the valley floor. The models take account of climate, soils, topography, management and supporting conservation practices. They are designed to run on a continuous simulation but can be operated for a single storm. A separate model, CLIGEN (Nicks 1985), is used to generate the climatic data on rainfall, temperature, solar radiation and wind speed for any location in the USA for input to WEPP.

The erosion model within WEPP uses the same version of the continuity equation (Eq. 5.16) as CREAMS and the same approach for determining soil particle detachment (Eq. 2.18) and deposition (Eqs. 5.18 and 5.19) by flow. No particle-size selectivity occurs in detachment but deposition is particle-size selective. The particle-size distribution of the sediment load is recalculated once deposition occurs. The model also assumes that all the material detached on the interrill areas is delivered to the rills. The model differs from CREAMS in no longer relying on factor values from the Universal Soil Loss Equation. Separate erodibility parameters are used for interrill and rill erosion and their values can be varied seasonally. Crop management effects are represented by the direct influence of plant canopy and surface cover on soil detachment and transport. The interrill erosion rate (Di) is given by:

$$Di = Ki\ I^2\ Ce\ Ge\ (Rs/w) \qquad (5.24)$$

where Ki is an expression of interrill erodibility of the soil, I is the effective rainfall intensity, Ce expresses the effect of the plant canopy, Ge expresses the effect of ground cover, Rs is the spacing of rills and w is the width of the rill computed as a function of the flow discharge. The canopy effect is estimated by:

$$Ce = 1 - F\ e^{-0.34PH} \qquad (5.25)$$

where F is the fraction of the soil protected by the canopy and PH is the canopy height. The ground cover effect is estimated by:

$$Ge = e^{-2.5gi} \qquad (5.26)$$

where gi is the fraction of the interrill surface covered by ground vegetation or crop residue. The detachment of soil particles by rill flow is given by:

$$Df = Dc\ (1 - Qs/Tc) \qquad (5.27)$$

where Dc is the detachment capacity, Qs is the sediment load in the flow and Tc is the sediment load at transport capacity. Detachment capacity is expressed as:

$$Dc = Kr\ (\tau - \tau_c) \qquad (5.28)$$

where Kr is the rill erodibility of the soil, τ is the flow shear stress acting on the soil and τ_c is the critical flow shear stress for detachment to occur. The transport capacity of the flow is obtained from:

$$Tc = k_t\ \tau^{3/2} \qquad (5.29)$$

where k_t is a transport coefficient and τ is the hydraulic shear acting on the soil. The model is now undergoing an extensive programme of testing and evaluation.

5.3.3 GUESS

GUESS (Griffith University Erosion Sedimentation System) is a mathematical model which simulates the processes of erosion and deposition along a hillslope (Rose *et al.* 1983). The model is designed as a guide to farmers, scientists and extension workers concerned with the evaluation and control of soil erosion by water. Soil erosion at any position on the slope and at any time during the storm is related to a sediment flux which depends on the hydrological conditions and the sediment concentration.

The model differs from CREAMS and WEPP in separating the surface soil into two parts: that which is the original soil and possesses a certain degree of cohesion and that comprising recently detached material with no cohesion. As in CREAMS and WEPP, no selectivity occurs in detachment but deposition is particle-size selective. GUESS, however, describes the soil in terms of 50 particle-size classes instead of eight, all of equal mass, and determined according to their settling velocity. The continuity equation is modified accordingly and takes the form:

$$\left(\frac{\partial Q_{si}}{\partial x}\right) + \frac{\partial(C_i h)}{\partial t} = e_i + e_{di} + r_i + r_{di} - d_i \quad (5.30)$$

where Q_{si} is the sediment load of sediment class i, C_i is the concentration of sediment class i in the flow, e_i is the rate of detachment of particles of sediment class i in the original soil by raindrop impact, e_{di} is the rate at which recently detached soil of sediment class i is redetached by raindrop impact, r_i is the rate of detachment of particles of sediment class i by flow, r_{di} is the rate at which recently detached soil of sediment class i is redetached by flow, and d_i is the rate of deposition.

The rate of detachment of soil particles of sediment class i by raindrop impact is given by:

$$e_i = a \, C_e \, I/N \quad (5.31)$$

where a is the detachability of the soil, C_e is the fraction of the soil surface exposed to raindrops, I is the rainfall intensity and N is the number of particle size classes. The same equation is used to calculate the rate of redetachment of soil particles by raindrop impact (e_{di}) except that a is replaced by a_d, an expression of redetachability of the soil. The detachment rate of soil particles by flow is modelled as a function of stream power (Ω), defined as the product of flow shear stress (τ) and flow velocity (v) above a critical value (Ω_c):

$$r_i = (1 - H) \, F \, (\Omega - \Omega_c) \, / \, I\mathcal{J} \quad (5.32)$$

where $(1 - H)$ is the fraction of the original soil surface exposed to runoff, F is the fraction of the excess stream power which is used in detachment, generally assumed to be ≈ 0.1, and $\mathcal{J}$ is the amount of unit stream power necessary to detach a unit mass of soil. For redetachment of soil particles, the rate is:

$$r_{di} = \left(\frac{\alpha_i HF}{g}\right) \left(\frac{\sigma}{\sigma - \rho}\right) \left(\frac{\Omega - \Omega_c}{h}\right) \frac{M_i}{M} \quad (5.33)$$

where α_i is a dimensionless parameter with a value dependent on the depth of flow (usually ≥ 1 but taken as 1 for shallow flow), H is the fraction of the surface soil covered by recently deposited material, σ is the submerged sediment density, ρ is the density of water, h is the depth of flow, M_i is the mass fraction of sediment class i to that of the total mass (M) of

material being redetached. The rate of deposition of sediment of class is given by:

$$d_i = \alpha_i \, v_{si} \, C_i \quad (5.34)$$

where v_{si} is the fall velocity of particles of sediment class i.

Unlike in the CREAMS and WEPP models, not all the sediment detached or redetached by raindrop impact is contributed to the flow. A limitation is placed so that:

$$C_{rain} = (H \, a \, I) \, / \, \Sigma v_{si}/N) \quad (5.35)$$

where C_{rain} is the contribution of detachment by raindrop impact to the sediment concentration in the flow, and H approximates 0.9. The transport capacity of the flow (C_{max}) is represented by (Rose and Hairsine 1988):

$$C_{max} = \left(\frac{F\rho}{\Sigma v_{si}/N}\right) \left(\frac{\sigma}{\sigma - \rho}\right) sV \quad (5.36)$$

This equation is applied either to the overland flow or, if rills are present, by assigning all the flow to the rills and, by dividing by the number of rills, calculating the water discharge and velocity per rill. The sediment flux in the rill can then be multiplied by the number of rills to give the total sediment flux for the slope.

The model needs to be calibrated to determine the values of $\mathcal{J}$ and Ω_c. Like WEPP it is currently undergoing trials. In its present form, it can only be applied to bare soil and there is no provision for simulating the effects of a plant cover. However, an earlier and simplified version of the model has been used successfully to predict erosion under rain forest, maize and soya bean in northern Thailand (Suddhapreda et al. 1988).

5.3.4 EUROSEM

EUROSEM (European Soil Erosion Model; Fig. 5.4) is an event-based model designed to compute the sediment transport, erosion and deposition over the land surface throughout a storm. It can be applied to either individual fields or small catchments (Morgan 1994; Morgan, Quinton and Rickson 1994). Compared with other models, EUROSEM simulates rill and interrill erosion explicitly, including the transport of water and sediment from interrill areas to rills, thereby allowing for deposition of material *en route*. This is considered more realistic than assigning all or a set proportion of the detached material to the rills. A more physically-based approach to simulating the effect of the vegetation or crop cover is used, taking account of the influence of leaf drainage. Soil conservation measures can be allowed for by choosing appropriate values of the soil, microtopography and plant cover parameters so as to describe the conditions associated with each practice. Unlike other models, however, EUROSEM does not describe the eroded sediment in terms of its particle size.

Equation 5.15 is used as the continuity equation in which the term (e), the net rate of pick-up of sediment, is defined as:

$$e = Di + Df \qquad (5.37)$$

The rate of detachment by raindrop impact (Di) is calculated from:

$$Di = k \cdot KE^{1.0} \cdot e^{-2h} \qquad (5.38)$$

where k is the detachability of the soil by raindrop impact and h is the depth of surface water. The kinetic energy (KE) of the rainfall is determined separately for the direct throughfall and the leaf drainage, using Eq. 3.2 for the energy of a millimetre of direct throughfall

and the following relationship for a millimetre of leaf drainage (Brandt 1990):

$$KE(LD) = (15.8 \cdot PH^{0.5}) - 5.87 \qquad (5.39)$$

where PH is the height of the plant canopy. The rate of detachment of soil particles by flow is modelled as a balance between detachment and deposition:

$$Df = \eta \, w \, v_s \, (C_{max} - C) \qquad (5.40)$$

where η is an expression of the efficiency of the detachment process, its value being dependent upon soil cohesion, and w is the width of flow. When the sediment concentration in the flow exceeds transport capacity, η assumes a value of 1.0 and the term, e, in Equation 5.15 becomes negative and represents a net deposition rate. Equation 2.23 is used to describe the transport capacity of the flow in terms of unit stream power, expressed as the product of slope and velocity. The model can be applied to overland flow over a relatively smooth surface or to a slope containing rills or other defined channels such as plough furrows. When rills are present the model places all the runoff into the rills and, taking account of the number of rills, calculates the flow velocity and depth per rill. If the rills overflow, a unified rill model developed by Dr Roger Smith (USDA), is used to calculate the hydraulic conditions of the flow as a function of the wetted perimeter. Extensive trials of EUROSEM are being undertaken in the UK, Denmark, Italy and Spain.

5.3.5 Wind erosion models

Although considerable research is taking place to develop a physically-based wind erosion model, the work is still in the developmental stage. Particular difficulties arise in simulating the high degree of complexity and randomness of the detachment and transport of soil particles by wind; in representing a three-dimensional sediment transport system as a simpler two-dimensional one; and in describing the physics of the system by mathematical equations which can be solved. Many of the present models apply only to rather limited conditions such as sediment movement at transport capacity or bare soil. The first state-of-art model with the ability to simulate the effects of soil, land cover and tillage is likely to be the Wind Erosion Prediction System (WEPS) being produced in the USA (Hagen 1991) as a daily simulation model, applicable to a field or at most a few adjacent fields.

5.4 Sensitivity Analysis

Sensitivity analysis indicates by how much the output of a model alters in relation to a unit change in the

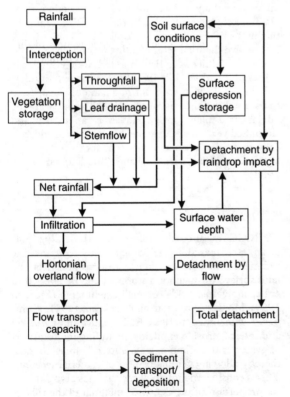

Figure 5.4 Flow chart for the European Soil Erosion Model (EUROSEM) (after Morgan, Quinton and Rickson 1994).

value of one or more of the inputs. It should be carried out on all models to show that the model behaves rationally and to indicate how accurately values of the inputs need to be measured or estimated. Rational behaviour is generally judged on whether the level of sensitivity of the factors in the model matches what is expected in reality and on whether the relationships between the output and the controlling factors accord with what is observed in the field or in laboratory experiments. The sensitivity analysis may be extended to evaluate whether the interaction between factors is correctly simulated and whether the model gives plausible results when operated for extreme conditions. It may also be used to identify which processes could be excluded without significant loss of information should a decision be made to simplify the model.

Clearly if a predictive model is to give satisfactory estimates of soil loss, a high level of accuracy is required in the values of the most sensitive inputs. The critical question is whether this accuracy can be obtained, particularly for inputs which are highly variable spatially, such as saturated hydraulic conductivity, or which are difficult to measure, such as soil detachability. Knowing the errors involved in the input data is important because they are a potential source of error in any predictions.

5.5 Model Validation

The success of any model must be judged by how well it meets its objectives or requirements. With a predictive model this means deciding on the time and space scale for which predictions are required and the level of accuracy. The failure of the Universal Soil Loss Equation and, under some circumstances, the erosion component of the CREAMS model to predict storm erosion does not make them bad models since, as emphasized above, they were not designed to make this kind of prediction. If they successfully meet their objective of evaluating mean annual soil loss, they must be classed as satisfactory models. Also, in one of the examples given above, the failure of the CREAMS model to predict erosion was attributed to poor quality input data rather than to deficiencies in the model itself. Thus, in making a judgement on the utility of a model, it is necessary to distinguish between failures due to misuse, those related to inadequate input data and those associated with the structure of the model or its operating functions. In the latter case, failure may result from poor conceptualization of the problem, omission of important factors or inaccurate representation of a particular element in the model by the operating function or equation employed. The solution is to modify or in some instances completely rethink the model.

The accuracy of model predictions is usually tested by comparing predicted with measured values and applying some measure of goodness-of-fit. The data used for validation should, of course, always be different from those used to develop the model. Criteria for validation are by no means clear-cut and need to be set for individual models in relation to their objectives. In many cases, a qualitative assessment is all that is required. Thus, when Meyer and Wischmeier (1969) produced their model to determine whether a process-based approach to simulating soil erosion down a hillside was feasible, they judged its success by how well it reproduced the patterns of sediment movement found in the field. For many applications, it may be sufficient to show that a model predicts the correct location of erosion and sedimentation.

Where the objective is to predict amounts of soil loss, it is necessary to compare the predictions with measured values. This can be achieved by dividing the predicted by the measured value to give a ratio. Ideally, the ratio should equal 1.0 but, since this is rarely the case, its value has to be related to some guideline in order to judge whether it is acceptable, for example, the ratio should be between 0.75 and 1.5 in value or, less stringently, between 0.5 and 2.0. The percentage of predictions that are acceptable can be used as a measure of a model's success. For some purposes, such as comparing the effectiveness of different conservation strategies, it may be enough for a model to predict realistic percentage differences in erosion between the strategies without giving absolute values. Although the use of ratios is a simple and effective method of validation, it cannot be applied when non-linear relationships are involved. In this case, the simplest measure is to examine the absolute differences between predicted and observed values. Wischmeier and Smith (1978) state that when tested against a data set of 2,300 plot years, the Universal Soil Loss Equation gave predictions of mean annual soil loss within $\pm$ 5 t/ha 84 per cent of the time when applied to agricultural land in the USA. Preliminary work in the UK on EUROSEM shows that some 50 per cent of soil loss predictions for individual storms are correct to $\pm$ 0.5 t/ha and 80 per cent to $\pm$ 2.5 t/ha for measured values within the range of 0.1 to 11 t/ha (Morgan, Quinton and Rickson 1994).

These types of tests assume that a given level of error is equally important over the full range of values whereas, for many purposes, higher levels may be acceptable at very low values of soil loss. A predicted value for mean annual soil loss of 0.2 t/ha may be acceptable in relation to a recorded value of 0.02 t/ha for evaluating erosion-control strategies since both values indicate that there is no on-site erosion problem. However, a predicted value of 200 t/ha against a recorded value of 20 t/ha would not be acceptable, since it would give a very misleading impression of the severity of the problem. Both predictions may be unsatisfactory when modelling off-site problems such as pollution. The validation criteria used for the Morgan, Morgan and Finney model were adjusted to allow for differences in acceptability of the prediction with its absolute value. The model was viewed as successful if (1) the annual predicted and observed values were both less than 1 t/ha, both indicating little or no erosion problem, or otherwise (2) the ratio of the predicted to observed value was between 0.5 and 2.0. The model successfully predicted annual soil loss for 47 out of 67 sites in 12 countries, a 70 per cent success rate. When considering only those sites, 31 in all, where data on soil properties were based on field measurements, the success rate rose to 90 per cent (Morgan, Morgan and Finney 1984).

Many workers prefer to use correlation and regression analyses in order to test the strength of the association between predicted and observed values. A high and significant value of the correlation coefficient, however, does not imply that a model is performing well, merely that the model is producing results with a degree of precision and with the right trend. Unless the predicted and observed values have a 1:1 relationship, the model predictions are inaccurate to a certain degree. Ideally, a regression equation for the relationship should have a slope of 1.0 and pass through zero. Statistical tests should be applied to determine whether particular relationships differ significantly from these two conditions. Use of a least-squares regression approach assumes that there will be some error in the predicted value but not in the observed value. This assumption is certainly questionable and it may be preferable to recognize this and use a reduced major axis line instead (Kermack and Haldane 1950; Till 1973) which will allow for errors in both sets of values.

Most validation trials on erosion models compare predicted and observed values of total soil loss for either set periods of time, such as a year, or individual storms. Such comparisons are sufficient when testing empirical models like the USLE and SLEMSA which predict only totals, but are inadequate when dealing with process-based models. With CREAMS, WEPP, EUROSEM and similar models, it is necessary to know whether the prediction has been arrived at in the right way. For example, if soil loss is limited by the detachment rate and the model predicts the right total but as a result of a limitation in transport capacity, it cannot be described as a successful model and it cannot be used for soil conservation since any strategy based on its predictions will reflect a wrong diagnosis. It is therefore essential that validation of process-based models be extended to cover all of their constituent modules.

5.6 Choosing an Acceptable Model

Broadly, it might be said that models are acceptable if they meet their objectives or design requirements. It is generally recognized that a good model should satisfy the requirements of reliability, universal applicability, easy use with a minimum of data, comprehensiveness in terms of the factors and erosion processes included, and the ability to take account of changes in land use and conservation practice. It should simulate the erosion system in a reasonably realistic way and also give plausible results for extreme conditions. For practical purposes, however, this requirement may be less important than its ability to perform well over a range of commonly experienced conditions in which case the limits of that range should be specified. An appropriate range for an erosion model is that it should deal with natural as well as disturbed land surfaces, bare and vegetated surfaces, areas close to divides as well as those at the bottom of slopes (Kirkby 1980b) and both typical erosive rains and extreme events. Unfortunately, many of the ideal characteristics of models conflict with each other. Ease of operation can mean that input data are readily available, for example from tables in an accompanying user's manual. Such data, however, are at best only guide values and there will be considerable uncertainty as to how well they describe the actual conditions at any one location in the field. In many instances, particularly with soil properties, it is almost impossible to describe the conditions by a single representative value. A model can only be as good as the data which are fed into it. If much of the input data contain errors of 10 per cent or more, it is unrealistic to expect a model to give predictions to accuracies of 5 per cent or less. Considerably more research is required to develop procedures for analysing uncertainty in model predictions and for expressing that uncertainty in a way that users can understand.

When selecting a model, care needs to be taken to avoid its misuse, extrapolating it to conditions beyond those of the database from which it was derived and being attracted to sophisticated schemes for which data input is difficult to obtain or which have not been properly validated. Despite the comments made above on the desirability of physically-based models, the present state of model development is such that a simple empirical model is often more successful in predicting soil erosion than a complex physically-based one which is difficult to operate and has been only partially evaluated.

The simplicity of models like the USLE and

SLEMSA makes them attractive to many users particularly for adding quantification to assessments of soil erosion risk. Their use in this way, however, is both unscientific and often misleading unless the models have first been validated against locally measured data. Altshul (1993) purposely applied the USLE and SLEMSA without local validation to parts of the Middle Veld of Swaziland to see how closely the predictions matched assessments of erosion based on erosion mapping. She found that both models explained the distribution of erosion severity less successfully than using factorial scoring because they did not take account of gully erosion. Further, the predictions made by the USLE were four to five times greater than those of SLEMSA. Without local validation there is no way of knowing which is right and no sensible judgements on the severity of the problem and the likely cost of conservation measures can be made. Similar discrepancies between the USLE and SLEMSA were found by Rydgren (1992) when testing both models in Lesotho against data from erosion plots. Even though SLEMSA gave results closer to reality, Chakela and Stocking (1988) found that the predictions of erosion in Lesotho based on SLEMSA could also be misleading since the intensively gullied lowlands around Maphutseng obtained a low erosion-risk rating. In addition, the predictions went against conventional wisdom that the mountain areas are less prone to erosion than the lowlands. The results do not imply that SLEMSA is wrong but more likely that its predictions of sheet and rill erosion are irrelevant to the situation in Lesotho.

As models become increasingly available for use on a personal computer, they are accessible to planners, consulting engineers, agricultural advisers and farmers. With this expansion in potential users, it is important that scientists develop and follow a code of practice to discourage model misuse. This means making clear the objectives of any particular model and indicating the kinds of problems, erosion processes and environmental conditions for which it is suitable so that users can make sensible choices about which model or models to use. It also means ensuring that models are properly validated, particularly where decisions involving expenditure on soil conservation are going to be made on the basis of their results. The success of all predictive erosion models depends on how closely the estimated rates of erosion compare with those measured in the field. The development and validation of models is therefore closely linked with data collection on rates of soil loss.

CHAPTER 6

Measurement of Soil Erosion

Data on soil erosion and its controlling factors can be collected in the field or, for simulated conditions, in the laboratory. Whether field or laboratory experiments are used depends on the objective. For realistic data on soil loss, field measurements are the most reliable, but because conditions vary in both time and space, it is often difficult to determine the chief causes of erosion or to understand the processes at work. Experiments, designed to lead to explanation, are best undertaken in the laboratory where the effects of many factors can be controlled. Because of the artificiality of laboratory experiments, however, some confirmation of their results in the field is desirable.

Experiments are usually carried out to assess the influence of one or more factors on the rate of erosion. In a simple experiment to study the effect of slope steepness, it is assumed and, as far as possible, built into the experimental design that all other factors likely to influence erosion can be held constant. The study is carried out by repeating an experiment for different slope steepnesses. A decision is necessary on the range of steepnesses, the specific steepnesses within the range and whether these should reflect some regular progression, e.g. $2°, 4°, 6° \ldots n°$, rising in intervals of $2°$, or $5° 48', 11° 36', 17° 30' \ldots n°$, rising in intervals of 0.1 sine values. The range can be selected to cover the most common slope steepnesses or extended to include extreme conditions. A decision is also needed on what levels to set for the factors being held constant, e.g. whether slope length should be 5 m or 50 m.

Measurements are subject to error. Since no single measurement of soil loss can be considered as the absolutely correct value, it is virtually impossible to quantify errors. However, they can be assessed in respect of variability. This requires replicating the experiment several times to determine the mean value of soil loss, for example for a given slope steepness in the above experiment, and the coefficient of variation of the data. This is frequently rather high. In a review of field and laboratory studies of soil erosion by rainsplash and overland flow, Luk (1981) found typical values of 13 to 40 per cent for the coefficient of variation. Extreme values of soil loss ranged from

± 39 to ± 120 per cent. To achieve a measurement of soil loss to an accuracy of ± 10 per cent with 95 per cent confidence for three different Canadian soils required six, 25 and 29 replications respectively of the experiment. A decision therefore needs to be made between carrying out between five and ten replications and accepting data with a coefficient of variation of generally between 20 and 30 per cent or completing 30 or more replications to obtain a more accurate measurement of erosion. Generally, the variability in field measurement is higher than that in laboratory studies (Luk and Morgan 1981). Roels and Jonker (1985) found coefficients of variation of 20 to 68 per cent in measurement of interrill erosion using unbounded runoff plots.

Systematic errors can be built into an experiment, for example, by starting a laboratory study on slope steepness effects on the gentlest slope, carrying out the rest of the study in a sequence of increasing slope steepness and following the same procedure for all the replications. In this way, the soil loss measured at a particular slope steepness can be influenced consistently by the amount of erosion which has taken place on the next lowest slope steepness in the sequence. Remaking the soil surface between the separate runs of the experiment may not entirely eliminate this effect. Strictly, the order of the runs should be randomized but there is often a need to balance randomization with expediency and conduct some runs in sequence where it is very time-consuming to keep switching operating conditions. For instance, in a study of slope steepness effects involving simulated rainfall of several intensities, it is expedient to complete all the runs at one intensity before proceeding to the next because of the problems encountered in continually altering the settings of the simulator. It must be appreciated, however, that failure to randomize an experiment may limit the use that can be made of statistical techniques in processing the data. Certain techniques, such as two-way analysis of variance, assume randomness in the data.

With the simple slope steepness experiment outlined above, the experimental design can be described as examining the effect of one variable (slope steepness)

with, say, five treatments (2°, 4°, 6°, 8° and 10°) and three replications, giving a total of 15 separate runs. These can be arranged in a random sequence, either in time if the experiment is carried out in the laboratory, or in space if the experiment is set up in the field. Many experiments, however, are designed to investigate the effects of more than one factor, for example examining the slope steepness effect in the laboratory at two different rainfall intensities. The design then becomes one of two variables, slope with five treatments and rainfall intensity with two treatments, giving a total of ten different treatments. With three replications, this gives a total of 30 separate runs. These could be organized in a completely random sequence or, for the reasons of expediency indicated above, in two blocks, one for each rainfall intensity, with the slope treatments randomized within each block. The resulting data can then be analysed to compare treatment means across the two blocks for all ten treatments and also to compare block means. Block randomized experiments are particularly useful in the field when, for example, an investigation of the effects of say four different land management systems might take up so much land that the results are affected by variability in soil conditions (Fig. 6.1). In this case, if three replications are used, they could be set up as three separate blocks with the treatments randomized in their position within each block. In many field situations, variations may occur in more than one factor across the experimental site, for example soil may change across the slope and slope steepness may change down the slope. In this situation, the four land management treatments could be organized on a Latin square design with each treatment located randomly, once in an across-slope direction and once in a downslope direction. Since each treatment would then occur four times, there would be four replications. The results would be analysed to compare across-slope means and downslope means as well as treatment means (Fig. 6.1; Mead and Curnow 1983).

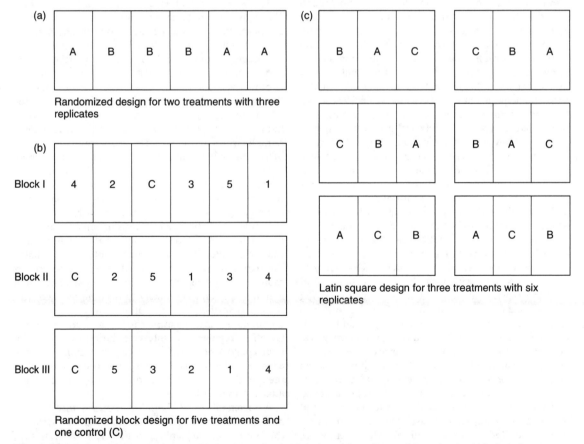

Figure 6.1 Layout of field experiments: (a) simple randomized design with replicates; (b) randomized block design; (c) Latin square design (after Mead and Curnow 1983).

The overall experimental design will vary according to the objective of the experiment and the constraints placed on it by the need to change the settings of equipment in the laboratory or variability in conditions in the field. In theory, no limits exist on the number of factors that may be incorporated in an experiment. For example, a laboratory experiment on the detachment of soil particles by rainsplash might include four soil types, five slope steepnesses, four rainfall intensities and four replications, giving 320 separate runs. In practice, the size and design of the experiment are usually limited by the amount of time and money available.

Experiments should be set up in such a way that they can be easily understood and repeated by other workers. Errors may arise due to different operators (Luk and Morgan 1981) or the use of slightly different equipment, for example, different rainfall simulators (Bryan and De Ploey 1983). Unfortunately, so few studies exist on these sources of error that their importance cannot be properly quantified at present.

A problem faced in many experiments is how much data to collect. For instance, if the objective of a study is to measure the total soil loss over a period of time, this can be achieved by collecting in bulk all the soil washed or blown from an area or by collecting the soil at regular shorter time intervals so as to obtain a total by summation of the loss in these time periods but also to learn about the pattern of loss over time. Restricting the measurement system to its bare essentials is usually cheaper and the data are generally easier to interpret. However, potentially useful information on whether most of the soil is eroded early or late in a storm is lost. This type of conflict is more apparent where the data are collected automatically, for example, by using an autographic rain gauge to measure daily rainfall. A decision then needs to be made on whether or not to analyse the shorter-period data.

There are no easy solutions to the problems of experimental design. Guidelines are to set up a study with clearly conceived objectives, define what needs to be measured and with what level of accuracy, and keep the experiment simple. This often means relying on one's experience and initiative. This is also necessary in respect of the choice of techniques and equipment. In much soil erosion and conservation work, equipment has been built by individual researchers and is not commercially available. Thus it needs to be constructed anew and in some cases modified; sometimes entirely novel equipment needs to be designed and built. The result is that there is a multiplicity of methods and equipment in use for erosion measurement and little standardization. The techniques described in this chapter are restricted to those in most common use.

6.1 Field Measurements and Experiments

Field measurements may be classified into two groups: those designed to determine soil loss from relatively small sample areas or erosion plots, often as part of an experiment, and those designed to assess erosion over a larger area such as a drainage basin.

6.1.1 Erosion plots

Bounded plots

Bounded plots are employed at permanent research or experimental stations to study the factors affecting erosion because the conditions on each plot can be controlled. Each plot is a physically isolated piece of land of known size, slope steepness, slope length and soil type from which both runoff and soil loss are monitored. The number of plots depends upon the purpose of the experiment but usually allows for at least two replications.

The standard plot is 22 m long and 1.8 m wide although other plot sizes are sometimes used. The plot edges are made of sheet metal, wood or any material which is stable, does not leak and is not liable to rust. The edges should extend 150–200 mm above the soil surface and be embedded in the soil to a sufficient depth so as not to be shifted by alternate wetting-and-drying or freezing-and-thawing of the soil. At the downslope end is positioned a collecting trough or gutter, covered with a lid to prevent the direct entry of rainfall, from which sediment and runoff are channelled into collecting tanks. For large plots or where runoff volumes are very high, the overflow from a first collecting tank is passed through a divisor which splits the flow into equal parts and passes one part, as a sample, into a second collecting tank (Fig. 6.2). Examples are the Geib multislot divisor and the Coshocton wheel. On some plots, prior to passing into the first collecting tank, the runoff is channelled through a flume where the discharge is automatically monitored. Normally an H-flume is chosen because it is non-silting and is unlikely to become blocked with debris. A further level of automation is to install an automatic sediment sampler to extract samples of the runoff at regular time intervals during the storm for later analysis of its sediment concentration; the time each sample is taken is also recorded. Rainfall is measured with both standard and autographic gauges adjacent to the plots.

A flocculating agent is added to the mixture of water and sediment collected in each tank. The soil settles to the bottom of the tank, and the clear water is then drawn off and measured. The volume of soil remaining in the tank is determined and a sample of known volume is taken for drying and weighing. The sample weight multiplied by the total volume gives the total weight of soil in the tank. If all the soil has been collected in the tank, this gives the total soil loss from the plot. For tanks below a divisor, the weight of the soil in the tank needs to be adjusted in accordance with the proportion of the total runoff and sediment passing into the tank. Thus for the layout shown in

EXPERIMENTAL PLOT LAYOUT

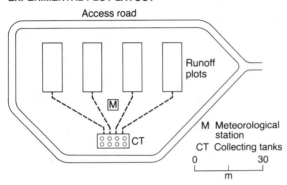

DETAIL OF COLLECTING APPARATUS

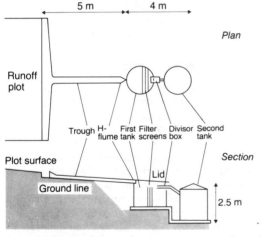

Figure 6.2 Typical layout of erosion plots at a soil erosion and conservation research station (after Hudson 1965).

Fig. 6.2, the total soil loss from the plot is the weight of the soil in the first tank plus, assuming one-fifth of the overflow from the first tank passes through the divisor into the second tank, five times the weight of soil in the second tank. Where automatic sediment sampling occurs, the sediment concentration is determined for each sample. Since the time that each sample was taken during the storm is known, the data can be integrated over time to give a sediment graph.

Full details of the equipment, manufacturing instructions for the divisors and operation of erosion plots are found in Agricultural Handbook No. 224 of the United States Department of Agriculture (1979).

Although the bounded runoff plot gives probably the most reliable data on soil loss per unit area, there are several sources of error involved with its use (Hudson 1957). These include silting of the collecting trough and pipes leading to the tanks, inadequate covering of the troughs against rainfall, and the maintenance of a constant level between the soil surface and the sill or lip of the trough. Other problems are that runoff may collect along the boundaries of the plot and form rills which would not otherwise develop, and that the plot itself is a partially closed system, being cut off from the input of sediment and water from upslope. The data obtained give a measure of the soil loss from the entire plot which may be reasonably realistic of losses from fields under similar conditions. The data do not give any indication of the redistribution of soil within a field or along a slope.

Small bounded plots of $4\,m^2$ and $8\,m^2$ have been used by Kellman (1969) in the Philippines to study erosion under a wide range of conditions at a large number of sites. With very small plots, however, it is not possible to study all the relevant processes. Plots of only $1\,m^2$ in size will allow investigations into infiltration and the effects of rainsplash but are too short for studies of overland flow except as a transporting agent for splashed particles. Thus Govers and Poesen (1988) used plots ranging from $0.5\,m^2$ to $0.66\,m^2$ in size to measure interrill erosion. Field studies of soil erodibility in Bedfordshire using small plots showed that erodibility increased with increasing fine sand content in the soils. This reflected the selectivity of fine sands to detachment by rainsplash but did not accord with erodibility assessments at the hillslope scale which incorporated runoff effects (Rickson 1987). Plots must be at least 10 m long for studies of rill erosion. Much larger plots are required for evaluating farming practices such as strip cropping and terracing. Although there is little uniformity on the size of plots for these types of experiment, they are generally in the range of 6 to 13 m wide and 15 to 32 m long.

Gerlach troughs

An alternative method of measuring sediment loss and runoff has been developed by Gerlach (1966) using simple metal gutters, 0.5 m long and 0.1 m broad, closed at the sides and fitted with a movable lid (Fig. 6.3). An outlet pipe runs from the base of the gutter to a collecting bottle. In a typical layout, two or three gutters are placed side-by-side across the slope and groups of gutters are installed at different slope lengths, arranged *en echelon* in plan to ensure a clear run to each gutter from the slope crest. Because no plot boundaries are used, edge effects are avoided. It is normal to express soil loss per unit width but if an areal assessment is required, it is necessary to assume a catchment area equal to the width of the gutter times the length of the slope. A further assumption is that any loss of water and sediment from this area during its passage downslope is balanced by inputs from adjacent areas. This assumption is reasonable if the slope is straight in plan. On slopes curved in plan, the catchment area must be accurately surveyed in the field. This disadvantage is offset by the flexibility of monitoring soil loss at different slope lengths and steepnesses within an open system. Because of their cheapness and simplicity, Gerlach troughs can be employed for sample measurements of soil loss at a large number of selected sites over a large area.

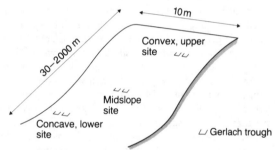

Siting Gerlach troughs on slope profile

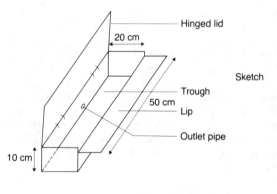

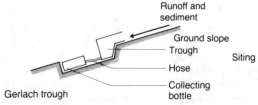

Figure 6.3 Gerlach troughs.

Splash erosion measurement

Erosion plots and Gerlach troughs provide information on water erosion by rainsplash, overland flow and rills combined. Attempts to assess the relative contributions of each are based on separate measurements of splash erosion and rill erosion with the balance being attributed to overland flow. Splash erosion has been measured in the field by means of splash boards (Ellison 1944) or small funnels or bottles (Sreenivas, Johnston and Hill 1947; Bollinne 1975). These are inserted in the soil to protrude 1 to 2 mm above the ground surface, thereby eliminating the entry of overland flow, and the material splashed into them is collected and weighed. An alternative approach is the field splash cup (Morgan 1981; Fig. 6.4) where a block of soil is isolated by enclosing it in a central cylinder and the material splashed out is collected in a surrounding catching tray. Because the quantity of splashed material measured per unit area depends upon the diameter of the funnels and cups, the following correction has to be applied to determine the real mass of particles detached by splash:

$$MSR = MS\, e^{0.054D} \qquad (6.1)$$

where MSR is the real mass of splashed material per unit area (g/cm^2), MS is the measured splash per unit area (g/cm^2) and D is the diameter of the cup or funnel (cm) (Poesen and Torri 1988).

Rill erosion measurement

The simplest method of assessing rill erosion is to establish a series of transects, 20 to 100 m long, across the slope and positioned one above another. The cross-sectional area of the rills is determined along two successive transects. The average of the two areas multiplied by the distance between the transects gives the volume of material removed. By knowing the bulk density of the soil, the volume is converted into the weight of soil loss and this, in turn, is related to an area defined by the length and distance apart of the transects (Bučko 1975). Since this method ignores the contribution of interrill erosion to the sediment carried in the rills and also depends upon being able to identify distinctly the edge of the rills, it is likely to underestimate rill erosion by 10 to 30 per cent (Zachar 1982).

6.1.2   Catchments

Investigations of sediment production in a catchment must be carried out on hillslopes and in rivers if a full picture of erosion is to be obtained. In many cases, the amount of sediment leaving a drainage basin is less than that removed from the steeper sections of the

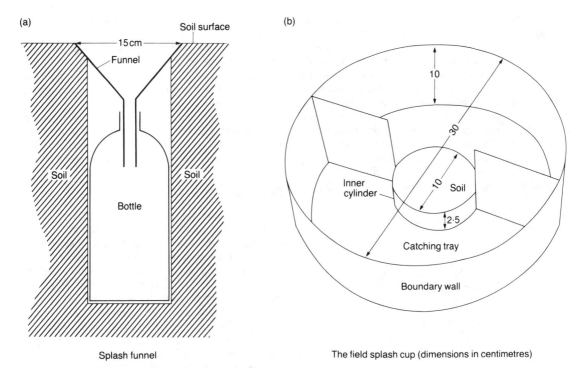

(a)

Soil surface

15cm

Funnel

Soil

Soil

Bottle

Splash funnel

(b)

10

30

Inner
cylinder

10

Soil

2·5

Catching tray

Boundary wall

The field splash cup (dimensions in centimetres)

Figure 6.4 Methods of measuring splash erosion in the field after (a) Bollinne (1975) and (b) Morgan (1981).

hillsides because of deposition of material at the base of the slopes and on the flood plains. Where more sediment leaves a basin than is contributed to the rivers from the hillslopes, the balance is derived from erosion of the river-bank.

The selection of measurement sites to establish the pattern of sediment movement poses a problem of sampling. One approach is to divide a large watershed into sub-basins and set up recording stations on the rivers at the mouth of each. Discharge is measured automatically using weirs and water depth recorders. Suspended sediment concentrations are determined from water samples taken at set times with buckets or specially-designed integrated sediment samplers, or they are monitored continuously by recording the turbidity of the water (Gregory and Walling 1973). Within each sub-basin, slope profiles are chosen at regular intervals or at random along the mid-slope line which lies half-way between the main river and the divide. Bounded erosion plots or Gerlach troughs are installed either randomly or in set positions on the profiles. A typical layout is shown in Fig. 6.5. Amphlett (1988) recommends that within such a nested arrangement, the respective sample catchments should differ in size by an order of magnitude. Thus, within a 100 km² catchment, the first level of sub-catchments should be 10 km² in size, the next

level 1 km², the next 0.1 km² and so on, down to the 10 m² or even 1 m² erosion plot.

Reservoir surveys

Sedimentation rates in lakes and reservoirs can indicate how much erosion has taken place in the catchment upstream, provided the efficiency of the reservoir as a sediment trap is known. Rapp *et al.* (1972b) used repeated surveys of designated transects across four reservoirs in the Dodoma region of Tanzania in relation to a set level or bench-mark. Using manual soundings of depth from a boat, a contour map was made of the bottom of the reservoir. The volume of the reservoir was obtained by adding up the partial volumes (V):

$$V = h \, (A + B) \, / \, 2 \qquad (6.2)$$

where V is the volume of sediment (m³), h is the contour interval (m), A is the area of the upper contour (m²) and B is the area of the lower contour (m²). To this total, the volume below the lowest contour was added, calculated as the product of the area of the lowest contour and the mean depth from that contour to the bottom. The reservoir volume was then compared with the initial volume. The reduction in volume represents the volume of sediment accumulated in the reservoir. Assuming an average dry bulk density of 1.5 Mg/m^3, the volume was converted into mass and divided by the area of the catchment upstream to give an erosion value in t/ha. This was further adjusted by assuming that the sediment in the reservoir represented about half of that eroded from the catchment. Dividing the adjusted total by the number of years since the reservoir was built gives a mean annual erosion rate. More rapid reservoir surveys can now be made using an echo-sounder to obtain depth readings, an electro-distance measuring theodolite or a laser theodolite to fix the position of the sounding, and a digital terrain model (DTM) to

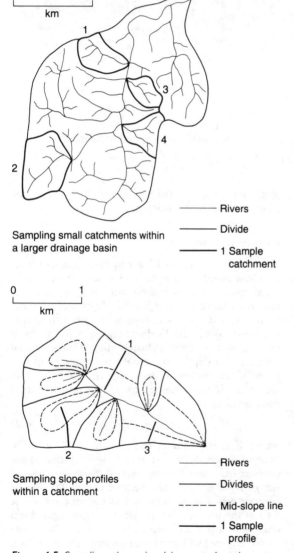

Sampling small catchments within a larger drainage basin

——— Rivers
——— Divide
——— 1 Sample catchment

Sampling slope profiles within a catchment

——— Rivers
——— Divides
----- Mid-slope line
——— 1 Sample profile

Figure 6.5 Sampling scheme involving nested catchments.

produce the contour map (Brabben, Bird and Bolton 1988).

Potential sources of error are quite high with reservoir surveys, the most important being the estimate of the trap efficiency of the reservoir, which requires knowledge of the frequency and sediment concentrations of flows carried over the spillway in times of flood, and errors arising in calculating the capacity of the reservoir. Even small errors in the latter may give rise to errors up to two orders of magnitude in determining the volume of deposited material.

Tracers

The most commonly used tracer in soil erosion measurement is the radioactive isotope, caesium-137. Caesium-137 was produced in the fall-out of atmospheric testing of nuclear weapons from the 1950s to 1970s. It was distributed globally in the stratosphere and deposited on the earth's surface in the rainfall. Regionally, the amount deposited varies with the amount of rain but within small areas, the deposition is reasonably uniform. The isotope is strongly and quickly adsorbed to clay particles within the soil. By analysing the isotope content of soil cores collected on a grid system varying in density from 10 × 10 m to 20 × 20 m, the spatial pattern of isotope loading is established. Figure 6.6 (Walling and Quine 1990) shows a typical situation. In the pasture land at the top of the slope, the isotope is concentrated at the surface; its presence in small amounts at depth is the result of earthworm activity in the soil. On the arable site, the isotope is more uniformly distributed with depth as a result of disturbance of the soil by ploughing. The decline in isotope loading by about 40 per cent on the steeper slope is a result of erosion. At the bottom of the slope, there is an increase in loading due to deposition of material. Since the deposition has been active for some years, some of the isotope lies below the present plough depth. The interpretation of spatial variations in isotope loading in comparison with those at a reference site, usually in either woodland or grassland, as indicating the patterns of erosion and deposition across a landscape was pioneered by Ritchie and McHenry (1975) and has been successfully applied in the USA (Mitchell *et al.* 1983), Canada (De Jong, Villar and Bettany 1982), Australia (Loughran, Campbell and Elliott 1988) and the UK (Walling and Quine 1990).

Where the changes in isotope loading can be correlated with measured sediment yields, the method can be used to estimate erosion rates. This can be done by taking samples on erosion plots and comparing the isotope loss, expressed as a percentage of the reference level, to the measured erosion rate or by applying a simple model which assumes that net soil loss is directly proportional to the percentage loss of caesium-137. The estimates of soil loss by the two methods may differ by as much as an order of magnitude. They should therefore be compared with other data since they may be subject to errors arising from uneven mixing of the isotope in the cultivated layer, changes in land use since 1955 and preferential adsorption of the isotope on to the clay particles during the processes of erosion, transport and deposition. Also the erosion expressed by a decline in isotope loading will represent the sum of all erosion processes whereas that measured in other ways, such as on an erosion plot, may be restricted to interrill and rill erosion.

6.1.3 Sand traps

The techniques for measuring wind erosion are less well established than those for monitoring water erosion. Various types of traps are used to catch sand moving in a band of unit width. Horizontal traps consist of troughs set in the ground, level with the surface and parallel to the direction of the wind. The trap is sometimes divided so that rolling and saltating particles fall into different compartments according to their length of hop. Alternatively, several traps of different lengths may be used. Horizontal traps have the advantage of minimum interference with the wind but a considerable length is required to collect a representative sample. Studies by Borsy (1972) on sand dunes in Hungary show that in a storm of 2 h 30 min, with a wind velocity of 8–10 m/s, the amount of sand collected in 10 cm wide horizontal traps varied from 218 g for 1 cm length, to 370 g for 50 cm length and 570 g for 1 m length. Vertical traps consist of a series of boxes placed one above the other so as to catch all of the particles moving at different heights. Many traps are unsatisfactory aerodynamically, however, because it is difficult to supply a sufficiently large exhaust to permit the free flow of air. The build-up of

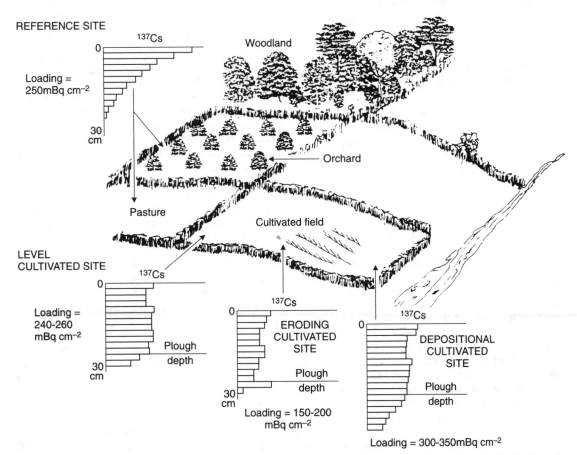

Figure 6.6 Schematic representation of the effect of erosion and deposition upon the loading and profile distribution of caesium-137 (after Walling and Quine 1990).

back pressure causes resistance to the wind which is deflected from the traps. By careful adjustment of the ratios between the sizes of inlet, outlet and collecting basin, a satisfactory trap can be produced. Examples are those designed by Horikawa and Shen (1960) and the Bagnold catcher. A problem with these horizontal

VERTICAL TRAP

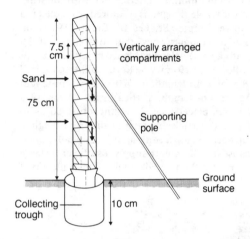

HORIZONTAL TRAPS

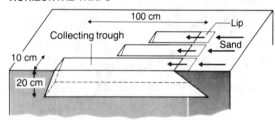

Figure 6.7 Sand traps.

and vertical traps (Fig. 6.7) is that they cannot be easily reoriented as wind direction changes. A trap was developed by Janssen and Tetzlaff (1991), comprising a 25-mm inlet or capture tube which leads into a 75-mm vertical tube where the wind-blown material settles out and falls on to a tray mounted on top of a balance. The weight of the material is recorded automatically. By fixing the top of the vertical tube to a wind vane and the base to a ball bearing, the apparatus rotates so that the capture tube always faces the wind. The traps can be placed at different heights to collect material throughout the wind profile. When tested in the field, they were found to be about 90 per cent efficient in collecting all the saltating particles (Janssen 1991).

An alternative automatically-recording sensor, the saltiphone, was developed at the Agricultural University, Wageningen, in The Netherlands (Spaan and van den Abeele 1991). It consists of a microphone installed in a 50–mm diameter and 130–mm long stainless steel tube. Two vanes at the back of the tube orient the apparatus into the wind. Saltating and suspended sediment passes into the tube and some of the particles bounce against the microphone. Only the saltating particles have enough energy to vibrate the membrane surrounding the microphone, producing a signal of high-frequency pulses which is then amplified and recorded automatically. A calibration curve has been developed to convert the number of pulses per second into particle impacts per second. Multiple saltiphones can be constructed, one above the other, to determine changes in the sediment concentration with height (van Dijk 1990). When tested in the field on the Dutch dune coast at Schiermonnikoog against the Janssen and Tetzlaff traps, the two instruments yielded very similar information (Spaan et al. 1990). The main disadvantage of the saltiphone is that it does not collect the material, so its particle-size distribution cannot be determined.

6.2 Laboratory Experiments

The key questions arising with laboratory studies concern the scale of the experiment, the greater influence of boundary effects and the extent to which field conditions are simulated. It is not usually possible to construct a scaled-down version of field conditions, for example by using a small plot to represent a large hillslope, because scale-equivalence cannot be maintained in raindrops and soil particles without affecting their basic properties or behaviour. It is therefore preferable to treat laboratory experiments as representing true-to-scale field simulation. Even so, many factors cannot be properly simulated and, unless the laboratory facilities are very large, nor can processes such as rill erosion and the saltation of soil particles by wind.

Many laboratory studies centre around the use of a rainfall simulator which is designed to produce a storm of known energy and drop-size characteristics which can be repeated on demand. The most important design requirements of a simulator are that it should reproduce the drop-size distribution, drop velocity at impact and intensity of natural rainfall with a uniform spatial distribution and that these conditions should be repeatable. The need to reproduce the energy of the natural rainfall for the intensity being simulated is generally regarded as less important (Bubenzer 1979).

Many simulators are available but none accurately recreate all the properties of natural rain (Hall 1970). There is insufficient height in most laboratories for water drops to achieve terminal velocity during fall so their kinetic energy is low. To overcome this, water is released from low heights under pressure. This results in too high an intensity and, because the increase in pressure produces small drop sizes, unrealistic drop-size distributions. The intensity can be brought down by reducing the frequency of rain striking the target area, either by oscillating the spray over the target or by intermittently shielding the target from the spray. It should be noted, however, that this only reduces the intensity measured as the amount of water applied over a given time period; the instantaneous intensity and impact energy of the simulated rain are not reduced.

Rainfall simulators are classified according to the drop-formers used. The most common are tubing tips, either hypodermic needles or capillary tubes (De Ploey, Savat and Moeyersons 1976), and nozzles (Morin, Goldberg and Seginer 1967). Meyer (1979) describes the various approaches to attaining the desired rainfall characteristics in simulators. Reviews of a large number of simulators produced by different researchers are made by Bubenzer (1979) and Hudson (1981).

Studies of splash erosion are made in the laboratory by filling containers with soil, weighing them dry, subjecting them to a simulated rain storm of pre-selected intensity and measuring the weight loss from the containers on drying. In this way different soils can be compared for their detachability. The standard container is the splash cup first used by Ellison (1944), a brass cylinder 77 mm in diameter and 50 mm deep with a wire mesh base. A thin layer of cotton wool or sponge rubber is placed in the bottom of the cup which is then filled with the soil, oven dried to a constant weight and weighed. The soil in the cup is brought to saturation prior to rainfall simulation. The soil level in the cup needs to be a few millimetres below the rim, otherwise the erosion is accelerated by material washing-off the surface during the early part of the storm (Mazurak and Mosher 1968); this effect can be enhanced if the initial raindrop impacts create a pitted or rough surface or if the surface is too loosely-packed (Kinnell 1974). Towards the end of the storm, the splash loss will be reduced because the level of the soil within the cup becomes too low and splashed particles are intercepted by the rim of the cup (Bisal 1950).

Where several soils are being compared it is important to treat them all the same. Moldenhauer (1965) recommends collecting them from the field under uniform moisture conditions, achieved by flooding the surrounding area with 100 mm of water and covering it with polythene sheet for 48 hours before taking the sample. This gives a condition close to field capacity. The use of disturbed samples is justified for studies of soil erosion on agricultural land because the soil is disturbed anyway by tillage. When the samples are brought back to the laboratory they should be dried to a constant weight and split into 200 g portions. These are poured into the splash cups uniformly, reversing the direction of pouring regularly so that all the large aggregates do not accumulate at one end. Differences in the moisture content and surface distribution of stones and aggregates are the main reasons for the high level of variability in measurements of splash erosion in laboratory experiments (Luk 1981). Although the soils are free draining, their condition does not represent the field situation because they are uncompacted and not subject to suction. Attempts to simulate soil suction on such a small scale have generally not been successful and it is difficult to achieve repeatable compaction. Fortunately, a soil in a loose saturated state is at its most erodible so laboratory experiments reproduce the worst conditions for erosion. Further processing of the soil, for example passing it through 4 mm or finer sieves to remove stones or break up aggregates, may reduce the variability in the splash measurements but can produce misleading results compared with small plot studies in the field (Rickson 1987).

Where the target is in the form of a small soil plot, the rainfall simulator may be supplemented by a device to supply a known quantity of runoff at the top of the plot, instead of relying solely on runoff resulting from the rainfall. This facility is helpful for studies of the hydraulics of overland flow during rain (Savat 1977). The set-up at the Experimental Geomorphology Laboratory, University of Leuven, consists of an aluminium and plexiglass flume, 4 m long and 0.4 m wide, which is fed by water at its upper end through a cylinder with ten openings. The discharge is controlled by a tap through which water is pumped to the cylinder from a container; a constant water level is maintained in the container (Savat 1975). The set-up at Silsoe College comprises a metal soil tray, 2.5 m long, 1.0 m wide and 0.5 m deep, with a wire-mesh base. A layer of foam rubber or a semi-permeable geotextile membrane is placed over the mesh to allow infiltration and the tray is filled with the test soil. Upslope of the tray is a 0.2-m long wooden approach slope to which runoff is supplied as overflow from a tank. The discharge is controlled and can be varied by the rate of inflow of water to the tank. The approach slope is covered with a thin layer of sandy soil mixed with an adhesive to allow soil roughness effects to be imparted to the flow before it reaches the test soil. With both the Silsoe and Leuven flumes, the slope can be adjusted to different steepnesses and rainfall simulators can be positioned over the top.

The main problem arising with these small flumes is

that edge effects are difficult to eliminate. They can be reduced, however, by collecting the runoff and sediment washing off a narrow strip down the centre of the plot instead of collecting for the whole of the plot width. The short length of the flumes makes it difficult to simulate rill erosion although the supply of runoff at the top of the slope can be adjusted to simulate the effect of different slope lengths. Sediment can also be added to the runoff upslope of the test soil.

A much longer combined rainfall and runoff simulator with adjustable slope has been built at the Centre for Geomorphology in Caen, France, to enable rill erosion to be investigated under controlled conditions (Govers *et al.* 1987). The flume is 20.4 m long, 1.4 m wide and 0.7 m deep. The base of the flume is covered with a 15-cm thick layer of gravel overlain by a perforated plastic sheet on top of which 40 cm thickness of the test soil is placed. A rainfall simulator, comprising 644 flexible capillary tubes, is mounted 1.7 m above the top of the flume. A fine wire-mesh is placed 70 cm below the capillaries to break up the 2.9 mm diameter drops into a drop-size distribution ranging from 0.6 to 4.6 mm with a median volume drop diameter of 2.8 mm. In a further development, the long flume in the Soil Erosion Laboratory, University of Toronto, Canada, has been constructed from ten flexible modules, each 2.45 m long, 0.85 m wide and 0.31 m deep which can be tilted and combined to produce slopes of varying profile shape (Bryan and Poesen 1989). A rainfall simulator comprising nine cone-jet spray nozzles is placed 8 m above the flume.

Almost all the basic studies on wind erosion have been carried out in the laboratory in wind tunnels. Wind is supplied by a fan which either sucks or blows air through the tunnel. The tunnel is shaped so that air enters through a honeycomb shield, serving as a flow straightener, into a convergence zone, where flow is constricted, passes through the test section, and leaves through a divergence zone, in which flow is diffused. A mesh screen at the outlet traps most of the sand particles whilst allowing the air to blow through without the build-up of back pressure. Realistic wind profiles are produced only in the test section and the value of the tunnel depends on the length of this. Small tunnels, such as the one described by De Ploey and Gabriels (1980) with a test section only 1.5 m long, do not permit a satisfactory sand flow to be attained. A section at least 15 m long is required for this though Bagnold (1937) was able to simulate sand flow in a 10-m long tunnel by feeding a stream of sand into its mouth.

The principles of wind tunnel design are described by Pope and Harper (1966). Most of the tunnels used for soil erosion research are of the open-circuit type where air is drawn in from outside. These are cheaper than closed-circuit tunnels and afford easier control over the air flow in the test section because there is less upstream contact between the air and the boundary walls and therefore less turbulence. In a closed-circuit

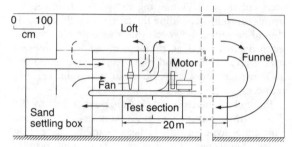

Figure 6.8 Details of a closed-circuit wind tunnel (after Knottnerus 1979).

(Fig. 6.8), the fan is located in a loft above the test section and dry centrally-heated air is forced downwards and into the test section through a funnel where unwanted turbulence is likely to be induced. Closed-circuit tunnels are less noisy and give good control over air humidity. The latter is important because it influences the critical wind velocity for particle movement. Studies with an erodible sandy soil in The Netherlands show that the critical velocity, as measured at 10 m height, is 8 m/s when the relative humidity of the air is 50 per cent, but 13 m/s when the humidity is 70 per cent (Knottnerus 1979). Soil trays are placed on the floor of the tunnel in the test section and these can be removed for weighing. Thus the amount of erosion can be determined by weight loss. Wind tunnels have been used to simulate the effects of soil roughness, especially that related to stones (Logie 1982) and low-growing plants (Knottnerus 1976) at full scale but have also been applied, using the principles of engineering similitude (Woodruff, Fryrear and Lyles 1963), to evaluating the effectiveness of shelterbelts in scaled-down experiments (Woodruff and Zingg 1955). They are also used for testing and calibrating sand traps.

6.3 Composite Experiments

In recent years greater use has been made of field experiments combined with rainfall simulation. They have now virtually replaced the natural runoff plot as the major research tool in the USA for studies of both soil erodibility (Laflen *et al.* 1991) and vegetation cover effects (Simanton and Renard 1992). The combination has the advantages of field conditions for soils, slope and plant cover, all of which cannot be easily reproduced in the laboratory, with the benefits of a repeatable storm.

For studies with small plots, the rainfall simulators used in the laboratory, like the one designed by Meyer and Harmon (1979) can be transferred directly to the field. Larger simulators can be built by joining together modules of the smaller ones. Modules of a simulator built by Bazzoffi, Torri and Zanchi (1980) have been combined to form a field simulator capable of covering a standard runoff plot (Zanchi *et al.* 1981). Alternatively, large-area simulators can be purpose-designed and built. The best example is the rotating-boom simulator (Swanson 1965) in which nozzles are mounted on a series of spray booms positioned radially from a central stem. The nozzles spray water from a height of 3 m and the booms rotate at four revolutions per minute to give a uniform distribution of rain over a circular area large enough to accommodate two plots, 10 m by 4.3 m.

Less use has been made of field wind tunnels partly because, as noted above, small ones, which would be the most portable, are of limited value. Since experimental field plots are not used in wind erosion research, there is little demand for large permanently-sited tunnels. However, some large wind tunnels have been developed for field use (Zingg 1951; Chepil, Woodruff and Zingg 1955; Fryrear 1984; Bocharov 1984).

CHAPTER 7

Strategies for Erosion Control

The aim of soil conservation is to obtain the maximum sustained level of production from a given area of land whilst maintaining soil loss below a threshold level which, theoretically, permits the natural rate of soil formation to keep pace with the rate of soil erosion. In addition, there may be a need to reduce erosion to control the loss of nutrients from agricultural land to prevent pollution of water bodies; to decrease rates of sedimentation in reservoirs, rivers, canals, ditches and harbours; and to limit crop damage by wind or by burial beneath water- and wind-transported sediments. In the longer term, erosion must be controlled to prevent land deteriorating in quality until it has to be abandoned and cannot be reclaimed, thereby limiting options for future use. Since erosion is a natural process, it cannot be prevented but it can be reduced to a maximum acceptable rate or soil loss tolerance.

7.1 Soil Loss Tolerance

Unfortunately, deciding the tolerance level theoretically as the rate at which soil loss equals the rate of soil formation is not possible in practice. It is difficult to recognize when this balance exists and, although rates of soil loss can be measured (see Chapter 6), rates of soil formation are so slow they cannot be easily determined. According to Buol, Hole and McCracken (1973), rates of soil formation throughout the world range from 0.01 to 7.7 mm/y. The fastest rates are exceptional, however, because the average is about 0.1 mm/y (Zachar 1982). For the USA, Kirkby (1980a) suggests rates of about 0.1 mm/y for the northeast, 0.2 mm/y for the loess soils of the Great Plains and 0.02 mm/y for the arid southwest. Evans (1981) considers 0.1 mm/y as a realistic rate for the UK. Dunne, Dietrich and Brunengo (1978) estimated rates in Kenya to range from 0.01 to 0.02 mm/y in the humid areas but to fall below 0.01 mm/y in the semi-arid areas. If it is assumed that all the soils in Iceland have formed in the last 10,000 years, the average thicknesses of silt loam profiles at the present time would indicate rates of formation of 0.05 to 0.15 mm/y (Jóhannesson 1960). Where deposition of wind-blown material has added to contemporary soils,

higher rates of formation have been measured. In eastern Iceland soils formed at rates of 0.12 mm/y between 1850 BC and 750 BC; 0.27 mm/y between AD 1362 and 1845; and 1.11 mm/y from AD 1875 to 1950 (Thórarinsson 1958).

An alternative approach which avoids the need to measure the rate of new soil formation directly, is to estimate the rate required to match the rate of soil removal by erosion and solution in areas where an equilibrium condition might be presumed to exist. Using data from small watersheds under forest and grassland vegetation, Alexander (1988) found the required rates to be between 0.3 and 2 t/ha with the majority being below 1 t/ha which, assuming a bulk density for the soil of 1.0 Mg/m^3, is equivalent to 0.1 mm/y. Such a rate, however, may be a rather conservative indicator for the development of an agriculturally-productive soil.

Bennett (1939) and Hall, Daniels and Foss (1979) suggest that in soils of medium to moderately-coarse texture on well-managed crop land the annual rates of formation of the A horizon can exceed 11.2 t/ha. This is because the subsoil can be improved by incorporating it with the top soil during tillage and by adding fertilizers and organic matter. It is against this background that values for soil loss tolerance are set to maintain an adequate rooting depth and avoid significant reductions in yield whilst the surface layer of soil is removed by erosion (McCormack and Young 1981). Soil loss tolerance is then defined as the maximum permissible rate of erosion at which soil fertility can be maintained over 20 to 25 years. A mean annual soil loss of 11 t/ha is generally accepted as appropriate but values as low as 2 t/ha are recommended for particularly sensitive areas where soils are thin or highly erodible (Hudson 1981). Where soils are deeper than 2 m, the subsoils capable of improvement and reductions in crop yield are unlikely to be brought about by erosion over the next 50 years or more, some scientists favour increasing the tolerance to 15 or 20 t/ha (Schertz 1983).

These recommendations on soil loss tolerance are based solely on agricultural considerations. In many parts of the world, problems of pollution and sedimentation that arise as plant nutrients and pesticides leave a field either in solution in the runoff or attached to sediment particles, outweigh those of lost farming potential. Particular concerns relate to the removal of nitrogen, phosphorus and organic matter but, in the context of erosion, most attention has been given to the removal of phosphorus which can occur in both soluble and particulate forms. Studies on small catchments in Denmark (Kronvang 1990; Hasholt 1991) show that between 45 and 82 per cent of the phosphorus contributed annually to water courses can be in particulate form amounting to some 0.05 to 0.47 kg/ha. Similar figures, ranging from 0.06 to 0.45, have been measured in southern Sweden (Ulén 1988; Alström and Bergman 1988). Much higher rates are recorded in the southern USA where particulate phosphorus can account for more than 80 per cent of the total phosphorus transported from agricultural land. Some 18–49 per cent may be bioavailable phosphorus which means that it is potentially available for uptake by algae (Sharpley and Smith 1990). Although no guidelines exist regarding acceptable rates of particulate phosphorus within runoff, the above figures match what would be expected if only the base content of phosphorus in the soil was being removed. According to Brady (1990), the phosphorus content of soil ranges between 200 and 2,000 mg/kg. Thus between 0.2 and 2 g of phosphorus would be eroded for every kg of soil, assuming all the erosion was in particulate form. At an annual erosion rate of 1 t/ha, this would give annual losses of particulate phosphorus of 0.2 to 2 kg/ha. Such simple calculations support the recommendation made by Moldenhauer and Onstad (1975) that a soil loss tolerance no greater than 1 t/ha may be required to reduce the effects of non-point source pollution from agricultural land to acceptable levels.

Although a soil loss tolerance of 11 t/ha may maintain a productive A-horizon over a 25- or even a 50-year period, the overall soil depth will still be reduced and, in 100 to 200 years, the surface material may become closer to gravel than a farming soil (Kirkby 1980a). A value as low as 2 t/ha may be inadequate to control the off-site effects of erosion, particularly because higher concentrations of particulate phosphorus are likely in events of low magnitude and high frequency when runoff volumes and velocities are low and the sediment load is made up of fine material. Current values for soil loss tolerance are therefore highly dubious. A better guideline is likely to be an assessment of the natural rate of erosion which, assuming that the environment is reasonably stable under natural conditions, will be close to the rate of new soil formation and the rate at which the base mineral status of the soil is maintained. This guideline would lead to tolerance levels of 1 to 2 t/ha but, in some areas, these would have to be reduced to 0.1 to 0.2 t/ha. Generally, soil loss tolerance should be an order of magnitude lower than most current recommendations.

7.2 Principles of Soil Conservation

From the discussion (Chapter 2) of the mechanics of the detachment and transport of soil particles by rainsplash, runoff and wind, it follows that the strategies for soil conservation must be based on covering the soil to protect it from raindrop impact; increasing the infiltration capacity of the soil to reduce runoff; improving the aggregate stability of the soil; and increasing surface roughness to reduce the velocity of runoff and wind. The various conservation techniques can be described under the headings of agronomic measures, soil management and mechanical methods. Agronomic measures utilize the role of vegetation to protect the soil against erosion. Soil management is concerned with ways of preparing the soil to promote plant growth and improve its structure so that it is more resistant to erosion. Mechanical or physical methods, often involving engineering structures, depend upon manipulating the surface topography, for example by installing terraces or wind breaks, to control the flow of water and air. Agronomic measures combined with good soil management can influence both the detachment and transport phases of erosion whereas mechanical methods are effective in controlling the transport phase but do little to prevent soil detachment (Table 7.1).

Generally, preference is always given to agronomic measures. These are less expensive and deal directly with reducing raindrop impact, increasing infiltration, reducing runoff volume and decreasing wind and water velocities. They are also more easily fitted into an existing farming system. Mechanical measures are largely ineffective on their own because they cannot prevent the detachment of soil particles. Their main role is in supplementing agronomic measures, being used to control the flow of any excess water and wind that arise. Many mechanical works are costly to install and maintain. Some, such as terraces, create difficulties for farmers. Unless the soils are deep, terrace construction exposes the less fertile subsoils and may therefore result in lower crop yields. On irregular slopes, terraces will vary in width, making for inefficient use of farm machinery, and only where slopes are straight in plan can this problem be overcome by parallel terrace layouts. Also, there is a risk of terrace failure in severe storms. When this occurs, the sudden release of water ponded up on the hillside can do more damage than if no terraces had been constructed. For

Table 7.1 Effect of various soil conservation practices on the detachment and transport phases of erosion

Practice	Rainsplash D	Rainsplash T	Runoff D	Runoff T	Wind D	Wind T
Agronomic measures						
Covering soil surface	*	*	*	*	*	*
Increasing surface roughness	–	–	*	*	*	*
Increasing surface depression storage	+	+	*	*	–	–
Increasing infiltration	–	–	+	*	–	–
Soil management						
Fertilizers, manures	+	+	+	*	+	*
Subsoiling, drainage	–	–	+	*	–	–
Mechanical measures						
Contouring, ridging	–	+	+	*	+	*
Terraces	–	+	+	*	–	–
Shelterbelts	–	–	–	–	*	*
Waterways	–	–	–	*	–	–

– no control; + moderate control; * strong control
D = detachment, T = transport

all these reasons, terracing is often unpopular with farmers.

7.3 Assessment Sequence

Soil conservation design most logically follows a sequence of events (Fig. 7.1) beginning with a thorough assessment of erosion risk using the techniques described in Chapter 4. This is followed by designing a sound land-use plan based on what the land is best suited for under present or proposed economic and social conditions, land tenure arrangements and production technology and what is compatible with the maintenance of environmental stability.

7.3.1 Land capability classification

By adopting land capability classification as a methodology for land-use planning a distinction can be made, as seen in Chapter 4, between areas where erosion is likely to occur through mismanagement and areas where erosion occurs through misuse. In the first case, erosion can be reduced to a tolerable level by suitable conservation practices but, in the second case, erosion control is expensive and difficult, if not impossible. The land is likely to become so degraded that eventually it will have to be abandoned. Much money can be spent attempting to reclaim land where soil degradation has become virtually irreversible but the low productivity of the land means that the investment is highly questionable since it can never be repaid.

7.3.2 Defining conservation needs

The ultimate success of soil conservation schemes depends on how well the erosion problem has been identified, the suitability of the conservation measures selected to deal with the problem and the willingness of farmers and others to implement them. These features of an erosion-control strategy are just as important as the ability of the engineer to design the conservation structures required. Correct identification of the major areas of erosion and therefore the main sources of sediment is also essential. Too often expensive conservation work has been carried out on areas of land which were thought to be important sediment sources but in fact were not (Perrens and Trustrum 1984). The conservation system must be closely related to the nature of the erosion problem.

Analysis of existing farming systems

The farming systems already operating in the area should be analysed to see how well the land is protected under current cropping or grazing regimes, particularly at the critical periods of the year when erosion risk is highest. This is best achieved by relating the farming calendar to the monthly pattern of erosivity. For example, in eastern Thailand (Prinz 1992), erosivity reaches its peak in the first third of the wet season when the crops of kenaf, groundnut and upland rice do not yet provide adequate cover. On the loamy soils in north-eastern France (Monnier and Boiffin 1986), the danger of erosion under winter-sown cereals is high, even when the soil is covered and rainfall is relatively low, because of the marked reduction in infiltration rates between December and March as a result of crusting of the soil in the autumn.

Perceptions of erosion

The relevance of conservation measures to a farming system depends in part on how farmers and others perceive the erosion problem and its consequences. Most farmers are aware of the problem and its effects and the notion of the peasant farmer damaging the land through ignorance is severely mistaken. The small-scale farmer is as much an experienced and efficient practitioner of land husbandry as the large-scale commercial farmer but with a different objective, namely that of survival rather than profit (Hudson 1981). If a farmer destroys land by overcropping or overgrazing, it is because there is no alternative employment from which to make a living. Farmers who work marginal land because there is no other land available are generally well aware of the damage they cause. Surveys of small-scale farmers in Sierra Leone

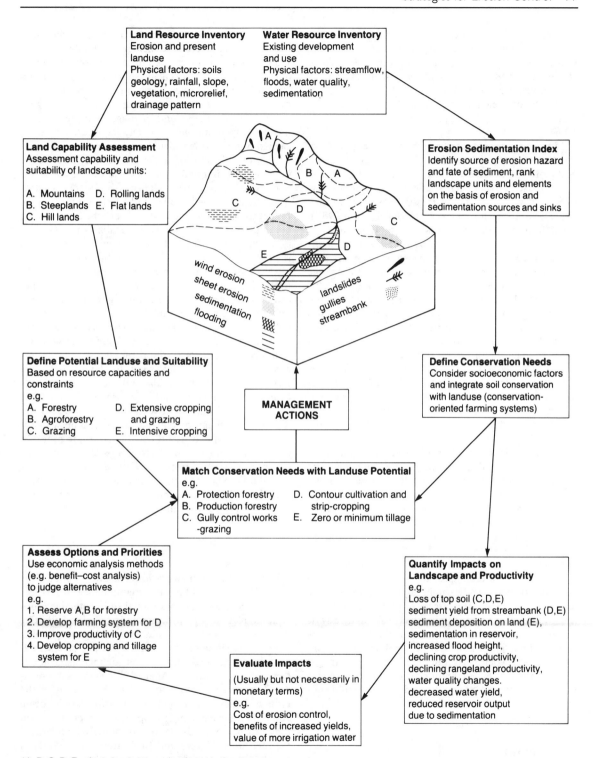

Land Resource Inventory
Erosion and present
landuse
Physical factors: soils
geology, rainfall, slope,
vegetation, microrelief,
drainage pattern

Water Resource Inventory
Existing development
and use
Physical factors: streamflow,
floods, water quality,
sedimentation

Land Capability Assessment
Assessment capability and
suitability of landscape units:

A. Mountains   D. Rolling lands
B. Steeplands E. Flat lands
C. Hill lands

Erosion Sedimentation Index
Identify source of erosion hazard
and fate of sediment, rank
landscape units and elements
on the basis of erosion and
sedimentation sources and sinks

wind erosion
sheet erosion
sedimentation
flooding

landslides
gullies
streambank

Define Potential Landuse and Suitability
Based on resource capacities and
constraints
e.g.
A. Forestry   D. Extensive cropping
B. Agroforestry   and grazing
C. Grazing   E. Intensive cropping

**MANAGEMENT
ACTIONS**

Define Conservation Needs
Consider socioeconomic factors
and integrate soil conservation
with landuse (conservation-
oriented farming systems)

Match Conservation Needs with Landuse Potential
e.g.
A. Protection forestry   D. Contour cultivation and
B. Production forestry   strip-cropping
C. Gully control works   E. Zero or minimum tillage
   -grazing

Assess Options and Priorities
Use economic analysis methods
(e.g. benefit–cost analysis)
to judge alternatives
e.g.
1. Reserve A,B for forestry
2. Develop farming system for D
3. Improve productivity of C
4. Develop cropping and tillage
   system for E

**Quantify Impacts on
Landscape and Productivity**
e.g.
Loss of top soil (C,D,E)
sediment yield from streambank (D,E)
sediment deposition on land (E),
sedimentation in reservoir,
increased flood height,
declining crop productivity,
declining rangeland productivity,
water quality changes.
decreased water yield,
reduced reservoir output
due to sedimentation

Evaluate Impacts

(Usually but not necessarily in
monetary terms)
e.g.
Cost of erosion control,
benefits of increased yields,
value of more irrigation water

*A, B, C, D, E refer to landscape units shown in the diagram.

Figure 7.1 Sequence of events in planning a soil conservation strategy (after Perrens and Trustrum 1984).

showed that the majority correctly perceived an erosion problem on their land and associated it with high rainfall, steep slopes and lack of vegetation cover (Millington 1987). Studies of large-scale farmers in erosion-prone areas of Ohio revealed that over 40 per cent of them knew they owned land on which erosion was severe enough to affect productivity (Napier 1990).

Most farmers are concerned with the effects of erosion on potential productivity and on increased costs of, for example, seeds for replanting a crop destroyed by erosion, fertilizers to maintain soil fertility, and water storage or irrigation facilities to provide additional water for crop growth (Shaxson 1987). Most farmer's decisions, however, tend to reflect a compromise between preventing long-term soil damage by erosion and maximizing short-term income. Generally, the farmer is not unwilling to change practices but will do so only if substantial benefits arise and the investment costs can be recovered. Where a land user does not perceive such benefits, soil conservation is unlikely to be adopted. For example, where land is cheap and readily available, the incentive is to make a rapid profit and then move on to new land when the soil becomes unproductive. For the subsistence farmer, a particular problem is whether or not a new practice increases the risk of crop failure. Taking-up higher yielding crop varieties which may also help to reduce erosion by increasing biomass will not be an acceptable gamble if, in one year out of ten, the crop yields less than the minimum required for subsistence, even though much higher yields will be obtained in the other nine years (Hudson 1981). These pressures invariably mean that, regardless of good intentions, the concept of a farmer managing the land according to an ethic of land stewardship, is an inappropriate base for soil conservation (Hudson 1988).

In addition to understanding the perception of the land user, it is often helpful to know how the whole community perceives erosion, especially the people experiencing the off-site effects of pollution and sedimentation. Their attitude towards the farmers may be influential in inducing change particularly in societies which adopt the principle of the 'polluter pays' and where land owners may pursue claims for damages through the courts. A community awareness may also promote collective action to address erosion problems.

Land tenure

The arrangements by which tenure is granted to the land user can influence attitudes towards soil con-

servation techniques. Where farmers own the land, they are more likely to consider the long-term consequences of their actions and adopt soil-protection measures unless the need for short-term survival dictates otherwise. Tenure systems based on short-term cultivation rights, share-cropping and collectives generally lead to poor land management because of uncertainty about whether any conservation work carried out on the land will be rewarded.

The overall size of a farm does not necessarily influence the frequency or the type of soil conservation measures employed. Hallsworth (1987) cites data from the Nyasa, Rift Valley and Western Provinces of Kenya which show that 51 per cent of the farmers with holdings of less than 1 ha use fertilizers to maintain productivity whereas on farms of over 10 ha the figure is still only 68 per cent. Equivalent figures for the use of terracing are 24 per cent and 21 per cent respectively. For mulching, however, the figures are 5 per cent and 17 per cent, the lower figure reflecting the difficulty of supplying mulch material on very small holdings. More important than farm size is the layout of the farm and, in particular, its degree of fragmentation since many conservation measures, such as terraces, become impractical when the land is held as a series of extremely small and scattered parcels. Mulching is more difficult because there may be a considerable distance between the strip of land from which a mulch can be taken and the strip on which it needs to be applied. Land consolidation programmes may not improve the erosion status, however, because, as seen in northern France, the reorganization of holdings into larger fields means the removal of field boundaries and increases in the length over which surface runoff can flow.

Labour

All soil conservation work implies extra labour. It is needed for the building and maintenance of terraces and for the growing of additional soil-protective crops either in rotation or by intercropping (Table 7.2). The likelihood of any soil conservation measure being adopted depends on whether the farmer and associated family can realistically meet the increased labour demands (Stocking and Abel 1992). Surveys in northern Thailand showed that whilst 76 per cent of the farmers believed that soil conservation would bring about an improvement in crop yields, 85 per cent said they did not have the labour resource to implement the measures (Harper and El-Swaify 1988). Whilst it may be thought that in countries with high population densities and high rates of unemployment, labour-intensive conservation schemes would be beneficial,

Table 7.2 Examples of labour requirements for soil conservation

Conservation practice	Labour (person-days/ha)
Construction of contour terraces	100–300
Construction of bench terraces	200–500
Construction of bench terraces with stone side slopes (Inca type)	1,000–1,800
Annual maintenance of terraces	40–60
Contour stone bunds	50–60
Contour grass barriers	40–45
Intercropping an additional crop with maize (extra labour over maize monoculture)	150–200
Construction of Katumani-type pits	100–120

Sources: Wenner (1981), Alfaro Moreno (1988), IFAD (1992), Stocking and Abel (1992), Gichangi *et al.* (1992).

this view ignores the fact that farmers may give a higher priority in their use of labour to other activities such as marketing produce, traditional craft-work, house maintenance and off-farm employment. Also, a poor small-scale farmer is often physically unable to put more hours of work into the land because of poor diet and poor health. If the family splits up to try to secure additional sources of income by the young men going to the towns or to mines, an inadequate workforce is left behind on the farm comprising grandparents, whose working life is over, young women, whose working time is limited by duties of housekeeping and child-bearing, and children.

The organization of labour within the family is frequently highly specialized by age and sex. Men may do the land clearing and ploughing and women the weeding and the collection of fuel wood; men may look after the cash crops and women the food crops. Knowing which groups perform which tasks is important so that extension workers can target the most appropriate sector of the labour force. When soil conservation introduces new tasks, these have to be assigned in acceptable ways without creating stresses elsewhere in the system. For example, replacing hand-hoeing with ox-ploughs may enable more land to be cultivated by the men but will result in much more weeding for the women (Milner and Douglas 1989).

The availability of labour can be improved if farmers have sufficient income to hire additional workers but, more often than not, serious constraints exist, particularly at times of land preparation and harvesting when the work has to be done in a very short time. Other ways of improving labour supply are through communal schemes where farmers and their families come together to build terraces, dig the soil or weed the crop for what is essentially a social occasion.

Access to soil conservation

The ability of farmers to adopt soil conservation measures will depend on their access to all appropriate resources, not just labour. These may vary from access to knowledge of new systems to an ability to afford the necessary inputs of capital to take them up. The numbers of extension workers with experience of soil conservation, the access of the farmers to extension staff and the perceived relevance of their recommendations will influence whether an extension service is successful or ineffective. Whether or not farmers have the cash to purchase the additional seeds, fertilizer or machinery required to support a more conservation-oriented farming system will clearly affect its uptake; many poor farmers have insufficient security to support loans and would consider the risk of borrowing money too high. Also, most credit agencies would view small-scale farmers as unacceptable risks and restrict access to credit to larger-scale and wealthier land users. Clearly it is pointless designing a soil conservation programme which requires levels of input to which the targeted farmers have no access. However, it should be recognized that many farmers use their own initiative, technical skill and labour to develop soil conservation measures where they benefit from so doing. Between 1948 and 1978, farmers in the Machakos District of Kenya (Tiffen, Mortimore and Gichuki 1994) made large investments in terracing, tree planting and hedging as well as improving cultivation techniques in order to grow coffee, cotton, oranges and papaya. Despite a rapidly increasing population density of 3 per cent per year, the cash value of farm output per hectare rose tenfold and soil loss was reduced by one-third. In these circumstances, the aim of the soil conservationist is to work with the farmer to enhance the performance of the measures and optimize their benefits.

External factors

The causes of soil erosion rarely relate only to the existing physical and socio-economic conditions of a farming system but arise also from its inability to adapt to changing circumstances brought about, for example, by drought, increases or decreases in population, fragmentation or enlargement of farms, economic pressures to grow cash crops rather than food crops, and new technologies. As shown by Franke and Chasin (1981) in Niger, the penetration of a cash economy into areas of traditional subsistence farmers and pastoralists frequently marginalizes those sections of the community who do not have sufficient economic resources to take advantage of the new income opportunities. Pastoralists are often displaced from the better quality land and forced to utilize and therefore overgraze poorer areas. Small-scale farmers who plant the new cash crops, which are often more soil-degrading, lose their food security and in years when the market price for the crop is low are unable to provide the necessary inputs to maintain soil fertility and conservation structures. To obtain money to buy food, they may have to relinquish control of their land and lose their security of tenure. With more of the land area taken up in cash crops, food crops are in short supply and more expensive. Similarly, surplus labour is taken-up by the larger-scale farmers, reducing its seasonal availability. As a result, many farmers can no longer adjust their farming system to meet the needs of soil conservation.

Erosion in much of Europe also reflects the breakdown of farming systems which, in the UK, France and Belgium, formerly allowed a rotation of cereals, root crops and grasses, and in the Mediterranean a combination of tree crops, cereals and pastures. With economic incentives to grow continuous cereals and roots, all essentially soil-degrading crops, the loss of grass has led to a decline in organic matter and a deterioration in the aggregate stability of the soils. Increasing mechanization over 15 to 30 years has led, on some soils, to the formation of a thick, compacted plough pan at 30 cm depth, reducing the infiltration rate and resulting in greater runoff. The increased use of harrows and vibro-driven cultivators has brought about a more pulverized and less resistant top soil (Roose and Masson 1985). Increasing the number of tractor passes over the land has also increased the compaction of the surface soil along wheelings.

If soil erosion as a response to changes in farming systems is to be avoided, soil conservation must be a dynamic process capable of adjusting to future change. It is insufficient merely to develop a strategy which solves an immediate problem.

Technological transfer

Though the principles of soil conservation are transferable, conservation strategies developed in one area will not necessarily work elsewhere. What is feasible on a 100 to 200 ha farm with a high level of mechanization in the Mid-West of the USA is unlikely to be applicable to a smallholding of less than 0.5 ha in the tropics. The non-transferability of soil conservation measures is perhaps best illustrated by the failure of channel terraces within the small-scale farming systems of Africa. Borrowed from the USA, where they were successfully developed and implemented in the 1930s and 1940s, such terraces formed the nub of soil conservation programmes proposed by the European colonial administrations in much of central and eastern Africa. Since the Crown Land made available to European settlers was not under pressure, there was little incentive to practise soil conservation and the colonial governments were forced to support the implementation of terraces and other structural measures by legal means to ensure that the works were maintained and appropriate soil management, such as contour-ploughing, followed. With the measures being generally effective on the large settler farms, the governments tried to impose them on the land reserved for the local population on which erosion was becoming particularly severe. Despite much exhortation and attempts at compulsion, the attempt was a failure. The measures were seen as too labour-intensive for the expected economic return, took too much scarce land out of cultivation, were not supported by the same system of subsidies and loans that were made available to the settlers, and were regarded as an illustration of what was increasingly seen as the unfairness of colonial rule (Temple 1972a; Wood 1992). Today, there are many examples throughout Africa where channel terraces and bench terraces implemented by governments, pre- and post-independence, have not worked whereas indigenous, often highly labour-intensive systems, have been successful (Roose 1992; Cheatle and Njoroge 1993; Tiffen, Mortimore and Gichuki 1994).

7.3.3 Impact assessment

The next stage is to quantify the impacts of the proposed land use and associated conservation strategy. Simple environmental impact assessments can be made using the predictive models described in Chapter 5. Over the last 15 years, models have also been developed for forecasting the impact of erosion on the productivity of the soil. Most of the models, such as the Productivity Index (PI) (Larson, Pierce and Dowdy 1985) and the Erosion Productivity Impact Calculator (EPIC) (Williams, Jones and Dyke 1984) are only valid in areas such as the USA and western Europe, where soil fertility is not a limiting factor because nutrients are being added to the soil through fertilizers. At present, the models can only serve as guides to the likely impacts of erosion because they have not been properly validated owing to a lack of adequate data. To obtain the necessary information on changes in the nutrient status and physical properties of a soil as a result of erosion requires a minimum period of 10 years under natural conditions. Trying to speed up the process by scalping the top soil to different depths results in an underestimation of the effect on nutrient removal by some five to ten times (Stocking 1985).

7.3.4 Economic evaluation

Ideally, the impact analysis should be followed by an economic evaluation but there are few examples of this in practice, particularly at the farm level. One problem of applying an economic analysis is that many aspects, such as the benefits related to minimizing adverse environmental impacts, are difficult to quantify. Although this may be of limited importance in a farm budget, it may be vital, at a project scale, for deciding on the size of financial incentives to farmers to promote soil conservation for the benefit of the community. An analysis directed at the community should take account of the costs of not controlling the erosion, including downstream or downwind sedimentation, and the benefits in terms of crop production and employment. However, at this level the analysis generally suffers from inadequate information on how to price labour costs when much of the labour is in the form of unpaid inputs from the farmers and their families; how to take account of the effects of different currency exchange rates where a black currency market exists alongside the official exchange; and how to cost the differences between a project with soil conservation and one without when the impact of continuing erosion on productivity cannot be quantified (Bojö 1992). There is also disagreement on whether projects should be evaluated in terms of their net present value (NPV), that is the sum of discounted benefits less the sum of discounted costs; their internal rate of return (IRR), that is the rate of interest that the project can afford to pay to cover the resources required; or their benefit–cost ratio, defined as the sum of discounted benefits divided by the sum of the discounted costs. Disagreement also exists on whether the evaluation should be made over 10, 20, 30 years or longer. Extending the time horizon to 50 years can have a detrimental effect on the outcome, especially if discount rates of 10 per cent or more are used (Bojö 1992).

If the analysis is applied to an individual farmer, costs and benefits can be more easily quantified and yearly cash flows calculated (Hedfors 1983). Benefit–cost ratios to farmers in the Acelhuate Basin, El Salvador from adopting a system of contour cultivation, contour grass strips and contour ditches to grow maize, sorghum and beans on steep slopes ranged from 1.0 to 2.8, depending on slope steepness and the depth of soil (Wiggins 1981). Greatest returns were predicted for the gentler-sloping land with deeper soil whereas benefits were very limited or neutral for farmers working slopes of 25–51° with no top soil and renting the land on annual leases. It is clear that these last farmers have no economic accessibility to soil conservation.

In contrast, economic analysis of farms in the Machakos District of Kenya shows that it pays to adopt soil conservation measures. Calculations for a 1.9 ha farm give a return of KSh 5 per man-day where no soil conservation is employed, KSh 11 where terracing and some tree- and grass-planting are used, and KSh 21 where all the terrace edges are planted with fodder grasses and fruit trees (Tjernström 1992). Respective labour inputs over a two-year period were 741, 863 and 934 person-days. Accessibility to soil conservation depends on farmers and their family being able to meet the higher labour requirements.

Studies in the USA (Ervin and Washburn 1981; Mueller, Klemme and Daniel 1985) indicate that investments in conservation are not economic to farmers in either the short- or long-term. With the effects of erosion on productivity being masked by other factors, like higher-yielding varieties and improved soil and water management, the cost of adopting a conservation technique is often higher than any benefit gained. With farmers coming under greater economic stress in recent years, existing soil conservation practices are frequently eliminated to reduce costs and avoid bankruptcy (Napier 1988). Clearly, under these conditions farmers are not going to undertake soil conservation work unless either compelled or paid to do so.

7.4 Approaches to Soil Conservation

7.4.1 Cultivated lands

A risk of erosion exists on cultivated land from the time trees, bushes and grasses are removed. Erosion is exacerbated by attempting to farm slopes that are too steep, cultivating up-and-down hill, continuous use of the land for the same crop without fallow or rotation, inadequate use of fertilizers and organic manures, compaction of the soil through the use of heavy machinery and pulverization of the soil when trying to create a seed-bed. Erosion control is dependent upon good management which implies establishing sufficient crop cover and selecting appropriate tillage practices. Thus soil conservation relies strongly on agronomic methods combined with sound soil management whilst mechanical measures play only a supporting role (Fig. 7.2).

Conservation strategies are aimed at establishing and maintaining good ground cover. The feasibility of this is determined by what crops are being grown and how quickly, under the local climatic and soil conditions, they attain 40 to 50 per cent canopy cover. From the results of the studies discussed in Chapter 3, it is clear that least protection of the soil is afforded by crops grown in rows, tall tree crops and low-growing

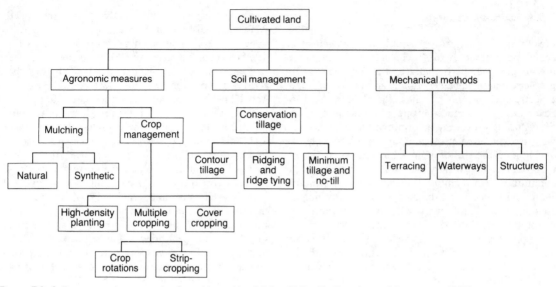

Figure 7.2 Soil conservation strategies for cultivated land (after El-Swaify, Dangler and Armstrong 1982).

crops with large leaves. Continuous production of cereals, rubber, oil palm, grape vines, maize, cassava, sugar beet, broccoli and Brussels sprouts can produce moderate to serious erosion problems. Rapid establishment of crops is important, particularly in those parts of the world where erosion risk is high at and immediately after planting, for example where the first rains of a wet season are highly erosive. With timely planting of sorghum early in the rainy season followed by a crop of good yield, 40 per cent canopy cover can be obtained by about 20 July in the Indore area of India. With late planting and a poor yield, 40 per cent cover is not attained until 30 September, thereby extending the period of erosion risk throughout the summer monsoon (Shaxson 1981). Aina, Lal and Taylor (1977) found that the time to establish 50 per cent canopy cover from the date of planting varied from 38 days for soya beans to 46 days for pigeon peas and 63 days for cassava. Soil loss under these crops was proportional to the time needed for the canopy to develop. Thus quick-growing crops may be viewed as soil-conserving crops. Unfortunately, the crops which present erosion problems are also those of considerable value either for industrial purposes or as food crops. The challenge is to develop soil conservation strategies that will allow these crops to be grown productively in the short term, to meet the immediate needs of the farmers and their families, and sustainably in the long term, so as not to deplete the soil resources for future generations.

As implied above, any strategy must be both technically sound and socially and economically acceptable to the farmers. One of the major complaints of farmers

in southern Mali in the late 1970s was the damage being done to their land by erosion. The solution proposed and tested on one village, Fonsébougou, was a system of graded channel terraces, diversion drains and grass waterways, supported by contour cultivation, and the use of mulches and organic manures. Although the measures successfully reduced the damage, the engineering works took up 10–14 per cent of the farm land and the terraces and the grass cover in the waterways were poorly maintained. Only a limited number of farmers were involved in decision making or the execution of the work. Farmers gained the impression that once the system was installed, their problems were solved and no further work was required. Most farmers viewed the diversion drain above the fields, which directly protected their land from runoff, as the most important part of the system and did not fully appreciate the role played by the other components. As a result of this experience, an alternative approach was tried in the village of Kaniko based on an understanding of the existing farming systems, and integrating the ideas of the farmers with those of the researchers and extension staff. Greater emphasis was placed on contour grass strips and the use of trees and hedges rather than terraces (Fig. 7.3; Kleene, Sanogo and Vierstra 1989; Hijkoop, van der Poel and Kaya 1991). Increasingly, it is recognized that strategies for soil conservation must rely on improving traditional systems instead of imposing entirely new techniques from outside (Roose 1992; Table 7.3) and on enhancing land husbandry (Hudson 1993; Hudson and Cheatle 1993). Even then, take-up can be slow. Whereas over 90 per cent of the

(a) Toposequence

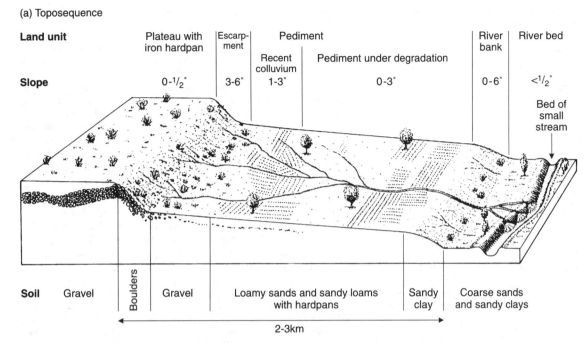

Land unit		Plateau with iron hardpan	Escarpment	Pediment			River bank	River bed
					Recent colluvium	Pediment under degradation		
Slope		$0\text{-}^{1}/_{2}°$	3-6°	1-3°		0-3°	0-6°	$<^{1}/_{2}°$
Soil	Gravel		Boulders	Gravel	Loamy sands and sandy loams with hardpans		Sandy clay	Coarse sands and sandy clays

Bed of small stream

2-3km

(b) Proposed land use and management

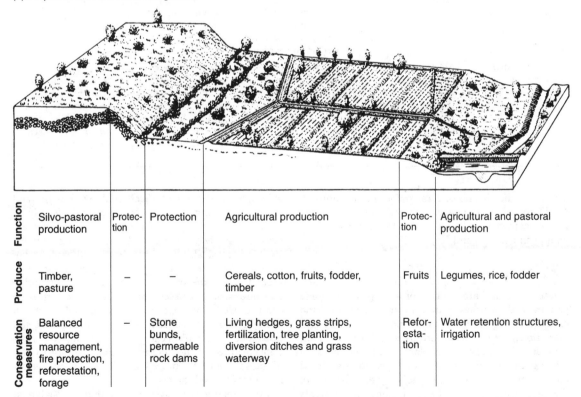

	Function	Silvo-pastoral production	Protection	Protection	Agricultural production	Protection	Agricultural and pastoral production
	Produce	Timber, pasture	–	–	Cereals, cotton, fruits, fodder, timber	Fruits	Legumes, rice, fodder
	Conservation measures	Balanced resource management, fire protection, reforestation, forage	–	Stone bunds, permeable rock dams	Living hedges, grass strips, fertilization, tree planting, diversion ditches and grass waterway	Reforestation	Water retention structures, irrigation

Figure 7.3 Proposed soil conservation scheme on cultivated land in southern Mali (after Hijkoop, van der Poel and Kaya 1991).

Table 7.3 Variation in traditional and proposed soil conservation systems with climate in West Africa (after Roose 1992)

Area	South Soudanian	North Soudanian	South Sahelian	North Sahelian
Rainfall	> 1,000 mm	700–1,000 mm	400–700 mm	150–400 mm
Conservation objectives	Dispose of excess rain whilst conserving soil	Hold rain *in situ* whilst conserving soil	Concentrate rain by water harvesting from a catchment area whilst conserving soil on cropped land	Concentrate rain by water harvesting from a catchment area whilst conserving soil on cropped land
Soils	Ferralitic	Leached ferruginous + vertisols + brown on basic rocks	Leached ferruginous + vertisols + brown on basic rocks	Sandy sheet on brown sub-arid soil
Vegetation	Tree savanna	Tree savanna	Bush savanna	Steppe, bush
Farming system	Drainage farming; drainage of excess water	Rainfed farming; drainage of hillside runoff	Runoff farming; trapping of rain and runoff	Valley farming; intense cropping limited to valleys
Traditional management	Yams on large mounds. Maize + intercropping in ridges. Millet + ground nut + various on ridges. Intercropping + agroforestry. Drainage between plots.	Superficial tillage. Cropping on the flat. Two weedings. Sorghum, cotton or millet + sorghum or ground nut + cowpeas. Hillslope runoff on waterways. Total infiltration or rain in the field. Stone bunds and walls. Grass, stone or brush barriers on contour.	Superficial tillage. Cropping on the flat. Two weedings. Sorghum or millet then ground nut or cowpeas. Mulching, stone bunds, grass or brush barriers on contour. Tied ridging on sandy soils (sometimes).	Sowing on the flat. Two weedings. Millet on sandy soils. Sorghum on clay soils on low ground. Grazing on hillslopes. Gardens in low-lying areas. Retreat flooding.
Proposed modern management	Afforestation of laterite screes. Grassed buffer strips. Gully restoration. Protection of rice fields.	Improvement of grassland. Living hedges + stone bunds or grassed buffer strips. Grass waterways.	Improvement of grassland. Pounds for cattle + supplement-ary irrigation of gardens. Stone bunds protected by grass. Gully management.	Shrub forage plantings in lunettes or ditches. Grassland management. Pounds for cattle. Trapping runoff on hillslopes. Stone bunds on pediment.

farmers in the villages of Kaniko and Try in southern Mali know about the benefits of hedges only just over 20 per cent have adopted them on their land (Hij-koop, van der Poel and Kaya 1991).

7.4.2 *Pasture lands*

These comprise areas of improved pasture where grasses and legumes suited to the local soil and climatic conditions are planted and managed by reg-ular applications of lime, fertilizers and organic man-ures, as well as areas of rangeland composed of native grasses and shrubs. Since grass provides a dense cover, close to the soil surface, it is a good protector of the land against erosion. Erosion problems arise only when this cover is removed through overgrazing,

although they can be exacerbated by drought and excessive burning. Erosion control depends largely on the use of agronomic measures (Fig. 7.4). These are directed at determining and maintaining suitable stocking rates, although this can be difficult if not impossible in areas where people attach cultural and social value to the size of their herds, and at planting erosion-resistant grasses and shrubs. The latter are characterized by vigorous growth, tolerance of drought and poor soils, palatability to livestock and resistance to the physical effects of trampling. Spe-cialized measures designed to increase the resistance of the soil may be required around field gates, water-ing points and salt boxes.

Traditional grazing systems are often well-adapted to the local conditions of climate, soils and vegetation,

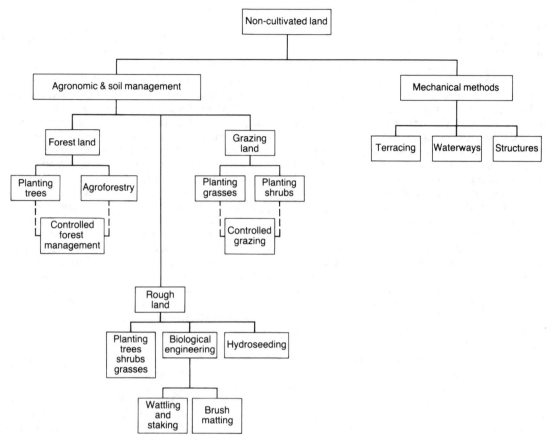

Figure 7.4 Soil conservation strategies for non-cultivated land (after El-Swaify, Dangler and Armstrong 1982).

making use of rotational grazing on a nomadic basis. The Turkana in Kenya have a carefully controlled and reasonably flexible system administered by the elders. Cattle are grazed in the lowlands immediately after the wet season, taking advantage of the annual flush of grass, and then moved to slightly wetter hilly areas during the dry season, towards the end of which they are kept around a permanent homestead in riverine woodland from which dry-season forage can be obtained (Barrow 1989). Sorghum is planted near the homestead during the rains, managed by the women, and used to supplement the pastoral diet. The dry-season hill-grazings are managed communally and include areas of reserved grazing which are used only in times of drought.

Virtually all traditional grazing systems are under pressure today. First, human population numbers, as a result of better health, and livestock numbers, as a result of better veterinary services, are both increasing. Second, there is a conflict between the individual and family ownership of livestock and the communal own-

ership of the pasture. The individual derives a positive utility of almost one for every additional animal owned and grazed on communal land and experiences a negative utility of only a small fraction of one as a result of ensuing overgrazing. The maximization of individual benefit at the expense of the community, termed by Hardin (1968), 'the tragedy of the commons', is one of the biggest challenges facing soil conservationists on rangeland. The conflict becomes most marked when incentives for commercial livestock production are so strong that individuals, usually those owning the larger numbers and better quality stock, break away and take over much of the land, enclosing it by fencing, displacing other members of the community and increasing the pressure on the remaining range. Third, the provision of additional watering points, often located without consideration of the traditional movement of stock, has meant that availability of forage rather than water has become the limiting factor on livestock numbers, creating additional pressure on the land. Fourth, the concentration of people in settlements to provide health, education and water more efficiently and to promote cash cropping with irrigation has caused people to abandon their nomadic tradition with the result that stock are kept all year on the pastures close to the homestead. Sometimes new settlement schemes will be sited on seasonal grazing areas, thereby removing them from the system. Fifth, education of the youth and changed political systems are gradually eroding the authority of the elders.

Grazing can be considered as the removal of biomass from rangeland. The rate of removal depends upon the number of animals and their daily intake of forage. Soil conservation is therefore aimed at controlling grazing numbers so that a sufficient vegetation cover is sustained over time to protect the soil. As seen in Chapter 3, loss of vegetation increases the rate of runoff and erosion and decreases the amount of water in the soil which, in turn, reduces the amount of vegetation growth. In reality, the vegetation–erosion–grazing interaction is more complex because of the need to consider soil fertility, loss of soil nutrients, production of litter, the palatability and digestive value to stock of the different species in the plant community, and the ability of the different species to survive under changing grazing, moisture and nutrient conditions. At present, the interaction is poorly understood. On the one hand, there is considerable evidence that many tropical rangelands are overstocked. It is suggested by Pratt and Gwynne (1977) that the threshold stocking rate on the semi-arid lands of northeast Kenya is about 1 Tropical Livestock Unit (TLU) per 25 ha whereas actual stocking rates range from 1 TLU per 4 ha to 1 TLU per 30 ha (Peden 1987). In the Middle Veld of Swaziland, present grazing density is about 1 TLU per 1.3 ha compared with estimated sustainable stocking rates of 1 TLU for every 2.5 to 3.5 ha (Nsibandze 1987; Dlamini and Maro 1988). On the other hand, application of a simple grazing model developed by Biot (1990) to the rangelands of Botswana suggests that at a grazing intensity of 1 TLU per 5 ha, the system would remain reasonably stable for the next 2,000 years. Clearly with such a discrepancy between conventional wisdom and model predictions, it is difficult to determine the safe level of grazing, particularly when short-term fluctuations in climate also need to be considered. Further, there may be a conflict between designing a strategy at which the sustained productivity of the rangeland and erosion remain in balance and which may therefore maximize economic benefit over the medium- to long-term but keeps erosion rates relatively high, and a strategy which maintains erosion below a much lower tolerable level to minimize environmental impacts.

7.4.3 Forest lands

Forests provide excellent protection of the soil against erosion. They maintain high rates of evapotranspiration, interception and infiltration and therefore generate only small quantities of runoff. Low runoff rates and the protective role of the litter layer on the surface of the soil produce low erosion rates. Increases in erosion occur where the land is permanently or, in the case of shifting cultivation, temporarily cleared for agriculture. Whilst the forest cover remains largely intact, the most important erosion problems are associated with cropping of the trees for firewood; destruction of the trees and surrounding shrub and ground cover by grazing; and logging operations.

An estimated 40 per cent of the world's population use wood as the primary fuel. In the rural areas of Tanzania and Thailand, consumption ranges from 1 to 2 tonnes per capita each year (Brown 1981). In Sudan, Colombia, Ethiopia, Nigeria and Indonesia, fuel wood accounts for 80 per cent or more of annual timber removals. This rate of removal can lead to large-scale deforestation, followed by the washing and blowing away of the soil. Afforestation schemes which include rapid-growing tree species that can be cropped for firewood are an important feature of erosion-control strategies (Fig. 7.4).

Livestock grazing is frequently detrimental to the survival of forests. The animals trample and compact the soil, injure roots close to the surface and browse on the tree seedlings. At the time of settlement between AD 875 and 930, about 65 per cent of the land surface of Iceland was covered with vegetation (Thorsteinsson, Ólafsson and van Dyne 1971). Between 25 and 40 per cent of the country was under low-growing birch forest with a luxuriant undergrowth of forbs and grasses. The productivity of the ground vegetation in these forests is about 70 per cent higher than the average for other grazing areas in the country (Thorsteinsson 1980a) so they provide the best grazing lands. Overgrazing and stripping of the bark from the trees has prevented the regeneration of the forest, much of which has also been cut for fuel. As a result the area under forest has been reduced to only 1 per cent of the area of the country and less than 25 per cent remains vegetated (Arnalds 1987).

Logging causes limited disturbance because erosion is confined to the area of land where the trees have been removed. With good management, the vegetation cover regenerates quickly so that high erosion rates are restricted to the first and sometimes the second and third years after felling. The level of disturbance is related to the method of clearance. Studies of three practices in Nigeria (Lal 1981) showed that more erosion followed mechanical clearance using crawler tractors with tree-pusher and root-rake attachments than manual clearance using chain saws. Least erosion followed manual clearance with machete and axe. Erosion immediately after clearance is generally associated with surface runoff. Another effect of forest removal, however, is the gradual loss of shear strength of the soil following the decay of the root systems (O'Loughlin 1974; O'Loughlin and Watson 1979). This induces a risk of landslides which is greatest about five years after clearance (Bishop and Stevens 1964) although some researchers suggest a slower deterioration with maximum slide hazard being reached 15 years after logging (Rice and Krammes 1970).

The main erosion problems in logged areas are associated with skid trails and roads which are frequently areas of bare compacted soil; also the cut and fill slopes on roadsides are often steeper than the angle of repose of the soil and are subject to landslides. As reported by Swanson and Dyrness (1975), who measured 2,300,000 t/ha on a roadside in western Oregon, USA, annual erosion rates from logging roads are very high. Roads, tracks and paths can be major contributors to total sediment yield. They account for between 15 and 35 per cent in catchments devoted to agriculture in the Aberdare Mountains, Kenya (Dunne 1979) but in a forested area where the sediment yield on the surrounding land is much lower, their contribution will be greater and may account for half the total (Reid, Dunne and Cederholm 1981). Generally, cut slopes are more erodible than fill slopes as evidenced by studies in the Karuah State Forest, New South Wales, where sediment losses from cut and fill slopes were respectively 100 and 10 times higher than those from natural slopes (Riley 1988). Erosion can be minimized by locating roads on ridges and gentle slopes to avoid the need for cut slopes. Another source of sediment, widely reported from afforestation schemes in Wales (Newson 1980; Robinson and Blyth 1982; Murgatroyd and Ternan 1983; Francis and Taylor 1989), occurs where the land has to be drained by surface ditches. The ditch banks erode as the channels adjust from their relatively deep and narrow cut-form to a shallower, wide one. The increased runoff from the ditches also causes bank erosion in the rivers immediately downstream.

7.4.4 Land clearance

The methods used for clearance of forest for agricultural development not only affect the amount of soil loss which takes place but also influence the subsequent crop yields. The mechanical clearance of forest on an alfisol in Nigeria (Lal and Cummings 1979) significantly increased the bulk density and

decreased the infiltration capacity of the soil compared with removal by slash-and-burn. Bulk densities were increased from $0.9 \, Mg/m^3$ on uncleared land to $1.12 \, Mg/m^3$ with slash-and-burn and $1.25 \, Mg/m^3$ with mechanical clearance. The respective infiltration capacities were 112, 52 and 19 mm/h. Clearance of forest on 30–40° slopes in Malaysia on an ultisol soil (Durian Series) reduced the infiltration capacity from 214 mm/h on uncleared land to 149 mm/h using slash-and-burn and 43 mm/h using crawler tractors with straight blades; the respective bulk densities were 1.06, 1.33 and $1.48 \, Mg/m^3$ (Jusoff and Majid 1988).

The response of crop yield to different types of land clearance is quite varied depending on the degree of disturbance and the subsequent method of cultivation. Potentially, the situation is worse with mechanical clearance because much of the litter layer is scraped away, removing the immediate protection of the soil against erosion and the source of organic material. Yet, on an ultisol soil on a 2° slope in northern Thailand, mechanical clearance of the forest by tractor with a tree-pusher blade followed by construction of contour bunds, no-tillage and inclusion of legumes in a rotation system led to higher rice yields compared with other treatments (Boonchee, Peukrai and Chinabutr 1988). In contrast, on ultisols in Peru (Seubert, Sanchez and Valverde 1977) and alfisols in Nigeria (Lal 1981), significantly higher yields were obtained using conventional tillage practices or chisel ploughing following mechanical clearance compared with using no tillage or mulching. The latter techniques initially resulted in rather low yields, even when fertilizer was added, because they did not overcome the increased compaction of the soil.

If disturbance of the surface is kept to a minimum and a system of land management imposed for eight to ten years after clearance to overcome any increases in compaction of the soil, forest land can be successfully cleared for agriculture. If the land cannot be cropped immediately because the disturbance has been too great, the planting of a cover crop will help restore the soil close to its original bulk density and infiltration capacity (Hulugalle, Lal and Terkuile 1984), after which a range of management practices, including no tillage, might be possible. Once under agricultural use, the appropriate soil conservation practices for cultivated land must be adopted.

7.4.5 Rough lands

Areas of rough ground remaining in their natural habitats because they are too marginal for other forms of land use include hilly and mountainous terrain with shallow stony soils and steep slopes, alpine grasslands, arctic tundras and sand dunes. Since they are often areas of spectacular scenery, they attract recreational use. Their marginality, however, means that they are sensitive to any disturbance, and their ability to withstand recreational impacts is low. The main problems of land degradation are associated with footpaths, tracks created by horses and motor cycles and off-road vehicles, and very intensive use as in car parks and on camp sites. Overuse of paths and tracks results in a reduction in overall vegetation cover, modifications to species composition because some plants resist trampling better than others, compaction of the soil and changes in soil moisture. The latter may increase in dry soils following compaction but decrease in wet areas.

Approaches to controlling erosion range from exclusion of people, appropriate in areas of special scientific interest or in nature reserves, to use of erosion-resistant plant species, supported where necessary by drainage to reduce soil moisture content and increase soil strength, and artificial strengthening or reinforcement of footpaths. Although few plants can withstand prolonged and heavy pressure from walkers and horses, studies of footpath erosion in the English Lake District show that grasses, including *Nardus stricta*, *Agrostis* spp. and *Festuca* spp. resist trampling much better than *Calluna* heathland (Coleman 1981). *Poa pratensis* and *Festuca idahoensis* grassland was found to suffer less from the effects of hikers, motor cyclists and horse riders than a *Pinus albicaulis* forest with shrubby understorey of *Vaccinium scoparium* in the northern Rocky Mountains, Montana, USA (Weaver and Dale 1978). One way of comparing the vulnerability of different habitats to trampling is to determine the number of passes by individual walkers they can withstand before the plant cover is reduced from about 100 to 50 per cent. Sand dune pasture can withstand between 1,100 and 1,800 passes, alpine plant communities less than 60 and arctic tundra only about eight passes (Liddle 1973). Trampling of a grass cover on a clay soil with a moisture content close to field capacity resulted in runoff rising rapidly between 0 and 50 passes, then levelling off before rising again between 700 and 900 passes. The rate of soil loss, however, increased linearly with increasing number of passes. Since patches of bare ground did not appear until 250 passes and even after 900 passes between 25 and 50 per cent of the cover remained, the critical period when increased runoff and erosion are initiated seems to occur before any decrease in the vegetation cover is seen (Quinn, Morgan and Smith 1980).

In addition to being able to survive in the climatic and soil environment, the plant species selected for revegetation of rough areas should ideally be local, so as not to introduce new species into the ecological system, and meet the following criteria: short or prostrate growth form, flexible rather than brittle or rigid stems and leaves, basal or underground growth points, rapid growth, ability to withstand burial by soil and rocks, and ability to withstand exposure of the root system (Coppin and Richards 1990). Often an ecological succession can be planned involving a mixture of species, both grasses and shrubs, to give an initial rapid growth and cover of the ground followed by longer-term development of erosion-resistant plants.

7.4.6 Urban areas

Urban development frequently results in intensive erosion. The exposure of bare soil during the construction phase results in higher volumes of peak runoff, shorter times to peak flow, higher and more frequent flood flows and rapid increases in erosion by overland flow, rills and gullies, producing high sediment concentrations. Residential development, often on a speculative basis, started in the Anak Ayer Batu catchment in Kuala Lumpur in the mid-1960s. By 1970 the river was choked with sediment, and erosion was widening and steepening the channel so that it could evacuate the more frequent flood flows. Flooding and deposition of sand occurred regularly on the flood plain, affecting particularly the University of Malaya campus. Suspended sediment concentrations ranged from 4 to 81,230 mg/l (Douglas 1978). By 1977 the situation had changed little with sediment concentrations ranging from 3.7 to 15,343 mg/l (Leigh 1982b). These high sediment concentrations compared with 7 to 1,080 mg/l in rain forest areas occurred with only 10 per cent of the catchment in bare or semi-bare condition; about 40 per cent was in secondary scrub forest or belukar. One area of 6.3 ha cleared in 1970 for housing development was abandoned by 1977 because gullies were up to 10 m deep and extending rapidly headwards (Leigh 1982b). Similar sediment concentrations with peaks between 15,000 and 49,000 mg/l have been recorded on the new campus of the University of Singapore on the Kent Ridge and at the old campus at Bukit Timah (Gupta 1982).

Strategies for erosion control in urban areas depend on scheduling developments to retain as much plant cover as possible, but since this is generally feasible only to a limited extent, a much greater reliance is placed on mechanical methods than is the case with other types of land use. Erosion control in the final phase when urban development is complete requires rapid establishment of plant cover and permanent use of purpose-designed waterways and embankment-stabilizing structures (Fig. 7.5). On-site erosion is very low at this stage because most of the land is covered in

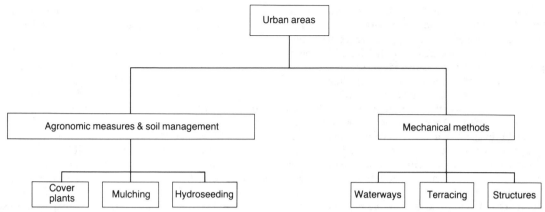

Figure 7.5 Soil conservation strategies for urban areas (after El-Swaify, Dangler and Armstrong 1982).

concrete but runoff from impervious surfaces of streets, pavements, roofs, gutters and sewers can cause bank erosion in rivers downstream of the urban area. The instability of roadside slopes often presents an additional problem.

7.4.7 Mining land

The main purpose of erosion control on land previously used for mining is to create a stable environment for vegetation establishment and growth, in order to promote reclamation of the land for agriculture or recreation, and to minimize off-site damage. Since mine-spoil banks are generally areas where erosion starts very quickly and, because of the infertile and often toxic nature of the material, vegetation grows very slowly, strategies involve a combination of mechanical and agronomic methods.

Initially, erosion is highest on the upper convex and upper mid-slope segments of a spoil mound and deposition occurs at the base of the mound, forming a concave segment. Over time, the reduction through erosion of the convexity causes the focus of erosion to move downslope on to the steep mid-slope segment (Haigh 1979). Where this becomes rilled or gullied, erosion rates can increase dramatically, reaching 100 to 500 t/ha annually (Porta, Poch and Boixadera 1989), especially if the gullies incise into unconsolidated deposits on the lower convexity. The evolution of the slope then depends on which of two conflicting processes becomes dominant. If the gully network becomes extensive, high erosion rates will be maintained and the overall slope will become more convex. If deposition on the concavity dominates, the overall gradient will decrease and the slope will become more stable (Toy 1989).

Revegetation of the spoil requires designing a plant

succession which will give adequate surface cover and increase the fertility of the soil. Ideally the succession should include rapid-growing grasses to give ground cover as quickly as possible and stabilize the surface; legumes, such as clovers and vetches, to fix nitrogen; and other grasses and shrubs to provide the long-term cover. Vegetation can also be planted in strips on or at a slight grade to the contour to form a barrier across the slope to reduce runoff and filter out moving sediment. Over time deposition upslope of the barriers will produce a cheap and simple system of terraces. Species selection will depend on local soil and climatic conditions but should aim to provide a uniform rather than clumpy pattern of vegetation to avoid concentrations of runoff and localized erosion. The ideal condition is difficult to achieve, however, where the soils have low water-retention capacity or are toxic, or where cold, drought or exposure inhibit plant growth. Choosing vegetation which has beneficial effects is important. In order to reduce the costs of reclamation of spoil mounds from tin mining on the Jos Plateau, Nigeria, the Mine Lands Reclamation Unit chose *Eucalyptus* spp. which could be cropped as a source of fuel and pole timber but these have resulted in significant declines in the base status and pH of the soil (Alexander 1990) to the detriment of other plant growth. Among the mechanical measures, terraces and waterways will be required to control runoff and remove excess water safely from the site, usually to silt-traps and soak-aways at the base of the slope, and revegetation may need to be supplemented by contour wattling, contour brush-layering or terrace construction.

7.4.8 Road banks

As indicated in Section 7.4.3, road banks on either cut or fill slopes are frequently subject to surface erosion and mass soil failures and are major sources of sediment. The methods used for erosion control range from engineering structures such as revetments and retaining walls to stabilization of the slope by vegetation. Ground vegetation, such as grasses, will protect the slope against erosion by raindrop impact and runoff, and also trap moving sediment, whilst shrubs and trees will increase the strength of the soil through root reinforcement. Also, trees will help support a slope by buttressing (Chapter 3). Vegetation increases the infiltration of water into the soil but this can cause problems where rainfall amounts and intensities are very high. Whilst the resulting reduction in runoff will help control surface erosion, the increased moisture content of the soil may exacerbate the risk of mass soil failure. Whether this occurs will depend on the extent to which root reinforcement has increased the cohesive strength of the soil and root penetration across the potential slide plane has anchored the soil mass to the underlying material. Where soil saturation is a danger, the vegetation may need to be supported by cut-off drains above the slope to prevent the influx of surface water, and trench or herringbone rubble drains to aid drainage of the slope itself. In the initial stages, until the vegetation cover is established, wattling, brush layering or geotextiles may be needed to provide temporary erosion control and stability. For surface erosion and for soil failures extending to depths of 2 m, vegetation has the advantages over engineering structures of being cheaper to install, aesthetically more pleasing, and, in the long term, self-repairing

with an indefinite life. It will, however, require regular maintenance and occasional repair. Although not yet adopted as a standard practice world-wide, civil engineers are showing increasing interest in both *bio-engineering*, in which vegetation is used on its own as an engineering material, and *biotechnical engineering*, in which the engineering properties of vegetation are combined with inert structures (Morgan and Rickson 1995).

Though road bank erosion adjacent to highways may be considered an engineering problem, its occurrence on unsurfaced roads in plantations, forests and agricultural areas is frequently the concern of the soil conservationist. Fortunately, these are areas where careful location of roads and vegetative solutions, either alone or with simple mechanical measures, will often be sufficient to control the problem. Even along highways, however, the civil engineer must take account of the nature of the land use on the slope above the road bank. In the Himalayas where the road banks are frequently oversteepened cut sections of the hillside, seepage from irrigation terraces on the slope above can increase instability. Inclusion of forestry, either as agroforestry or as community-managed forests, may help to stabilize both the slope above and the road bank itself and be more satisfactory than an engineering solution. If the forest can provide fodder, fuel and fruits, its long-term management may be vested in the individual farmers and the community as a whole.

CHAPTER 8

Crop and Vegetation Management

Agronomic measures for soil conservation use the protective effect of plant covers to reduce erosion (Section 3.4). Because of differences in their density and morphology, plants differ in their ability to protect the soil. Generally, row crops are the least effective and give rise to more serious erosion problems (Section 7.4.1). This is because of the high percentage of bare ground, particularly in the early stages of crop growth, and the need to prepare a seed bed. In designing a conservation strategy based on agronomic measures, row crops must be combined with protection-effective crops.

8.1 Rotation

The simplest way to combine different crops is to grow them consecutively in rotation. The frequency with which row crops are grown depends upon the severity of erosion. Where erosion rates are low, they may be grown every other year but, in very erodible areas, they may be permissible only once in five or seven years. A high rate of soil loss under the row crop is counteracted by low rates under the other crops so that, averaged over a six- or seven-year period, the annual erosion rate remains low.

Suitable crops for use in rotations are legumes and grasses. These provide good ground cover, help maintain or even improve the organic status of the soil, thereby contributing to soil fertility, and enable a more stable aggregate structure to develop in the soil. The effects are often sufficient to reduce erosion and increase yield during the first year of row-crop cultivation, but they rarely extend into the second year. For this reason, two continuous years of planting with a row crop should be avoided. Hudson (1981) shows that a rotation of tobacco–grass–grass–tobacco–grass–grass is more effective in Zimbabwe than one of two consecutive years of tobacco followed by four years of grass. The respective mean annual soil loss rates are 12 and 15 t/ha. The same effect is illustrated more dramatically by Kellman (1969) for shifting cultivation in the Philippines where, during the cropping period, soil loss from land under upland rice averages $0.38 \, g/m^2$ per day on a new clearing but rises

to $14.91 \, g/m^2$ per day on a clearing in its twelfth year of cultivation.

8.1.1 Shifting cultivation

Shifting cultivation is a traditional method of reducing soil erosion in the tropics by rotating the location of the fields. An area of forest is cleared by slash-and-burn, the soil loosened by hand hoeing and a crop planted. Where two crops a year can be obtained, a second crop is grown after the harvest of the first, otherwise the land is left in a weed fallow. The same area may be cropped for a second year before being allowed to revert to scrub and secondary forest. Typical crops grown are cassava, maize, upland rice and yam. The residual effect of the forest on the organic content and aggregate stability of the soil generally lasts for the first year of farming so that soil loss remains low. The clearance of an area of secondary forest, the growing and harvesting of a crop of upland rice and reversion of the land to bush fallow on a shallow loamy soil on a 25° slope in Sarawak, Malaysia, produced 0.033 t/ha of soil loss for the year compared with 0.038 t/ha under forest in the year prior to clearance (Hatch personal communication). As already indicated, erosion rates rise rapidly if the land continues to be cropped in subsequent years. In Sabah, Malaysia (Sinajin 1987) soil loss from plots cropped to cabbage was 0.05 t/ha for the period from transplanting to mature growth in the first year but increased to 2.02 t/ha in the third year of cropping, even though terracing was installed.

The practice of shifting cultivation will maintain soil fertility and reduce erosion to acceptable levels provided the associated conditions of low crop yields and low ratio of population to land area remain socially and economically desirable. The critical factor in the system is the length of the fallow period. This is traditionally between 7 and 20 years in West Africa (Okigbo 1977) but increasing population densities and the desire of people to raise their standard of living by changing from subsistence to cash cropping are putting pressure on the land, resulting in a reduction and sometimes the elimination of the fallow period. Serious erosion problems are created when the fallow period is reduced and alternative farming practices to the shifting cultivation system are required to solve them, using intercropping, mulching and agroforestry.

8.1.2 Row-crop cultivation

Particular problems are associated with maize which, when grown as a row crop with conventional tillage and clean weeding, results in an annual soil loss on 2–5° slopes of between 10 and 120 t/ha taking data from Zimbabwe (Hudson 1981), Malaysia (Sulaiman, Maene and Mokhtaruddin 1981) and India (Singh, Bhardwaj and Singh 1979). These rates are well above most soil loss tolerance levels. Where maize is grown in a more traditional manner with cultivation being restricted to ploughing and with no weeding, it presents fewer difficulties but the yields are rather low, often only 1 t/ha compared with over 5 t/ha regularly achieved in the USA. There is an urgent need to find a way of increasing the yield without increasing the erosion.

When considering the use of crop rotation, attention must be paid to the order in which the crops are grown, which affects both the degree of erosion control and the yield. Lal (1976) found that in 1973, two crops of maize grown on a 6° slope near Ibadan, Nigeria, resulted in a soil loss of 7.2 t/ha whereas a crop of maize followed by one of cowpeas with no tillage produced a loss of only 0.2 t/ha. Growing cowpeas followed by maize but with tillage gave a soil loss of 6.2 t/ha. Although it is difficult to isolate the effects of zero tillage, it seems that a maize–cowpeas sequence produces less erosion than cowpeas–maize because maize is a soil-depleting crop and, when grown second, is planted into an already partially-exhausted soil. Also, soil loss is greater under maize than cowpea (Lal 1977a) and the difference between the two crops is likely to be enhanced when maize is the second crop. Crop yields were 4.9 t/ha for maize and 4.3 t/ha for cowpea in the maize–cowpeas sequence but 5.0 t/ha for cowpea and only 2.1 t/ha for maize in the cowpeas–maize sequence. Thus it is always preferable to grow maize as the first crop in the rotation. A suitable four-year rotation for the Ibadan area is maize–cowpeas; pigeon peas–sorghum; maize interplanted with cassava; and fallow crop (Okigbo and Lal 1977).

Soya beans are often recommended as an alternative to cowpeas because of their apparent ability to reduce soil loss by intercepting a high percentage of the rainfall. At 90 per cent canopy cover, soya beans intercept 58 per cent of the rainfall, compared with 40 per cent for maize interplanted with cassava and only 28 per cent for cassava on its own. Respective annual soil losses are 4.0, 6.9 and 11.0 t/ha in the Ibadan area (Aina, Lal and Taylor 1979). On a 2° slope near Dehra Dun, India, mean annual soil losses measured over a three-year period were 15.8 t/ha under maize and only 5.5 t/ha under soya bean (Singh, Bhardwaj and Singh 1979). Despite soya bean being a legume, studies in the Mid-West of the USA indicate that it can result in as much if not more erosion than maize (Laflen and Moldenhauer 1979). The average annual soil loss on a 4° slope with a silt-loam soil over seven years was 7.0 t/ha for continuous maize, 6.5 t/ha for maize following soya beans and 9.6 t/ha for soya beans following maize. The main reasons for the increased erosion under soya beans are that the crop affords less protection than maize in the stage between 75 per cent canopy cover and harvest, and produces less residue after harvest.

In addition to controlling erosion, the inclusion of grasses and legumes in a rotation can increase yields of the main crop. Yields of maize greater than 5 t/ha were obtained at Ibadan following a year with the land under *Centrosema pubescens*, *Setaria splendida* or *Stylosanthes gracilis* (Okigbo and Lal 1977). Unless the fallow crops can be used for grazing or fodder, however, they have no immediate value to the farmer. Thus, crop rotations with grasses or legumes are rarely practised in the main cereal-growing areas of the world and are unlikely to be acceptable anywhere if their inclusion gives no income to the farmer. Under these conditions, an alternative approach to erosion control is required based on minimizing the period of bare ground, for example, by leaving crop residue on the land after harvest and delaying ploughing until the following spring, or, where winter cereals are grown, sowing as early as possible to ensure a good cover before winter temperatures inhibit plant growth. Late sowing of winter cereal has been a major cause of soil erosion in recent years over much of southern and eastern England (Boardman and Evans 1994) and northern France (Monnier and Boiffin 1986).

8.1.3 Grazing land management

Rotation is commonly practised on grazing land, moving the stock from one pasture to another in turn, to give time for the grass to recover. Generally, rangelands should not be exploited to more than 40 to 50 per cent of the annual production of their most palatable species (Fournier 1972; Thorsteinsson 1980b), a condition which is reached when about 10 to 20 per cent of the annual growth of the woody species has been removed. The vegetation should be allowed to regenerate sufficiently to provide a 70 per cent ground cover at the times of erosion risk. Grazing has to be very carefully managed. Whilst overgrazing can lead to deterioration of the rangeland and the onset of erosion, undergrazing can result in the loss of nutritious grasses, many of which regenerate more rapidly when grazed.

Overgrazing of the summer pastures in Iceland has led to a decline in the proportion of grasses within the plant community from 30 to 12 per cent and rises in the proportions of sedges and rushes, from 18 to 26 per cent, and mosses, from 15 to 24 per cent (Thorsteinsson 1980b). The annual yield of woody species on overgrazed land is only 150 kg DM/ha compared with 800 kg DM/ha on protected land. The plant density is often low with open scars in the vegetation cover. An increase in the numbers of sheep and a decrease in the length of the growing season from 240 to 210 days between 1944 and 1968 created pressure on the better pastures of the Peak District in the UK, leading to a breakdown in the turf mat, soil erosion and a change in sward type from *Agrostis–Festuca* to *Nardetum* (Evans 1977) with a consequent lowering of the sheep-carrying capacity. This change in vegetation type and also that from heather to grassland has been observed generally in the UK following overgrazing of upland pastures over periods of five to ten years (Fenton 1937; Jeffers 1986). Where undergrazing occurs, grassland and heather give way to birch scrub and woodland over a period of 10 to 50 years but are also in danger of invasion by bracken, in which case the succession to trees may be delayed and take centuries to occur (Jeffers 1986).

Controlled burning, preferably practised on a rotational basis, is essential for the removal of undesirable species whereas uncontrolled burning can prevent plants from re-establishing and, by increasing the extent of bare ground, result in serious erosion. Soil erosion on the North Yorkshire Moors is attributed to burning of the heather every two or three years which is too frequent considering the six-year period necessary for *Calluna* to develop from seed to a height of 30 to 40 cm and provide a complete canopy to the ground surface (Imeson 1971). Generally, soil loss is highest in the first year after fire (Shakesby *et al.* 1993) and returns to background levels within three to four years. The absolute increase in erosion may be quite low, however, on soils with a stable granular structure (Kutiel 1994). On other soils, a high soil loss occurs because vaporization and readsorption of stable organic substances during the burn makes the soil hydrophobic (De Bano, Mann and Hamilton 1970), which decreases the infiltration rate, increases the runoff and also makes the soil more easily detachable by raindrop impact (Terry and Shakesby 1993). The severity of the effect depends upon the temperature of the burn. In an experimental study under Medi-

terranean *maquis* vegetation, annual erosion following a fire characterized by a mean soil temperature of 180°C was only five times that of an unburned area but was 50 times greater following a burn with a mean soil surface temperature of 475°C (Giovannini, Lucchesi and Giachetti 1988). The higher temperature burn causes the finer particles in the subsoil to form sand-size cement-like aggregates with low porosity (Giovannini 1994) which encourages the generation of more runoff. Once the top soil layer has been removed, however, this hard layer is more resistant to erosion so that soil loss declines over time. Also, although the hydrophobicity can last for up to five years (Dyrness 1976), the regrowth of the vegetation cover helps to protect the soil after the first year. Depending upon soil, rainfall, temperature of burn and the degree of insulation provided by duff (Dimitrakopoulos, Martin and Papamichos 1994), it takes some three years for the surface soil to be restored to pre-burn condition and some seven to twelve years for the plant community to re-establish following a fire (Naveh 1975; Trabaud and Lepart 1980) which implies that burning every seven to ten years is probably required to control invasion of the rangeland by undesirable species.

Prescribed burning modifies the density, stature and composition of brush stands within a plant community but kills only a few species outright. Most brush species sprout vigorously after fire to the detriment of grass. Grasses, however, fill the gaps between the brush and reach their maximum development in the second and third years following a fire. Thus, fire can be used to topkill and open up mature stands of brush so that a seed bed can be created for the establishment of grasses. Once the brush resprouts, however, it has a competitive advantage over the young grass and needs to be controlled by herbicides until the grass is strong enough to carry fire. After this stage has been reached, burning can be used periodically to control brush and to maintain a more diverse plant community than would occur if the ecological succession were allowed to develop uninterrupted (Kutiel 1994).

A critical factor in a rotational grazing system is the quality of the poorest rangeland. The traditional sheep-grazing system in Iceland relies on cultivated pastures close to the farm which are grazed in the spring and autumn and on communal upland pastures which are grazed in the summer while two crops of hay are cut from the cultivated lands. During the 1960s the farmers reduced the time that sheep spent in the uplands with the result that only one crop of hay was obtained and winter feed had to be imported. However, with imported feed, more stock could be kept during the winter. With increasing sheep numbers, the pressure on the summer grazings intensified, leading to overgrazing and erosion. The resulting deterioration of the upland pastures is important because, whilst the productivity of the lowland pastures is greater, giving an annual biomass production of 835 kg DM/ha compared with 666 kg DM/ha for the uplands (Thorsteinsson 1980a), lambs grazed on the uplands have higher rates of growth than those grazed on the cultivated pastures (Guðmundsson 1980). The availability of summer pasture thus becomes the limiting factor in sheep production (Thorsteinsson, Ólafsson and van Dyne 1971). Avoiding overgrazing and erosion in the uplands through the adoption of appropriate stocking rates is therefore vital.

There is no universal formula for determining stocking rates and the carrying capacity of pastures is usually imprecisely defined. The determination is more difficult in regions with high variability in rainfall from year to year so that overgrazing is almost inevitable when several years of drought follow in succession. The sheep-grazing capacity of the pastures in Iceland has been assessed by mapping the vegetation of the country from aerial photography and a supporting field survey and classifying it into 94 different types (Steindórsson 1980). The average annual production of dry matter has been determined for each type (Thorsteinsson 1980a). Based on this work, a knowledge of the palatability of the plant species of each type to sheep and limiting the annual removal to 50 per cent of the annual growth of the palatable species, the carrying capacity can be calculated (Thorsteinsson 1980b; 1980c). In reality, such surveys can only determine an average carrying capacity. The actual carrying capacity at a given moment will depend upon local soil and vegetation conditions, variations in climate and the extent to which trees and shrubs are cut for fuel and timber. Regular inspection and judgement of the quality of the pastures are thus needed to determine when stock should be moved from one area to another in a rotation system. The carrying capacity is also affected by the browsing of wildlife. The migration of wildlife on to a ranch in the Narok area of Kenya in the rainy season causes the stocking intensity to rise to 1.23 TLU/ha compared with a wet-season carrying capacity of 0.9 TLU/ha. When the wildlife move away in the dry season, the stocking rate falls to 0.5 TLU/ha which is very close to the dry-season carrying capacity (Mwichabe 1992). Clearly, if the land user is to obtain optimal stocking, wildlife must be kept out of the pastures.

Rotational grazing and burning are most appropriate for large-scale ranches of 4,000 ha or more since this size allows for greater flexibility in the design of suitable systems with respect to variability in land

quality, water supply, levels of degradation and climate. Figure 8.1 shows examples of rotational grazing systems used in the Arid and Semi-Arid Lands (ASAL) project in Kenya (Pratt and Gwynne 1977; Mututho 1989) to manage communal grazing.

8.1.4 Forest management

The principles of rotation can be applied to other forms of land use. The timber resources of forests are often exploited commercially by patch-cutting, also termed clear-felling, on a rotational basis. Erosion rates are highest in the years immediately following logging but decline in subsequent years with either regrowth of the natural vegetation or replanting so that, averaged over 12 or more years, they may be little different from those of undisturbed land. Leaf (1970) shows that in Fool Creek, Colorado, in the Rocky Mountains, annual soil loss averaged 0.22 t/ha whilst logging and road construction were in progress in the 1950s but, since 1958, the year after operations ceased, has averaged only 0.10 t/ha. The latter figure compares with soil loss of 0.05 t/ha from uncut forest land. Further north in western Oregon, annual sediment yield from Deer Creek catchment averaged 0.97 t/ha for the period 1959 to 1965 (Beschta 1978). Roads were constructed in late 1965 for the patch-cutting of 25 per cent of the catchment in late 1966. The sediment yield in 1966 increased by 150 per cent but in 1967 it was only 40 per cent above the pre-cutting average. Sediment production in subsequent years was similar to the pre-cutting average until 1972 when it was 170 per cent above because of several mass failures along roadsides in the patch-cut area (Section 7.3.3). The 1973 sediment yield was again similar to the pre-cutting average. Thus, provided erosion on roadside embankments and cuttings can be controlled, an area can recover quickly from the effects of a small amount of patch-cutting. Clearance of much larger areas can create longer-term problems. In the neighbouring Needle Branch watershed, annual pre-cutting sediment yield averaged 0.53 t/ha. Following patch-cutting of 82 per cent of the catchment in 1966, sediment yield was 50 per cent above the pre-cutting average in 1966, 260 per cent above in 1967 and did not return to the pre-cutting levels until 1973. The average annual sediment yield for 1966 to 1973 was 1.46 t/ha.

Forest fires represent an additional hazard once an area has been cut because, as shown by observations following wildfire in Pine Creek, Boise National Forest, Idaho, the erosion directly attributable to burning is greater in cut than in uncut areas (Megahan and Molitor 1975). Responses to fire in terms of hydrophobicity of the soil, generation of runoff and erosion

(a)

Block	Year 1			Year 2			Year 3			Year 4		
	J-A	M-A	S-D	J-A	M-A	S-D	J-A	M-A	S-D	J-A	M-A	S-D
I	G				G				G			
II		G				G				G		
III			G				G				G	
IV				G				G				G

Scheme for a four-block balanced rotational grazing system, as used for the rehabilitation of overgrazed land in Kenya, showing the movement of livestock as each block is grazed in turn (G) for a period of four months. The complete grazing cycle lasts four years, at the end of which the cycle is repeated. In the areas where this system has been used, the wet season lasts from April to September, as shown by the shaded area.

(b)

Block	Year 1				Year 2				Year 3				Year 4			
	J-M	A-J	J-S	O-D	J-M	A-J	J-S	O-D	J-M	A-J	J-S	O-D	J-M	A-J	J-S	O-D
I	G				≡			G			G			G		
II		G			G		≡					G		G		
III			G			G			G		≡					G
IV	≡			G		G			G				G			

A variation of the 4-block grazing system. The grazing is still balanced between blocks, but grazing periods last 3 months and rest periods are of different lengths. Provision is included for burning each block once in 4 years, with a full growing season's rest before and after burning, but one of the other rest periods during the cycle does not include part of a growing season. However, under a bimodal rainfall pattern, with rains in April-June and October-December, the latter objection would not arise and there would be more opportunities for burning.

(c)

Block	Year 1				Year 2				Year 3				Year 4			
	J-M	A-J	J-S	O-D	J-M	A-J	J-S	O-D	J-M	A-J	J-S	O-D	J-M	A-J	J-S	O-D
I		G			≡				≡				G			
II						G			≡				≡			G
III	≡				≡		G						G			
IV	≡		G						G				≡			

A second variant of the 4-block grazing system with longer grazing periods, each of 6-months' duration, which eliminate the short rests which appear in Fig. (b). Two opportunities for burning are shown in each block, though the one that is preceded by a short rest period should only be used in years of favourable rainfall. Alternatively, grass not burnt might be used as reserve grazing.

(d)

Block	Year 1			Year 2			Year 3			Year 4		
	J-A	M-A	S-D	J-A	M-A	S-D	J-A	M-A	S-D	J-A	M-A	S-D
I	G			≡			≡	G				
II					G		≡			≡	G	
III	≡	G					G			≡		
IV	≡			≡	G					≡		

A balanced 4-block grazing system, in which the grazing applied to each block comprises one period of 4 months and one of 8 months. This is an alternative to Fig. (c), though the second opportunity for burning during the long rest period would seldom be used since it is followed soon after by 8 months grazing.

Figure 8.1 Rotational grazing system for semi-arid lands in Kenya (after Pratt and Gwynne 1977; Mututho 1989).

are similar to those described above for burning on rangeland (Section 8.1.3). Replanting needs to take place quickly after clear felling before the loss of top soil and plant nutrients through erosion reduces the quality of the land. If planting is delayed, it is desirable to establish ground cover using some of the cover crops referred to in Section 8.2.

Although with careful management and limiting clearance to small areas at a time patch-cutting may be an acceptable management technique, in very erodible areas soil loss may be too severe during logging to permit its use. In mountainous areas of Peninsular Malaysia, slides and gullies develop rapidly when forest is cleared, particularly along log landings, skid trails and roads (Berry 1956; Burgess 1971), and increased erosion and runoff may result in problems of sedimentation and flooding downstream. In these areas, selective felling, whereby only the mature trees are removed and the other trees remain to provide a plant cover, is a better conservation practice. A selective management system is proposed by Wan Yusoff (1988) in which only marked trees, identified during an inventory of the forest, are removed, using directional felling to avoid damage to seedlings, young trees and rivers. A filter strip, with a minimum width of 20 m, from which no trees can be removed should be maintained on the banks of perennial streams. Harvesting by tractors fitted with shear blades and using skid trails should be permitted only on slopes up to 30°. On steeper slopes, only high-lead yarding or skyline cables should be used. Similar practices, but with a 15° upper limit on slopes for harvesting operations, are recommended for the timber plantations operated by the Kapunda Development Co. Pty. Ltd., at Bombala, New South Wales (Marshall and de Fegely 1987).

8.1.5 Recreation management

In recent years, as the problem of erosion has become more severe, interest has focused on whether rotational practices can be adapted to recreational areas, for example, by periodically fencing off sections of land to close certain footpaths and access routes so that the vegetation cover has a chance to regenerate. In many upland areas, where grazing, forestry and recreation are complementary but sometimes competing activities, the prospect of applying a common conservation measure is extremely attractive.

8.2 Cover Crops

Cover crops are grown as a conservation measure either during the off-season or as ground protection under trees. In the USA they are grown as winter

annuals and, after harvest are ploughed in to form a green manure. Typical crops used are rye, oats, hairy vetch, sweet clover and lucerne in the north, and Austrian winter peas, crimson clover, crotalaria and lespedeza in the south. To be effective, the cover crop must be quick to establish, provide an early canopy cover, aggressive enough to suppress weeds and possess a deep-root system to improve the macroporosity of the soil. The broadcast sowing of winter rye at a rate. of 120–130 kg/ha as early as possible in the autumn is practised to control wind erosion successfully on the sandy soils of the northern Netherlands where sugar beet, potatoes and maize are grown (Eppink and Spaan 1989). Early sowing is essential because a good crop cover must be obtained by the end of December since the climate is too cold for growth in February and March, the period when the soil surface is dry and strong winds with low humidity are most common. The crop is spray-killed with paraquat at a suitable time in the spring, usually before the drilling of sugar beet or before the emergence of potatoes and maize. Wind tunnel studies have shown that winter sown rye can prevent blowing of the soil with windspeeds up to 21 m/s as measured at a height of 10 m (Knottnerus 1976). Winter rye is now being adopted by farmers in Limbourg, The Netherlands, to control water erosion (Kwaad and van Mulligen 1991). Maintaining a cover crop over winter instead of a bare soil is also beneficial in reducing the leaching of nitrogen to the groundwater.

German scientists have shown that undersowing maize or sugar beet with clover offers sufficient protection of the soil with no reductions in yield. Compared with conventional cultivation, undersowing reduced soil loss on loess soils in Rheinland under sugar beet in 1984 from 9.2 t/ha to 0.1 t/ha (Buchner 1988). On sandy soils in Schleswig-Holstein, undersowing of maize reduced the mean annual soil loss for 1985–87 from 3.4 t/ha to 0.4 t/ha (Goeck and Geisler 1989).

Ground covers are grown under tree crops to protect the soil from the impact of water drops falling from the canopy. They are particularly important with tall crops such as rubber where the height of fall is sufficient to cause the drops to approach their terminal velocity (Section 3.4). With bare soil underneath the trees, erosion rates greater than 20 t/ha have been recorded under oil palm (Lim 1988) in Malaysia, even with bench terracing. The erosion can be reduced to less than 0.05 t/ha with cover crops, although the harvesting paths still remain vulnerable (Maene et al. 1979). The most common cover crops used are *Pueraria phaseoloides*, *Calapogonium mucunoides* and *Centrosema pubescens*. Although they grow rapidly and retain nutrients in the soil which would otherwise be removed by leaching, their use can sometimes give problems. First, there is a risk that a satisfactory cover will not be attained. This is because, as the main crop becomes established, ground conditions change from strong, open sunshine to shade, causing some plants to die out. Second, the cost of growing cover crops may outweigh the benefits an individual farmer receives. Most covers give no income and this restricts their use on smallholdings where farmers do not have sufficient cash reserves to wait for the tree crop to mature. Research is required to find suitable cover crops, particularly varieties of beans and peas, which the smallholder can grow. Third, ground covers compete for the available moisture and, in dry areas, may adversely affect the growth of the main crop. Studies in rubber plantations in eastern Java show that cover crops may reduce the soil moisture by up to 50 per cent during the dry season compared with clean-weeding (Williams and Joseph 1970). An alternative conservation measure is required in these circumstances.

When used in vineyards in the Beaujolais region of France a permanent grass cover, managed by mowing, reduced erosion to less than 7 per cent of that in vineyards with bare soil and did not compete with the vines for moisture despite the hot dry summer period. It is suggested that increased infiltration promoted by the grass limited the water loss by runoff during storms and thus helped to counterbalance any potential water deficit (Gril, Canler and Carsoulle 1989). In the Mosel Valley, Germany, grass reduces the available water in the top 40 cm of soil in the summer when rainfalls are about average but improves the moisture status during dry summers, with no significant difference in yield compared with maintaining a bare tilled soil under the vines (Husse 1991). Nevertheless, a permanent grass cover is rarely maintained in vineyards because of the fear of moisture competition. A grass cover reduces annual soil loss from vineyards in the Ruwer Valley, Germany to 0.31 t/ha, compared with 2.77 t/ha without grass, but appears to cause delays in the blooming and ripening of the grapes and a reduction in harvestable yield from 16.5 t/ha to 4.23 t/ha, a performance explained by a nitrogen deficiency, however, rather than moisture competition (Bamberger 1991).

8.3 Strip-cropping

With strip-cropping, row crops and protection-effective crops are grown in alternating strips aligned on the contour or perpendicular to the wind (Fig. 8.2a). Erosion is largely limited to the row-crop strips and soil removed from these is trapped within and behind the next strip downslope or downwind which is gen-

erally planted with a leguminous or grass crop. The strip-cropping of maize with soyabean on a 2° slope near Dehra Dun, India, gave an annual soil loss of 9.5 t/ha compared with 15.7 t/ha for maize alone (Singh, Bhardwaj and Singh 1979). Crop rotation is applied to the strips which vary in width according to the degree of erosion hazard. Guidelines for widths, which are generally between 15 and 45 m, are given in Table 8.1.

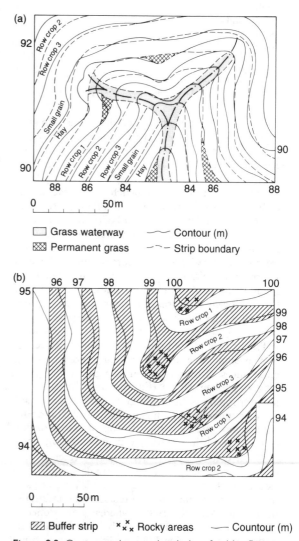

Figure 8.2 Contour strip-cropping designs for (a) a 5-year crop rotation and (b) the use of buffer strips. The contour lines are in metres above an arbitrary datum. Note the use of a grass waterway to evacuate excess runoff from the strip-cropped area and the inclusion of rocky areas which cannot be farmed within the buffer strips (after Troeh, Hobbs and Donahue 1980).

Table 8.1 Recommended strip widths for strip-cropping

Water erosion (soils with fairly high water intake)

2–5 per cent slope	30 m
6–9 per cent	25 m
10–24 per cent	20 m
15–20 per cent	15 m

Wind erosion (strips perpendicular to wind direction)

Sandy soil	6 m
Loamy sand	7 m
Sandy loam	30 m
Loam	75 m
Silt loam	85 m
Clay loam	105 m

After FAO (1965); Cooke and Doornkamp (1974).

On steep slopes or on very erodible soils, it may be necessary to retain some strips in permanent vegetation. These buffer strips are usually 2 to 4 m wide and are placed at 10 to 20 m intervals (Fig. 8.2b). The gradual build-up of soil behind each buffer strip leads, in time, to the formation of bench-type terraces. On a contour grass strip system with a tufted grass, *Setaria anceps*, on a clay nitisol soil near Nairobi, Kenya, up to 12 cm depth of soil had accumulated above the strips within two years and the local slope had been reduced from 6° to 5° (Wolde and Thomas 1989). However, sediment deposition also leads to a rise in height of the ground surface above the strip relative to that below, thereby locally increasing the slope steepness. Unless the soil within the strip is properly protected by the vegetation, water leaving the strip, having had its sediment load filtered out and now flowing over the steeper slope, may initiate erosion. Under these circumstances, erosion rates can be higher with contour grass strips than without (Emama and Morgan 1995). Strip-cropping is best suited to well-drained soils because the reduction in runoff velocity, if combined with a low rate of infiltration on a poorly-drained soil, can result in waterlogging and standing water.

When used to protect land against water erosion, contour grass strips are not normally required on slopes less than 3°. On slopes around 5°, the strips act mainly by retarding flow and encouraging infiltration of runoff (Melville 1992) whereas, on slopes of 12–16°, they control erosion by filtering out the sediment from the flow but have little effect on runoff (Boubakari 1992). The maximum slope at which grass strips remain effective has not been determined. Although they may be inadequate as the sole conservation measure on slopes above 8.5° (FAO 1965), they have been adopted by farmers on slopes up to 30° (Wolde and Thomas 1989).

The main disadvantage of strip-cropping is the need to farm small areas which limits the kind of machinery that can be operated. The technique is not therefore compatible with highly-mechanized agriculture. Although this is a less relevant consideration on small-holdings, the difficulty here is that much land is taken up with protection-effective crops of limited value. Contour grass strips were introduced into Swaziland by a decree of King Sobhuza II in 1948 and by the 1950s some 112,000 km of strips were laid out, protecting virtually all the arable land surrounding individual homesteads. The strips were originally 2 m wide but by the 1970s there was evidence that farmers were gradually allowing the cropped areas to encroach on the strips. However, the system appears to be relatively flexible regarding strip width and there may be little additional benefit in having strips wider than 1.5 m. Wolde and Thomas (1979) found that the effectiveness of grass strips increased exponentially. Strips of 0.5 and 1.0 m wide reduced soil loss to 36 per cent of that from an unprotected bare plot but a 1.5 m wide strip reduced soil loss by only a further 18 per cent.

The plants chosen to form the buffer strips are usually grasses. They should be perennial, quick to establish and able to withstand periods of both flood and drought. They should have deep-rooted systems to reinforce the soil and reduce scouring, a uniform density of top growth to provide a filter for sediment and reduce flow velocity; their growth points should be close to the ground or below the soil so that they are not grazed out and can recover from damage after fire; and they should either be sterile or propagate very slowly so that they do not become weeds in the adjacent cropped strips. For this last reason, rhizomous species should be avoided since they spread very rapidly on to surrounding land. Tussocky grasses, such as Kikuyu grass (*Pennisetum clandestinum*), should be avoided because they concentrate the flow. Even tufted grasses like *Setaria anceps* are not ideal though, as seen above, they have successfully reduced erosion in Kenya. Grasses with an erect growth habit of interwoven stems which act like a porous filter are more effective in reducing erosion than those with slender creeping rhizomes and a more horizontal growth form (Lakew 1991; Melville 1992). Among the more promising grasses are Napier grass (*Pennisetum purpureum*), Guatemala grass (*Tripsacum laxum*), Makarikari grass (*Panicum coloratum*), oat grass (*Hyparrhenia* spp.), wheat grass (*Agropyron* spp.) and *Elymus* spp.

Recently, considerable publicity has been given to Vetiver grass (*Vetiveria zizanoides*) as a result of a vigorous campaign by The World Bank in India (World Bank 1990; National Research Council

1993). Vetiver grass establishes quickly from splits and does not compete with adjacent crops. On experimental plots on 9–12° slopes in Venezuela, 0.5-m wide strips of Vetiver grass reduced annual erosion under horticultural crops (beet, carrot, cabbage and cauliflower) to less than 1 t/ha compared with 7–33 t/ha for other management practices (Rodriguez and Fernandez de la Paz 1992). Over an 11-month period on an oxisol under cassava at the Santander de Quilichao Research Station, near Cali, Colombia, soil loss was 1.3 t/ha with Vetiver grass contour barriers, 4.0 t/ha with Napier grass strips and 8.3 t/ha with no conservation measures. Whilst the Napier grass strips took 25 per cent of the land out of production, the Vetiver grass took only 12.5 per cent (Laing 1992). This is because Vetiver grass strips are only 0.5-1.0 m wide. Over a three-year period on the Punjabrao Krishi Vidyapeeth University Farm, Manoli, India, annual soil loss averaged 3.3 t/ha with Vetiver grass strips on the contour compared with 11.4 t/ha using only across slope cultivation for a rotation of green gram–pigeon pea–safflower; pearl millet–safflower; and pearl millet (Bharad and Bathkal 1991). Whilst Vetiver grass has the advantage that it is not palatable to livestock and therefore any strips formed from it are likely to remain undamaged, it provides no income to the farmer, whereas Napier grass can be cut and fed to cattle. Clearly, the ideal situation is where the economic value of the grass strips equals or exceeds that lost by taking the land out of agricultural production.

Strip systems can also be used to provide in-field shelter for wind-sensitive crops and protect the soil against erosion by wind. In the Fens of eastern England, live barley strips are used to protect onions, sugar beet and carrots from wind damage and to control erosion on the lowland peat soils. The barley is sown in February or March, allowing time for it to emerge before the main crop is drilled in mid to late April. The barley thus provides cover at the critical period for erosion. Once the main crop has sufficient biomass to perform a protective role, the barley is killed with a selective herbicide. Care is required with the amount and timing of the herbicide to ensure that the barley is properly removed without destroying or stunting the growth of the main crop. Live barley meets the design requirements of a plant for infield shelter in having bladed rather than ovate leaf forms to minimize the risk of streamlining in strong winds, rigid leaves to reduce flutter and thereby minimize the 'wall effect' (Section 3.4) and an upright growth habit with at least 0.65 kg DM/m^3 of uniform biomass in the lowest 5 cm to provide an adequate filter (Morgan 1989). The barley strips should be spaced at distances of eight to ten times their height which is every 2 m if

the barley is 20 cm tall at the time when the risk of erosion is highest.

8.4 Multiple Cropping

The aim of multiple cropping is to increase the production from the land whilst providing protection of the soil from erosion. The method involves either sequential cropping, growing two or more crops a year in sequence, or intercropping, growing two or more crops on the same piece of land at the same time. Many schemes involve a mixture of the two. Multiple cropping has been traditionally practised in the West Indies in the kitchen gardens which, averaging 0.2 ha in size, provide the subsistence component of fruits and vegetables for many families in the rural areas. Those in Grenada are dominated visually by fruit trees with banana the most ubiquitous but also including coconut, cocoa, mango and breadfruit. About two-thirds of the tilled land in the garden are devoted to vegetables, especially root crops such as sweet potatoes and yams. Pigeon peas, groundnuts, dasheen, okra, pepper and tomatoes are also grown. The home garden thus mimics the natural, multi-layered ecosystem (Hoogerbrugge and Fresco 1993). Although the

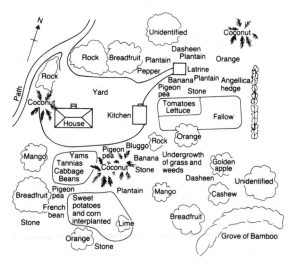

Figure 8.3 Land use in kitchen garden, near Concord, St John's, Grenada, showing an example of multiple cropping (after Brierley 1976).

garden has a confused appearance (Fig. 8.3), the organization of the cropped land is systematic and the better farmers carry out regular weeding, selective application of fertilizers and irrigation (Brierley 1976).

The intercropping of maize with cassava offers the advantages of a two-storey canopy, giving a higher interception capacity and reducing detachment of soil particles by raindrop impact to 35 and 60 per cent of the respective values from cassava or maize alone (Lal 1987). On a 6° slope, mixed maize-cassava reduced annual soil loss to 86 t/ha compared with 125 t/ha for cassava as a monoculture (Lal 1977a). Both values, however, are well above most soil loss tolerance levels as are those found in similar experiments involving cassava with groundnuts, maize, cowpeas and peppers (16.2 t/ha) near Benin City, Nigeria (Odemerho and Avwunudiogba 1993). Comparable results apply to maize intercropping systems, for example maize with sesame, soya bean and pigeon pea (83.8 t/ha) in the East Khasi Hills, Meghalaya State, India (Awasthi, Prasad and Chatterjee 1987). Intercropping maize with leguminous ground covers is more effective. On an oxic aridisol soil at Molokai, Hawaii, soil loss during the 135-day cropping season was 4 t/ha from maize alone but 2.0 and 2.5 t/ha respectively when the maize was intercropped with rose clover (*Trifolium hortum*) and kalo (*Lotus corniculatis*) (El-Swaify *et al.* 1988) but these plants have the disadvantage of limited economic value to the farmer. Generally, multi-cropping needs to be combined with other practices.

8.5 High Density Planting

High density planting is used to try to obtain the same effect for a monoculture that multiple cropping achieves with two or more crops. Again the technique may need to be supplemented by other practices. Hudson (1981) showed in Zimbabwe that increasing the planting density of maize from 25,000 to 37,000 plants per hectare and using a trash mulch at the higher density reduced the annual soil loss from 12.3 to 0.7 t/ha. Respective crop yields were 5 t/ha and 10 t/ha. Over one cropping season in Trinidad, Mohammad and Gumbs (1982) found that increasing plant density from 41,500 to 62,000 plants per hectare reduced the soil loss under maize from 32 to 21 t/ha. No other practices, like mulching, were adopted which may explain why the crop yield increase, though significant, was less dramatic from 4.2 t/ha with the low plant density to 5.1 t/ha with the higher density. The arrangement of the plants can also have an effect. With a plant population of 55,000 plants per hectare, the lowest soil loss from maize in the Doon Valley, northern India, was obtained by increasing the distance between the rows and increasing the plant density within the rows. Thus changing from a row spacing of 45 cm and a plant spacing of 40 cm to a row spacing of 90 cm and a plant spacing of 20 cm reduced

annual soil loss under maize from 22.6 to 13.9 t/ha (Bhardwaj *et al.* 1985).

8.6 Mulching

Mulching is the covering of the soil with crop residues such as straw, maize stalks, palm fronds or standing stubble. The cover protects the soil from raindrop impact and reduces the velocity of runoff and wind. From the conservation viewpoint, a mulch simulates the effect of a plant cover. It is most useful as an alternative to cover crops in dry areas where insufficient rain prevents the establishment of a ground cover before the onset of heavy rain or strong winds, or where a cover crop competes for moisture with the main crop.

In the semi-humid tropical areas, the side-effects of a mulch in the forms of lower soil temperatures and increased soil moisture are beneficial and may increase yields. Elsewhere the effects of mulching can be detrimental. In cool climates, the reduction in soil temperature shortens the growing season whilst in wet areas, higher soil moisture may induce gleying and anaerobic conditions. Crop yields may also be reduced because the mulch competes with the main crop for nitrogen as it decomposes.

The effectiveness of mulching in reducing erosion is demonstrated by the field experiments of Borst and Woodburn (1942) who found that, on a silt-loam soil on a 7° slope, annual soil loss was 24.6 t/ha from uncultivated bare land but only 1.1 t/ha from land covered with a straw mulch applied at 5 t/ha. Lal (1976) found that covering an alfisol on a 6° slope with 6 t/ha of straw mulch resulted in an annual soil loss of 0.2 t/ha compared with 23.3 t/ha recorded on bare soil. Applying a mulch of lalang grass (*Imperata cylindrica*) at a rate of 3 t/ha to maize grown on a sandy loam soil on a 4° slope on the Experimental Farm of the Universiti Pertanian Malaysia, near Serdang, Selangor, reduced the soil loss between October 1978 and July 1979 to 0.5 t/ha compared with 7.5 t/ha recorded for maize grown without a mulch (Mokhtaruddin and Maene 1979).

Although such low soil losses are not always achieved, the extent to which a mulch will reduce erosion is generally impressive. A grass mulch at 4 t/ha applied to a silt-loam soil on a 4.5° slope near Dehra Dun, India, reduced annual soil loss under maize to 5.0 t/ha compared with 22.4 t/ha for maize tilled conventionally up-and-down slope (Khybri 1989). A grass mulch at 15 t/ha on a sandy loam soil with a 3° slope near Pakhribas in eastern Nepal gave an annual soil loss under maize of 16.9 t/ha compared with 32.8 t/ha for the traditional practice of maize planting followed by two hoeings (Sherchan *et al.* 1990).

Mulches can also be used under tree crops. Using pruned fronds to cover harvesting paths in an oil palm plantation in Johor, Malaysia, reduced annual soil loss to 4.2 t/ha from 14.9 t/ha recorded on unprotected paths (Maene *et al.* 1979). A grass mulch applied between rows of young tea on a 6° slope on an ultisol near Kericho, Kenya reduced average annual soil loss over three years to 0.68 t/ha compared with 210.8 t/ha for bare hand-tilled soil between the rows (Othieno 1978).

Most applications of mulching involve spreading crop residues on the surface of the soil but this method creates problems in drilling or planting through the mulch, unless specialized equipment is used. The mulch may also delay plant emergence and, unless the mulch cover is 90 per cent or more, encourage the growth of weeds. An alternative approach is to incorporate the mulch in the soil. Although this reduces the overall surface cover, the mulch elements help to bind the soil, simulating the effect of plant roots, and increase infiltration rates. Duley and Russel (1943) found that incorporated crop residue reduced annual soil loss from 35.7 t/ha on a bare soil to 9.9 t/ha. The effect of incorporated mulch depends upon the material used. Wheat straw is particularly effective because it produces a large number of individual mulch elements, resulting in a more uniform pattern of incorporation, good contact between the mulch and the soil, and sufficient material on the surface to form miniature dams, behind which water ponds, and protect the soil against crusting. In contrast, maize stalks perform badly because their large size means that the surface is sparsely covered, the contact with the soil is poor and the individual elements are easily washed out and transported downslope (Abrahim and Rickson 1989). Overall, incorporated mulches are less effective than surface mulches but where erosion rates are only just above the soil loss tolerance, they may be sufficient to provide the necessary control with the advantage that they can be used with conventional tillage equipment.

When using mulches to control wind erosion, standing stubble is required because of the danger of spread mulches blowing away. A mulch of 1 t/ha of standing wheat stubble or 2 t/ha of flattened wheat straw will reduce annual wind erosion rates to a tolerable level of 0.2 t/ha. To achieve the same effect with sorghum stubble requires a mulch of 6.7 t/ha (Chepil and Woodruff 1963).

There is considerable experimental evidence (Wischmeier 1973; Lal 1977b; Laflen and Colvin 1981; Norton, Cogo and Moldenhauer 1985; Bekele and Thomas 1992) to show that the rate of soil loss decreases exponentially with an increase in the percentage area covered by a mulch. The mulch factor

(*MF*), defined as the ratio of soil loss with a mulch to that without, is related to the percentage residue or mulch cover (*RC*) by the expression (Laflen and Colvin 1981):

$$MF = e^{-a.RC} \qquad (8.1)$$

where a ranges in value from 0.01 to 0.07 with 0.05 being generally applicable. The value is affected by the degree of soil disturbance by tillage and ranges from 0.03 when the mouldboard plough is used to 0.06 with no tillage (Norton, Cogo and Moldenhauer 1985). Hussein and Laflen (1982) found that the exponential relationship applied only to rill erosion and that the rate of interrill erosion decreased linearly with increasing residue cover. Since for most soils the contribution of interrill erosion to total soil loss is quite small for slope lengths in excess of 25 m, Eq. 8.1 can be used to obtain mulch factor values which, when multiplied by the *C*-factor values, allow the effects of mulching to be included in the Universal Soil Loss Equation (Section 5.1.1).

A mulch should cover 70 to 75 per cent of the soil surface. With straw, an application rate of 5 t/ha is sufficient to achieve this. A lesser covering does not adequately protect the soil whilst a greater covering suppresses plant growth. Denser mulches are sometimes used under tree crops to control weeds. An estimate of the required application rate to control erosion can be made for a preselected set of conditions using the Manning equation for flow velocity (Eq. 2.13) and the relationship between Manning's n and mulch rate for maize straw determined by Foster, Johnson and Moldenhauer (1982):

$$n_m = 0.071\, M^{1.12} \qquad \text{interrill erosion} \quad (8.2)$$
$$n_m = 0.105\, M^{0.84} \qquad \text{rill erosion} \qquad (8.3)$$

where n_m is the value of n due to the mulch and M is the mulch rate (kg/m^2) (Table 8.2). No similar procedure exists for determining mulch rates to control wind erosion, but the effectiveness of standing crop residues depends upon the number of stalks per unit area and their size.

As indicated above, controlling erosion with a mulch poses special problems for the arable farmer because tillage tools become clogged with the residue, weed control and pest control are more difficult, planting under the residue is not always successful and crop yields, especially in humid and semi-humid areas, are sometimes lower. Mulching on its own is not always an appropriate technique but where it is combined with conservation tillage, many of these problems can be overcome, and it has tremendous potential as a method of erosion control.

Many workers have proposed that rock fragments on the surface act in the same way as a mulch and that stone mulches might therefore be appropriate for erosion control on non-agricultural areas such as road banks and construction sites. Relationships between the rate of erosion and percentage stone cover frequently follow that described by Eq. 8.1 (Poesen 1992). However, where the stones are embedded in a sealed or crusted soil, interrill erosion can be enhanced and will increase with percentage stone cover because the runoff generated on the impermeable stone elements is unable to infiltrate the surrounding soil (Poesen and Ingelmo-Sanchez 1992). On an

Table 8.2 Estimating the required density of a maize stalk mulch to control water erosion

---

Estimation is made for the following conditions:
Sandy soil, desired maximum flow velocity	$= 0.75$ m/s
Flow depth in small channels	$= 100$ mm
Slope	$= 5°$

Estimating required value for total Manning's n
From the Manning equation (Eq. 2.13)

$$n = \frac{r^{0.67} s^{0.5}}{v}$$

For simplicity, assume the hydraulic radius (r) is approximated by the depth of flow and that slope (s) can be represented by the tangent of the slope angle. Thus

$$n = \frac{0.1^{0.67}\ 0.087^{0.5}}{0.75}$$

$$n = 0.0843$$

Estimating required value for Manning's n due to mulch (n_m)
According to Foster, Johnson and Moldenhauer (1982)

$$n = (n^{3/2} - n_s^{3/2})^{2/3}$$

Taking a value of Manning's n due to the soil (n_s) = 0.02 gives

$$n_m = (0.0843^{3/2} - 0.02^{3/2})^{2/3}$$

$$n_m = 0.0767$$

Estimating required mulch application rate (M)
Rearranging Eq. 8.3 gives

$$M = \left(\frac{n_m}{0.0105}\right)^{1.06}$$

$$M = \left(\frac{0.0767}{0.105}\right)^{1.06}$$

$$M = 0.72\ \text{kg/m}^2$$

Note: Theoretically this procedure can be used to determine application rates for other mulches but in practice the relationships between M and n_m have not been established. Equation 8.3 cannot therefore be used for wheat straw, soya bean residue, palm fronds or other mulch materials.

erodible soil, rock fragments can also induce local turbulence leading to scouring on the upstream side of the stones and deposition on the downstream side. With small stones or pebbles around 1.5 cm in size, scouring can cause sediment yield to increase as cover increases from zero to 20 per cent but with further increases in percentage cover, the protective effect prevails. On covers of 60 per cent and above, scour is almost impossible. For cobbles around 8.6 cm in size, however, no reduction in sediment production occurs and the soil loss increases exponentially with increasing percentage cover (Bunte and Poesen 1994). Since rock fragments can both decrease and increase erosion according to local circumstances, their use as a mulch to control erosion cannot be recommended.

8.7 Revegetation

Vegetation plays the major role in the process of erosion control on gullied areas, landslides, sand dunes, road embankments, construction sites and mine spoils. Rapid revegetation is also necessary for replanting forest in areas cleared by patch-cutting and for covering land cleared of forest in favour of agriculture. The first-listed cases represent marginal environments for plant growth where the risk of vegetation failing to re-establish is high. The second-listed cases are less marginal and the objective must be to minimize damage during clearance and to establish cover quickly before the environment has time to deteriorate.

When developing a plan for revegetating an area, a soil test should be carried out to establish pH, nutrient levels, moisture status, salinity levels and the presence of toxic ions, all of which will influence the range of species which will grow. Climatic conditions should also be studied, including the frequency of drought and waterlogging. Topographic influences on the local climate are important, for example, differences in temperature and moisture between sunny and shady slopes, and frost hollows. Topography also determines the location of dry and wet sites through its effect on movement of water through the soil. Plant species should be selected for their properties of rapid growth, toughness in respect of diseases and pests, ability to compete with less desirable species and adaptability to the local soil and climatic conditions. Wherever possible, native species should be chosen. A study of neighbouring sites often gives a good indication of what species are most likely to survive and thrive. The use of introduced or exotic species should not be ruled out, however, especially where the local environment has deteriorated beyond that of adjacent sites or where numbers of local species are limited. The revegetation plan should allow for plant succession which will take place naturally. In many cases, the objective is to establish pioneer species which will give immediate cover and improve the soil, permitting native species to come in and take over as the pioneer plants decline. Generally, a mix of plant species is required because it is impossible to predict the success of any one species in marginal environments where the vegetation is going to receive little or no maintenance. A monoculture is also more susceptible to disease. The species mix should include grasses, forbs and woody species, both bushes and trees, except where specific requirements make such a mix undesirable, as with certain types of gully reclamation.

The most common way of planting grasses and forbs is by broadcast seeding. On small areas this can be carried out by hand but over large areas, aerial seeding is used. Hydroseeding is the application of the seed in a water slurry, usually with fertilizer and sometimes with a mulching material. Woody species are planted either by spot seeding or as cuttings. Grasses, forbs and woody plants may all be established by transplanting which provides a quick method of obtaining ground cover but needs to be undertaken when there is adequate moisture in the soil. The limited availability of water is a frequent cause of transplant failure. More detailed coverage of the methods of vegetation establishment is given by Gray and Leiser (1982) and Coppin and Richards (1990).

8.7.1 Restoration of gullied lands

Revegetation is used in gully erosion control as a method of increasing infiltration and reducing surface runoff. The area around the gullies should be treated with grasses, legumes, shrubs and trees or combinations thereof, aided in the early stages by mulching or the use of geotextiles if necessary. Research carried out at the Suide Soil and Water Conservation Experiment Station of the Huang He Conservancy Commission shows that afforestation can reduce runoff in the gullied loess areas by 65 to 80 per cent and soil loss by 75 to 90 per cent. Growing grass reduces runoff by 50 to 60 per cent and soil loss by 60 to 80 per cent (Gong and Jiang 1977). By planting trees and herbs on the steep slopes of the gully sides and raising crops on bench terraces on the gentler slopes of the divides (Fig. 8.4), the land can be stabilized and sediment prevented from entering the gullies. By supporting these measures with check dams and reservoirs along the gully beds, the volume of sediment entering the Huang He from gullies in the Wuding Valley was reduced by 44 per cent over the period 1971 to 1978 (Jiang, Qi and Tan 1981). The main types of trees and herbs used are *Caragana* sp., locust, poplar, Chinese pine, jujube, pear, apple, alfalfa, sweet clover and

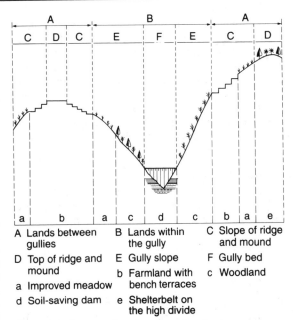

A	Lands between gullies	B	Lands within the gully	C	Slope of ridge and mound
D	Top of ridge and mound	E	Gully slope	F	Gully bed
a	Improved meadow	b	Farmland with bench terraces	c	Woodland
d	Soil-saving dam	e	Shelterbelt on the high divide		

Figure 8.4 Cross-section showing soil and water conservation measures adopted in the gullied loess area of China (after Jiang, Qi and Tan 1981).

Astragalus sp. The effectiveness of the *Astragalus adsurgens* grass and a seven-year old cover of black locust forest is shown by trials on 22–28° slopes in the Xingzihe Watershed within the Yanhe basin where, in 1980, the respective soil losses were 1.47 t/ha and 0.37 t/ha compared with 44.17 t/ha on farmland under millet and 10.44 t/ha on overgrazed land. Provided about 30 to 35 per cent vegetation cover remains on the land, closing an area on the gullied loess in the Xingzihe Basin by fencing and prohibiting its use for grazing can lead to a natural increase in cover to 70 to 90 per cent within three to four years. With supplementary planting of grasses and shrubs, the same effect can be achieved within two years (Tang *et al.* 1987). In order to provide niches into which aerially-sown seeds can drop and germinate successfully, a rough surface is created by breaking up the ground to a depth of 5 cm. *Astragalus adsurgens*, *Caragana microphylla* and *Hippophae rhamnoides* are the most suitable species for aerial sowing because of their small seeds, rapid germination, resistance to strong sunlight and high temperatures, and resistance to winter kill. The inclusion of *H. rhamnoides* in the seed mixture has solved the problem of vegetation succession following the decline of *A. adsurgens* after eight or nine years (Wu *et al.* 1988). With this approach, some 4,700 hectares of gullied loess were successfully reclaimed between 1975 and 1984.

In the studies mentioned above, the main objective was to increase infiltration. Achieving this will help reduce gullying when surface processes are involved but where the gullies are being fed by subsurface pipes or tunnels, it is necessary also to promote infiltration in a uniform pattern. Since trees and grasses have different root densities and their root networks extend to different depths, their mixture may result in more infiltration under the trees which may, in turn, feed water into a pipe system. Even if the pipe network has been previously broken up by subsurface ripping, concentrations of water in the soil may encourage pipes to reform. This is more likely to occur if tree species with long tap roots are planted. Thus, the best vegetative treatment for tunnelled areas is to establish a dense, uniform grass cover. Where conditions are too marginal for grass to grow, shrubs and trees will have to be used but species with a good system of lateral rather than vertical roots should be chosen.

8.7.2 Restoration of landslide scars

Tree planting is recommended as a method of stabilizing slopes prone to mass movement. Although the addition of trees to a hillside may sometimes induce sliding because of the increase in weight (De Ploey 1981b), this effect is generally offset by an increase in cohesion associated with the binding of the soil within the root network and by the tensile strength of the roots themselves. Also, the surcharge effect can be minimized by placing trees at the bottom rather than the top of the slope. Living tree roots can contribute up to 20 kPa to the soil shear strength (O'Loughlin and Watson 1979). It is believed that the lateral roots contribute most to binding because of their greater density whereas the vertical roots add most of the tensile strength and, where they cross a potential slide plane, help anchor the soil on to the slope. Deep-rooted species are thus preferred for stabilizing the slope and increasing its resistance to sliding. Bishop and Stevens (1964) show that large trees can increase the shearing stress required for sliding by 2.5 kPa which is why, as seen in Section 7.3.3, their removal can promote landslides.

Closing off an area, particularly from livestock but also from wild animals, will allow vegetation to colonize landslide scars naturally. The rate of recovery is slow, however. Herbs come in first followed by grasses but only about four years after the slide do perennial grasses dominate the cover and shrubs start to appear. After seven years the cover on scars in the Mgeta Valley, Tanzania, was only 25 per cent (Lundgren 1978). Recovery of soil slip scars in the Wairarapa hill country, New Zealand, takes 20 years but even then the productivity of the pasture is only 80 per cent of that on uneroded land. Despite recolonization, the quality of the land deteriorates (Trustrum, Thomas and Lam-

bert 1984). Tree planting was attempted on the slide scars in the Mgeta Valley but the species used, *Acacia mearnsii*, *Cupressus lusitanica* and *Eucalyptus maidenni*, proved unsuitable. All the *Cupressus lusitanica* seedlings died and the survival of the others was threatened by gully erosion (Lundgren 1978). This example emphasizes the need for careful selection of species in relation to the environmental conditions.

8.7.3 Afforestation

Many countries now have afforestation programmes aimed at arresting erosion and regulating floods. Most schemes involve closing the land to other uses. It should be recognized, however, that the success of such schemes depends on the methods used to prepare the land for replanting and that it takes some years before the all-important litter layer develops on the soil surface. As seen in Section 7.3.3, erosion often increases in the early years of reforestation work.

In the Vasad and Kota areas of Gujarat, India, closing of gullied lands to grazing allowed the establishment first of a good grass cover of desirable species and then an increase in tree numbers through natural colonization. Afforestation trials were also successfully implemented, with bamboo, teak, sissoo and eucalyptus as the most promising species (Tejwani 1981). The principal species in the forest plantations in Kenya, *Cupressus lusitanica*, *Pinus patula*, *Pinus radiata* and *Eucalyptus saligna*, are selected because of their rapid growth rather than their value in soil and water conservation (Konuche 1983). Although the pines are generally satisfactory, the cypress results in a bare forest floor liable to erosion by overland flow unless the stands are pruned and thinned. The role of eucalypts is unclear because of evidence that when planted in wet areas they result in a reduction of water supply in springs and rivers (Gosh, Kaul and Subba-Rao 1978) yet when grown in drier areas they do not consume large quantities of water (Konuche 1983). Since eucalypts develop their root systems rapidly and promote infiltration and subsurface drainage, they may induce mass movements when planted on steep slopes with shallow soils. This is because of the impedance to subsurface water movement at the soil–rock interface and the reduction in the shear strength of the soil as the moisture content increases to saturation. This mechanism has been invoked as a cause of landslides in the forested areas of the Serra do Mar, Brazil (De Ploey and Cruz 1979).

8.7.4 Restoration for pasture

Revegetation of lands for pasture is the major activity of Landgraeðsla Ríkisins, the State Soil Conservation

Service of Iceland. Work is concentrated on restoration of bare moving sands and gravels using aerial seeding. The land is first fenced to keep out livestock. *Elymus arenarius* is then planted on the moving dunes in strips at right-angles to the erosive winds and fertilized annually until serious sand movement has been halted. This plant thrives well in drifting sand and collects and fixes its own dunes. Once the sand has been stabilized, the *Elymus* dies out and *Festuca rubra*, *Poa pratensis* and *Phleum pratense* are aerially seeded. Fertilizer is applied aerially each year for the next two to four years by which time a reasonable vegetation cover has been obtained. The next step depends on whether the land is upland rough grazing or is lowland to be used for cultivated pastures and hay production. In the first case, the land remains fenced but no further work is carried out. In the absence of continued fertilizer application, the seeded grasses die out and there is a period of about seven to ten years after the start of reclamation when the vegetation cover becomes poor. Sufficient stability and organic matter have been achieved, however, to allow native vegetation to colonize. The land remains protected for 30 years (Runólfsson 1978; Arnalds, Aradóttir and Thorsteinsson 1987). In the second case, the land continues to be fertilized and the grass is cut annually for hay. Reseeding sometimes becomes necessary if the grass suffers from winter kill. At present the reclamation relies on imported seeds because the local climate is too severe to provide a reservoir of locally available seed. Experiments are in progress on the use of *Lupinus nootkatensis* which seems well-adapted to Icelandic conditions, provides good ground cover as well as a supply of seed and fixes nitrogen which is then available to later species in the plant succession (Arnalds 1988).

In many countries of the world, revegetation is limited not only by the infertility of the soil but also by lack of water. In addition to supplying fertilizers in the early stages of reclamation, it is necessary to provide water-conserving structures. Mututho (1989) modified a traditional pitting practice of the WaMatengo people in southern Tanzania (Pike 1938) to form semi-circular pits, 5–12 m^2 in area surrounded by a 15–30-cm high bank to trap water on rangeland reclamation schemes in the Kitui District, Kenya. The sites were excluded from grazing and the pits planted with indigenous perennial grasses. Good recovery was achieved within two years. The pits, however, require considerable inputs of labour for their construction. A modified version, known as the Katumani pit, has been developed (Gichangi *et al.* 1992) which is only 1.5–3.0 m^2 in area. These pits are also planted with indigenous grasses which recover well within two years but, in the first year after construction, a crop of beans and cowpeas can be obtained to give a short-term income until the grazing land has been rehabilitated.

8.7.5 Embankment and cut slopes

Rapid establishment of a grass or legume cover is essential on embankment and cut slopes to minimize surface erosion and enhance slope stability. This is commonly achieved by hydroseeding a seed and fertilizer mix. Where immediate erosion control is necessary, mulches or geotextiles should be used to protect the soil and prevent the seeds from being washed away. A straw mulch, applied at 4 t/ha at the time of seeding and fertilizing the soil in the autumn, was found to reduce erosion to acceptable levels on 45° roadside slopes in western Oregon (Dyrness 1975) and allow the establishment of more than 70 per cent vegetation cover by the end of the following summer. Maximum cover, mainly rye grass, bent and fescue, was achieved after some three years but then lack of nitrogen caused the vegetation to decline to only 10 per cent cover after eight years. By this time the slope was protected by a dense litter of dead grass. Addition of more fertilizer quickly revived the vegetation which developed to 90 per cent cover within one year. Most erosion occurred in the first year of the treatment after which soil loss was virtually zero. An alternative way of establishing cover is to plant natural turf over the slope as is standard practice in the urban areas of Singapore (Ramaswamy, Aziz and Narayanan 1981). Prior to seeding, the slopes should be covered with 100 mm of top soil and a seed bed prepared. Once the soil has been stabilized, ornamental trees and shrubs may be planted or the land can be left to be colonized by native vegetation.

Grasses, mainly *Cynodon dactylon* and *Pennisetum clandestinum*, are also used to control surface erosion on 30–65° road banks in eastern Nepal (Howell *et al.* 1991) but the increased infiltration that results enhances the risk of shallow slides. In order to reduce the risk, deeper-rooted grasses, such as *Saccharum spontaneum*, *Neyraudia arundinacea* and *Pennisetum purpureum*, are recommended to try to anchor the soil and root mat to the underlying weathered material. However, these grasses are clumpy in habit and cause runoff to concentrate. Also, their roots do not always adhere well to the coarse debris and pull-out when the material starts to move. Under these extreme conditions it is often necessary to supplement the vegetative measures with structural contols.

8.7.6 Sand dune restoration

The main plants used in stabilizing sand dunes in coastal areas of Europe and the USA are marram grass

(*Ammophila arenaria*) and American beach grass (*Ammophila breviligulata*). These have strong extensive root networks in both lateral and vertical directions which enable them to bind the sand whilst the grass acts as a sediment trap (Hesp 1979). They thrive well in moving sand with low nutrient availability. Grasses should be planted as 10–15-cm wide culms rather than seeded because of the risk of the seeds being blown away and the young plants being damaged by sand blasting. Once stabilized, the dunes can be planted with shrubs and trees. Scots pine, Corsican pine and Lodgepole pine were successfully used in the stabilization of the Culbin Sands near Nairn, Scotland.

Similar approaches are used to stabilize desert dunes whereby the establishment of pioneer plants binds the sand and leads to an improvement in soil conditions. Within eight years of establishing the fast-growing *Tamarix aphylla* on sands near the Al-Hasa oasis in Saudi Arabia, the depth of the organic horizon had increased from zero to 10 mm and the calcium carbonate content decreased from 30 to 15 per cent (Stevens 1974). The long-term effect of such schemes is not clear, however, because, as the trees mature, the soil moisture within the dune is depleted. Survival of the vegetation then depends on the ability of the tree roots to seek out groundwater. If this turns out to be salty, the trees may die and the dunes be remobilized (Gupta 1979). Generally, where precipitation annually exceeds 250 mm, dunes can be stabilized by vegetation within five to ten years. Where annual precipitation is between 100 and 250 mm, vegetational stabilization is possible but may take 20 to 30 years; also woody species cannot grow and only shrubs and grasses can be used. Where precipitation is less than 100 mm per year, a vegetation-based approach is not feasible except with irrigation over very small areas.

8.7.7 Restoration of recreational areas

Closing the land and replanting with trees, shrubs and grasses is frequently adopted to renovate areas eroded through recreational use. Examples are the Tarn Hows project in the English Lake District (Barrow, Brotherton and Maurice 1973), the reclamation of gullies in the Box Hill area of the North Downs in southeast England (Streeter 1977) and the footpath restoration work in the Three Peaks area of the Yorkshire Dales National Park in England (Rose 1989). In addition to ensuring that the species used for replanting are compatible with the physical environment and resistant to trampling, plant selection may be influenced by aesthetic considerations so as to enhance landscape quality and by the need to create varied and interesting wildlife habitats.

8.8 Agroforestry

Trees can be incorporated within a farming system by planting them on land which is not suitable for crop production. Where trees are deliberately integrated with crops or animals or both to exploit expected positive interactions between the trees and other land uses, the practice is defined as agroforestry (Lundgren and Nair 1985). Trees help to preserve the fertility of the soil through the return of organic matter and the fixation of nitrogen. They improve the soil's structure and help to maintain high infiltration rates and greater water-holding capacity. As a result less runoff is generated and erosion is better controlled. Trees are also attractive to the farmer where they provide additional needs, especially fuel, fodder and fruits. Multipurpose trees and shrubs are thus fundamental to agroforestry.

Agroforestry is being encouraged in many countries as a way of modifying existing farming systems to promote soil fertility, erosion control and a diversified source of income. Three types of practice may be defined. Agrisilviculture involves the combination of trees and crops. Silvopastoralism is the combination of trees and animals. Agrosilvopastoralism combines trees with crops and animals. Within these systems, trees can be used to supplement existing erosion-control measures, for example by being added to contour grass strips and terraces or they can be used to control erosion direct, for example in mutiple-cropping systems such as kitchen gardens or with contour-aligned hedgerows in intercropping. Much attention has been given to alley cropping systems whereby multipurpose trees are grown as contour hedges separated by strips of cropland. On a 16° slope at the Butare Research Station in Rwanda, the annual soil loss over four years from cassava in an alley cropping system with 5-m wide strips of the leguminous shrub *Calliandra calothyrsus* grown on microterraces was 12.5 t/ha compared with 111 t/ha for traditional cultivation of cassava. Mixed cropping of cassava in the alleys with beans, maize and sweet potatoes, alternating with leguminous cover crops, reduced soil loss to 1 t/ha (König 1992). In a range of alley systems on an alfisol on a 4° slope near Ibadan, Nigeria, annual soil loss over two years for a maize–cowpea rotation was 1.6 t/ha with *Leucaena* hedges at 4 m spacing, 0.15 t/ha with *Leucaena* hedges at 2 m spacing, 0.88 t/ha with *Gliricidia* hedges at 4 m spacing and 1.7 t/ha with *Gliricidia* hedges at 2 m spacing compared with 8.7 t/ha with conventional cultivation and 0.025 t/ha with no tillage (Lal 1988). Crop yields of the maize were similar in all systems.

Agroforestry systems require careful selection of both crops and tree species if a beneficial interaction is to be obtained. Indeed, several studies bring into question whether such interaction can be achieved. With alley cropping systems at Bellary, India, yields of sorghum, safflower and Bengal gram were reduced by 16, 53 and 53 per cent respectively when grown between *Leucaena leucocephala* alleys at 4.5 m spacing (Srivastva and Rama Mohan Rao 1988), mainly as a result of soil moisture depletion on land close to the hedgerows. Sorghum yields were also reduced when grown in combination with *Acacia nilotica*, *Azadirachta indica* and *Eucalyptus* hybrid. Yield reductions in sorghum and pearl millet were observed in agrisilvicultural systems near Karnal, northern India (Kumar *et al.* 1990). Clearly, with these adverse effects, the

acceptability of the system will depend on whether the grain yields are still sufficient for survival and whether additional income can be obtained from the trees.

The most important tree species are *Leucaena leucocephala* which is a quick-growing fodder tree but also provides timber for fuel and pulpwood; *Prosopis juliflora* and *Prosopis chilensis* which are drought-resistant and provide wood for fuel and poles; *Acacia albida* which is well-adapted to sandy soils and produces good fodder, *Acacia nilotica* and *Sesbania grandiflora*. Alternative systems, however, which might generate more income, could include fruit trees and more shade-loving crops.

CHAPTER 9

Soil Management

The aim of sound soil management is to maintain the fertility and structure of the soil. Highly fertile soils result in high crop yields, good plant cover and, therefore, in conditions which minimize the erosive effects of raindrops, runoff and wind. These soils have a stable, usually granular, structure which does not break down under cultivation, and a high infiltration capacity. Soil fertility can thus be seen as the key to soil conservation.

9.1 Organic Content

One way of achieving and maintaining a fertile soil is to apply organic matter. This improves the cohesiveness of the soil, increases its water retention capacity and promotes a stable aggregate structure. Organic material may be added as green manures, straw or as a manure which has already undergone a high degree of fermentation. The effectiveness of the material varies with the isohumic factor, which is the quantity of humus produced per unit of organic matter (Table 9.1; Kolenbrander 1974). Green manures, which are

Table 9.1 Farmyard manure equivalents of some organic materials (after Kolenbrander 1974)

Material	Isohumic factor	FYM equivalent
Plant foliage	0.20	0.25
Green manures	0.25	0.35
Cereal straw	0.30	0.45
Roots of crops	0.35	0.55
Farmyard manure	0.50	1.00
Deciduous tree litter	0.60	1.40
Coniferous tree litter	0.65	1.60
Peat moss	0.85	2.50

normally leguminous crops ploughed in, have a high rate of fermentation and yield a rapid increase in soil stability. The increase is short-lived, however, because of a low isohumic factor. Straw decomposes less rapidly and so takes longer to affect soil stability but has a higher isohumic factor. Previously fermented manures require still longer to influence soil stability

but their effect is longer lasting because these have a still higher isohumic factor (Fournier 1972).

Ekwue, Ohu and Wakawa (1993) found that ploughing in groundnut haulm on a range of soil types in northern Nigeria was more effective than cow dung in immediately reducing soil detachment by raindrop impact. Bonsu (1985) also found that cow dung was not fully effective in the savanna regions of West Africa because the extremely high temperatures bring about volatilization of nitrogen and desiccation of the manure. He found that it was better to combine cow dung with a straw mulch to increase the protection of the surface soil against erosion and enhance the moisture content of the soil. A combined application of cow dung at 5 t/ha and wheat straw mulch at 4 t/ha gave the lowest soil loss and the highest yields of grain sorghum compared with a range of other practices on experimental plots in northern Ghana. The dung and the mulch have to be applied every year because termites attack the straw mulch during the dry season. Combined cow dung and maize straw mulch were similarly effective in reducing erosion in the humid tropical region of Ghana. By reducing soil temperatures, the mulch helped to ensure a more gradual mineralization of the dung and the release of nutrients over a longer period of time.

To increase the resistance of an erodible soil by building up organic matter is a lengthy process. Before any effect on stability is observed, the organic content must be raised above 2 per cent (Section 3.2). On soils with less than 1 per cent organic content, a large supply of organic material is required. Ploughing in maize residue at 5 to 10 t/ha was found in Nigeria to increase the organic carbon content of the soil in absolute terms by only 0.004 to 0.017 per cent whilst application of farmyard manure at 10 t/ha was sufficient only to maintain but not increase an existing level of organic content (Jones 1971). A three-year grass ley, however, was equivalent to an annual application of farmyard manure at 12 t/ha. In the UK, Ekwue (1992) found that grass leys were more effective in reducing soil detachment by raindrop impact on a sandy loam soil than either farmyard manure or straw. Numerous field experiments world-wide show that grass leys are the only effective way of building up the organic content. One reason for this is that the stability of the larger soil aggregates depends on the density of the roots and hyphae and this is increased by pasture and decreased by arable cropping (Tisdall and Oades 1982). Unfortunately, with the trend in many parts of the world towards larger mechanized arable farms where there is no demand for grass, clover or alfalfa, the addition of organic matter becomes uneconomic.

Newbould (1982) refers to studies in Sweden which show a fall in the organic carbon content of soils from 3.5 to 2.5 per cent over 60 years of cereal production except where straw residue was returned to the soil and nitrogen added regularly. In the Great Plains of Canada the organic content has been reduced from 1.7 per cent in 1900 to 0.9 per cent in the 1980s under continuous cereal cropping up to 1935 and alternating cereals and fallow since. On the silty clay loams at Rothamsted Experimental Station, Harpenden, England, however, organic carbon contents have remained virtually unchanged in trials started in 1852 on unmanured and chemically-fertilized plots and have increased slightly on plots treated annually with 35 t/ha of farmyard manure. In contrast, on the sandy loam soils of the Woburn Experimental Farm, Bedfordshire, England, organic carbon contents have fallen from 1.5 to 0.76 per cent over 100 years and rotational cropping has not halted this decline. Three-year leys in a five-year rotation with the addition of farmyard manure to the first arable crop every five years is the only treatment to bring about an increase in organic carbon, from 1.02 to 1.44 per cent (Johnston 1982).

The value of organic matter is enhanced by the presence of base minerals in the soil as these bond chemically with the organic materials to form the compounds of clay and humus which make up the soil aggregates. The base minerals are thus retained in the soil rather than removed by leaching or subsurface flow. Where these minerals, which provide the essential nutrients for plant growth, are absent, they should be added to the soil as fertilizer in the amounts normally recommended for the crops being grown. However, mineral fertilizers cannot improve the aggregate structure of a soil on their own; they need organic support. The continual use of mineral fertilizers without organic manures may lead to the structural deterioration of the soil and increased erodibility.

9.2 Tillage Practices

When managed so as to maintain their fertility, most soils retain their stability and are not adversely affec-ted by standard tillage operations. Indeed, tillage is an essential management technique: it provides a suitable seed bed for plant growth and helps to control weeds. The effect of wheeled traffic and tillage implements on a soil depends upon its shear strength, the nature of the confining stresses and the direction in which the force is applied. The main effect of driving a tractor across a field is to apply force from above and compact the soil. This may result in an increase in shear strength through an increase in bulk density and in the number of clods larger than 30 mm in diameter, but these effects are often more than offset by decreased infiltration and increased runoff so that wheelings are frequently zones of concentrated erosion. The pattern of compaction depends upon tyre pressure, the width of the wheels and the speed of the tractor, the latter controlling the contact time between the wheel and the soil. Compaction generally extends to the depth of the previous tillage, up to 300 mm for deep ploughing, 180 mm for normal ploughing and 60 mm with zero tillage (Pidgeon and Soane 1978).

The tillage tools pulled by the tractor are designed to apply an upward force to cut and loosen the compacted soil, sometimes to invert it and mix it, and to smooth and shape its surface. When the moisture content of the soil is below the plastic limit (Section 3.2), the soil fails by cracking with the soil aggregates sliding over each other but remaining unbroken. The soil ahead of the loosening tool moves forwards and upwards over the entire working depth. A distinct shear plane is formed at the base of the tool which is crescentic in shape in the case of tines but is modified where the tool turns and inverts the soil as with a ploughshare. Soil loosening is effective where the confining stresses resisting upward movement are less than those resisting sideways movement of the soil. Vertical confining stress is obviously zero at the surface and increases with depth until, at a critical depth, it equals the lateral confining stresses and crescentic failure ceases to occur. Below this depth, the soil moves only forwards and sideways, no distinct shear plane is formed and lateral failure occurs with a risk of compaction (Godwin and Spoor 1977).

9.2.1 Conventional tillage

Over the years a reasonably standard or conventional system of tillage, involving ploughing, secondary cultivation with one or more disc harrowings, and planting, has been found suitable for a wide range of soils. Ploughing is carried out with the mouldboard plough although, on stony land or with soils which do not fall cleanly off mouldboards, the disc plough is often used. Ploughs invert the plough furrow and lift and move all the soil in the plough layer, usually to a depth of 100 to 200 mm. Secondary cultivation to form the seed bed and remove weeds is carried out by either disc or tine cultivators. With disc cultivators the soil is broken up by the passage of saucer-shaped metal discs mounted on axles. The most common tine cultivator is the chisel type which consists of a series of metal blades mounted on a frame. The blades vary in width from narrow, 50 mm, to wide, 75 mm, but may be up to 300 mm wide. Another tine cultivator is the sweep or blade type which has V-shaped blades between 0.5 and 2 m wide. Ploughing produces a rough cloddy surface with local variations in height of 120 to 160 mm. Secondary cultivation reduces the roughness to 30 to 40 mm whilst drilling and rolling decrease it still further (see Section 2.1). Roughness is also reduced over time by raindrop impact and water and wind erosion. Soil loss (SL) by water erosion decreases with increasing roughness (R) according to the relationship (Cogo, Moldenhauer and Foster 1984):

$$SL \propto e^{-0.5R} \qquad (9.1)$$

which means that small increases in roughness from a virtually smooth surface can have substantial effects on reducing erosion but much larger increases will be needed to have the same effect with an already rough surface. This explains why tillage can often be successfully used to roughen the surface to control wind erosion in an emergency (Woodruff, Chepil and Lynch 1957). Usually a chisel is used to produce ridges and furrows across the path of the prevailing wind.

9.2.2 Conservation tillage for problem soils

Conventional tillage causes problems on dusty, fine sandy soils, particularly when dry; on very heavy, sticky soils; and on structureless soils, especially those with a high sodium content. In the first case, conventional tillage tends to produce a large number of failure planes, pulverize the soil near the surface and create a compacted layer at plough depth which reduces infiltration and results in increased runoff. The soil is then readily eroded by water and, on drying into a fine dust, by wind. Thus, whilst tillage can improve the coarse structures on heavy soils, it can destroy the structure of non-cohesive soils. To overcome these effects, tillage operations are reduced and at least 30 per cent cover of crop residue is maintained on the soil surface (Table 9.2).

Numerous studies have been undertaken in recent years to examine the effects of different types of conservation tillage on soil erosion rates, soil conditions and crop yields. The results show that the success of the systems is highly soil specific and also dependent upon how well weeds, pests and diseases

Table 9.2 Tillage practices used for soil conservation

Practice	Description
Conventional	Standard practice of ploughing with disc or mouldboard plough, one or more disc harrowings, a spike-tooth harrowing, and surface planting.
No tillage	Soil undisturbed prior to planting which takes place in a narrow, 2.5–7.5 cm wide seed-bed. Crop residue covers of 50–100% retained on surface. Weed control by herbicides.
Strip tillage	Soil undisturbed prior to planting which is done in narrow strips using rotary tiller or in-row chisel, plough-plant, wheel-track planting or listing. Intervening areas of soil untilled. Weed control by herbicides and cultivation.
Mulch tillage	Soil surface disturbed by tillage prior to planting using chisels, field cultivators, discs or sweeps. At least 30% residue cover left on surface as a protective mulch. Weed control by herbicides and cultivation.
Reduced or minimum tillage	Any other tillage practice which retains at least 30% residue cover.

are controlled. Generally the better-drained, coarse and medium-textured soils with low organic content respond best and the systems are not successful on poorly-drained soils with high organic contents or on heavy soils where the use of the mouldboard plough is essential. Since the effectiveness of all the techniques depends on the amount of crop residue left on the surface at the time of greatest erosion risk, it is difficult to isolate the role of tillage *per se* in controlling the erosion from that of the residue which acts as a mulch (Section 8.6). The systems are also often integrated with crop rotations using leguminous or grass species as off-season cover crops.

No tillage

No tillage describes the system whereby tillage is restricted to that necessary for planting the seed. Drilling takes place directly into the stubble of the previous crop and weeds are controlled by herbicides. Generally between 50 and 100 per cent of the surface remains covered with residue. The technique has been found to increase the percentage of water-stable aggregates in the soil compared with tine or disc cultivation and ploughing (Aina 1979; Douglas and Goss 1982). It is not suitable, however, on soils which compact and seal easily because it can lead to lower crop yields and greater runoff.

No tillage reduced erosion rates under maize (Bonsu and Obeng 1979) and millet (Bonsu 1981) in Ghana to levels comparable with those achieved by multiple cropping but generally not to the levels obtained with surface mulching. Also, no tillage was not always effective in the first year of its operation because of the low percentage of crop residues on the surface. At Ibadan, Nigeria, the technique reduced annual soil loss under maize with two crops per year to 0.07 t/ha compared with 5.6 t/ha for hoe and cutlass, 8.3 t/ha for a mouldboard plough and 9.1 t/ha for a mouldboard plough followed by harrowing (Osuji, Babalola and Aboaba 1980).

No tillage is viewed as the leading technology to control erosion on the ultisol soils of the Southern Piedmont of the USA which have become severely eroded after 150 years of continuous cropping. Trials with strip-zone tillage in the 1950s were not successful because the methods of cutting the residue and opening seed furrows were inadequate and weeds could not be controlled. Since the 1960s, suitable herbicides have become available and fluted coulters and chisel ploughs have been developed which makes secondary tillage unnecessary. Using these techniques combined with winter cover crops, erosion can be substantially reduced. As alternatives to the conventional system of growing soya bean with a bare fallow over winter, the following no tillage systems are proposed: soya bean with wheat as a winter cover and using in-row chisel tillage; soya bean with barley as winter cover and using fluted coulters; and soya bean with rye as a green manure and using fluted coulters. The respective mean annual soil losses for the four systems are 26.2, 0.1, 0.1 and 3.4 t/ha (Langdale, Mills and Thomas 1992). Soya bean yields are also increased (Langdale, Thomas and Hargrove 1987) and the organic carbon content of the top centimetre of the soil has risen within three to four years from 0.7 to 3 or 4 per cent (Langdale, Thomas and Mills 1988).

The use of no tillage and herbicides as a practice to control wind erosion at North Platte (Wicks and Smika 1973) and Sidney (Fenster and Wicks 1977), Nebraska, resulted in less weed growth, higher soil moisture storage and higher grain yields than conventional tillage.

Strip tillage

With strip tillage, the soil is prepared for planting along narrow strips with the intervening areas left undisturbed. Typically, up to one-third of the soil is tilled with a single plough–plant operation. When used for maize cultivation on research plots of the University of Science and Technology, Kumasi, Ghana, the technique reduced soil loss from 23 storms totalling 452 mm of rain to 0.2 t/ha compared with 0.9 t/ha with a plough–harrow–plant sequence and a surprisingly high 1.4 t/ha with traditional tillage using a hoe and cutlass (Baffoe-Bonnie and Quansah 1975). The plough–plant system caused the least soil compaction, conserved the most soil moisture and reduced losses of organic matter, nitrogen, phosphorus and potassium (Quansah and Baffoe-Bonnie 1981). Plough–plant systems have not become popular, however, because of problems of weed control and the slow speed of planting.

Mulch tillage

Difficulties with weed control, operating with large amounts of residue and, in many cases, lower yields have also prevented the widespread take-up of stubble–mulch tillage. Nevertheless, this system can be used successfully to control wind erosion and conserve moisture in drier wheat-growing areas. Average annual soil loss over eight years was 2 t/ha on stubble–mulched land compared with 6.5 t/ha on conventionally-tilled land (Fenster and McCalla 1970).

Minimum tillage

Minimum tillage or reduced tillage refers to practices using chiselling or discing to prepare the soil whilst retaining a 15 to 25 per cent residue cover. With one disc cultivation prior to planting on a silty clay soil under continuous wheat production near Pisa, Italy, runoff was increased over that from conventional tillage. This was because the minimum tilled plots retained moisture and this in turn reduced the cracking which plays a major role in promoting infiltration of water in these soils. Despite the higher runoff, annual soil loss was lower under minimum tillage at 1.6 t/ha compared with 4.1 t/ha from conventional tillage (Chisci and Zanchi 1981).

Chiselling instead of ploughing with a mouldboard in the autumn to produce a rough surface but retain residue cover followed by disc cultivation in the spring to smooth the seed bed and cover the residue is now widely practised in maize–soya bean agriculture in the Corn Belt of the USA. It can reduce soil loss by an order of magnitude over that recorded with conventional tillage (Siemens and Oschwald 1978; Johnson and Moldenhauer 1979). The technique works well when soya bean is planted in the chiselled maize residue but is not satisfactory when planting maize in the chiselled soya bean residue because the latter deteriorates very rapidly and by the following spring there is insufficient cover to protect the soil. Alternative practices for this year of the rotation include growing a winter cover crop, killing it in the spring with a contact herbicide and planting maize in the residue of the cover, or planting maize with no tillage in order to minimize the disturbance of the soil.

Long-term effects of conservation tillage

Since conservation tillage does not break up the soil surface, there is concern that in the long term it may lead to a less porous surface with a resulting increase in runoff and erosion. Soane and Pidgeon (1975) found that chisel ploughing produced a less porous surface than the mouldboard plough on a loamy soil in Scotland and, as noted above, Chisci and Zanchi (1981) found that no tillage led to an increase in runoff. Some researchers, however, believe that these effects are short-lived and that after using conservation tillage with a chisel plough for three or four years the porosity of the top 150 mm of soil is the same as that of a mouldboard-ploughed soil. After seven years, the porosities of the top 300 mm of soil are similar. These effects were observed on a silty clay loam (Voorhees and Lindstrom 1984). Kemper and Derpsch (1981) suggest that no tillage can be effective in restoring the porosity of oxisols and alfisols where it has been reduced by the development of a plough pan but that it will take ten to twelve years.

9.2.3 Alternatives to conservation tillage

An alternative approach to conservation tillage is to attempt, through careful timing of operations in relation to soil conditions, to use tillage to produce an erosion-resistant surface. Several farmers on sandy soils in the Midlands counties of England have adopted the Glassford system of ploughing and pressing the soil to produce a cloddy surface to control wind erosion on land devoted to sugar beet. When the soil is moist but not wet, a chisel is used to break up the crust and produce ridges and furrows at right-angles to the direction of erosive winds. The land in the furrows is then rolled either in the same operation, by modifying the chisel plough to incorporate a roll press, or as soon as possible in a second operation before the soil dries out. This tillage is preferably carried out in January and the resulting surface of hills and valleys remains stable throughout the spring blowing period even after it has been broken up by drilling which is carried out transverse to the press ridges. The Glassford system can thus be practised with standard farm equipment. Since the stability of soil clods to raindrop impact and slumping on wetting depends upon the soil conditions at the time of clod formation and these have a much greater effect than subsequent changes in soil moisture and the influence of weathering, it may be feasible to devise a similar tillage system to produce a surface

which is resistant to water erosion (Stuttard 1984). Some attempts have already been made to do this but they rely on specially-designed conservation ploughs for imprinting the soil surface with complex geometric patterns (Dixon and Simanton 1980).

Heavy sticky soils with a moisture content above the plastic limit will fail compressively during tillage, produce few or no fissures and become smeared (Spoor and Godwin 1979). The type of clay mineral in the soil determines its behaviour under these conditions (Spoor, Leeds-Harrison and Godwin 1982). Unconfined swelling is likely to take place with smectitic or micaceous clays and this may cause either the disappearance of aggregates or the formation of new but less stable ones. If drainage cannot occur and the failure is therefore undrained, an already smeared soil may become puddled or even turn into a slurry. Reductions in runoff and therefore erosion on heavy soils are best achieved by increasing the rate of subsurface water movement by drainage. Erodible soils with more than 20 per cent clay content will benefit from the installation of mole drains and from the break-up of compacted layers at depth by subsoiling. Pipe drainage has been found particularly effective in reducing erosion on clay soils derived from Pliocene marine sediments in central Italy (Chisci, Zanchi and Biagi 1978; Zanchi, 1989).

Deep tillage using a crawler tractor to pull two chisels through the ground to open up furrows about 100 mm wide and 500 to 700 mm deep is a recommended practice to break up subsurface pipes and tunnels (Colclough 1965; Crouch 1978). This treatment aids the establishment of grasses, forbs and legumes and is therefore carried out prior to reseeding the land for pasture. Control of tunnel erosion is dependent upon the success of the revegetation because the effects of ripping decline after three to five years (Aldon 1976).

Conservation tillage has not proved appropriate for the management of the highly plastic, sticky, calcareous black clay soils of the tropics and subtropics. These vertisols have a high percentage of smectitic clays which undergo pronounced shrinking when dry, resulting in deep cracks which close only after prolonged wetting. Drainage is a problem when the soils are wet because the infiltration rate is very low; even rains of light intensity cause runoff and erosion. Stubble mulching on vertisols in Australia devoted to continuous wheat production failed to increase the aggregate stability of the soil compared with conventional tillage (Marston and Hird 1978) and did not reduce erosion to an acceptable level (Marston and Perrens 1981). However, crop residue retention accompanied by no tillage did reduce annual erosion on vertisols in northern Queensland to 2 t/ha compared with 4–5 t/ha using stubble-mulching alone and 30–60 t/ha with conventional practice involving a bare soil fallow (Freebairn *et al.* 1986).

In India, sowing has to be carried out in the dry soil in advance of the rains in order to obtain a good crop cover to protect the soil from erosion in the rainy season. This poses a problem on vertisols of how to prepare a good seed bed in the hard soil. Experiments by ICRISAT showed that by preparing a surface of broad-based beds (95 cm wide) and furrows (55 cm wide and graded at 1:150) as soon in the rainy season as the soil becomes workable, annual soil loss was reduced over a six-year period to 1.2 t/ha compared with 6.6 t/ha with conventional tillage practice (Pathak, Miranda and El-Swaify 1985). With better moisture control due to surface drainage along the graded furrows, crop yields were also increased (El-Swaify *et al.* 1985). However, the take-up of the technique by farmers has yet to be fully tested and this may be hindered by the need for specialized equipment and increased labour requirements.

9.3 Soil Stabilizers

Improvements in soil structure can be achieved by applying soil conditioners. These may take the form of organic by-products, polyvalent salts and various synthetic polymers. Polyvalent salts such as gypsum bring about flocculation of the clay particles while organic by-products and synthetic polymers bind the soil particles into aggregates.

Gypsum has been used successfully to improve the structure of sodic soils in southeast Australia (Davidson and Quirk 1961; Rosewell 1970). Sodic soils are highly erodible because excess sodium results in the dispersal of the clay minerals when in contact with water, with consequent structural deterioration. Such soils appear to be particularly susceptible to tunnel erosion. The most effective treatment is to apply gypsum as a cation to replace the sodium. A good drainage system is also necessary to help wash out the sodium from the soil. The treatment is extremely expensive over large areas and, unless accompanied by ripping to break up the tunnels and the sowing of grass, gives only temporary relief. Gypsum has also been used to reduce surface crusting and runoff on red-brown earths in the wheat-growing region of South Australia where the soils are unstable because of high contents of exchangeable magnesium (Grierson 1978).

Temporary stability, lasting from two weeks to six months, can be obtained on most soils by using oil- or rubber-based stabilizers or conditioners containing poly-functional polymers which develop chemical bonds with the minerals in the soil. They are normally applied with water as a spray. Although too expensive for general agricultural use, where the cost is warranted they are helpful on special sites like sand dunes, road cuttings, embankments and stream banks, to provide a short period of stability prior to the establishment of a plant cover.

Soil conditioners fall into two groups, those that render the soil hydrophobic and therefore decrease infiltration and increase runoff, and those that make the soil hydrophyllic, increase infiltration and decrease runoff. Hydrophobic conditioners based on bitumen are generally effective in controlling erosion for only a few storms and are not always suitable for soil conservation purposes. They can, however, be employed to increase water yield; for example, to supply farm ponds (Laing 1978). Asphalt and latex emulsions will also seal the surface and increase runoff, but they are effective in stabilizing the soil and preventing crusting and erosion at least until the seal is broken. When applied to agricultural soils, for example, subsequent discing to a depth of 200 mm can partially destroy the seal and promote aggregate destruction (Gabriels and De Boodt 1978). This problem can be alleviated to some extent by incorporating the emulsion in the top 100 to 200 mm of the soil (Gabriels *et al.* 1977). The critical factor in this case is the size of the aggregates which are produced: if they are too small, infiltration rates remain low. For effective infiltration with hydrophobic conditioners, the aggregates should be at least 2 mm in size and ideally larger than 5 mm (Pla 1977).

Experiments with polyacrylamide conditioners which are hydrophyllic show that high infiltration rates can be obtained regardless of aggregate size. Small plot studies with rainfall simulation show that, for best results, the conditioners should be sprayed directly on to the surface rather than mixed into the top soil (Wallace and Wallace 1986). In contrast, on severely degraded fluvisols near Lake Baringo, Kenya, better results were achieved by applying the polyacrylamide conditioner to a tilled surface and then raking it into the top 20 mm of the soil (Fox and Bryan 1992). Considerable interest has been shown in combining polyacrylamide conditioners with polysaccharides which are biodegradable and which swell in the presence of water and can therefore increase the water-holding capacity of the soil. A combination of 10 t/ha polysaccharide, extracted from guar beans, and 20 kg/ha polyacrylamide reduced erosion over one winter period on 18 to 31° slopes on a range of soil types in Israel to less than 20 t/ha compared with 80 to 120 t/ha on untreated soils. However, its performance was not significantly different from that of a combination of 10 t/ha phosphogypsum and 70 kg/ha polysaccharide which was found easier to apply. The application of polyacrylamide was problematic because of its low dissolution rate and its high viscosity in water (Agassi and Ben-Hur 1992).

Polyurea polymers contain a mixture of hydrophyllic ethylene oxide and hydrophobic propylene oxide in proportions determined according to the degree of hydrophyllicity or hydrophobicity required. They have been used successfully to stabilize sand dunes at Oulled Dhifallah, Tunisia, where they provided a resistant skin, 5 mm thick, to the soil which was resistant to water erosion. *Acacia cyanophylla* plants, used in the revegetation programme, had a higher survival rate and made faster growth on the soil-stabilized areas (De Kesel and De Vleeschauwer 1981). When soil conditioners were applied uniformly to the windward side of dunes in Saudi Arabia, they stabilized the sand but enhanced saltation transport along the slope, resulting in rapid deposition of sand in the lee of the dune. A more effective method was to apply the soil conditioner in 2.0 m strips, spaced 4 m apart. Initially, sand was removed from the untreated bands but eventually, except where the treated strips were undercut, scouring ceased and the slope was stabilized. It is therefore important to select a conditioner such as asphalt which is not easily destroyed (Watson 1990). Stanley and Watt (1990), however, found that soil conditioners were inferior to using brush matting for stabilizing and revegetating coastal dunes at Grassy Head, New South Wales, Australia.

Numerous materials have been tested for use in soil stabilization. Most are too expensive for agricultural use except with high-value crops such as vegetables or, like the slurries and sludges, are unpleasant to handle, create odours and without specialized soil incorporation equipment can only be applied in small quantities. The most promising are the organic by-products of crops which are often available locally as wastes in large quantities. They can be successfully used to give temporary stability on road embankments and on construction sites in urban areas.

CHAPTER 10

Mechanical Methods of Erosion Control

Mechanical field practices are used to control the movement of water and wind over the soil surface. A range of techniques is available and the decision as to which to adopt depends on whether the objective is to reduce the velocity of runoff and wind, increase the surface water storage capacity or safely dispose of excess water. Mechanical methods are normally employed in conjunction with agronomic measures.

10.1 Contouring

Carrying out ploughing, planting and cultivation on the contour can reduce soil loss from sloping land compared with cultivation up-and-down the slope. The effectiveness of contour farming varies with the length and steepness of the slope. It is inadequate as the sole conservation measure for lengths greater than 180 m at 1° steepness. The allowable length declines with increasing steepness to 30 m at 5.5° and 20 m at 8.5°. Moreover the technique is only effective during storms of low rainfall intensity. Protection against more extreme storms is improved by supplementing contour farming with strip cropping (Section 8.3).

On silty and fine sandy soils, erosion may be further reduced by storing water on the surface rather than allowing it to runoff. Limited increases in storage capacity can be obtained by forming ridges, usually at a slight grade of about 1 : 400 to the contour, at regular intervals determined by the slope steepness. Contour ridging is generally ineffective on its own as a soil conservation measure on slopes steeper than 4.5°. Contour ridges reduced soil loss under cabbage and cauliflower on trial plots over two years near Maracay, Venezuela, to 9.9 t/ha compared with 15 t/ha without ridges (Rodriguez and Fernandez de la Paz 1992). Ridge cultivation of cassava on 3° slopes near Benin City, Nigeria, reduced soil loss over 20 storms in 1990 to 16.8 t/ha compared with 29.5 t/ha for flat-bed cultivation and 22.2 t/ha for mound cultivation (Odemerho and Avwunudiogba 1993). Experiments on ways of reducing erosion under vines in the Tokaj area of Hungary showed that on an 18° slope the construction of a ridge every tenth row gave an annual soil loss

of 3 t/ha compared with 0.9 t/ha with a ridge every fifth row, 0.06 t/ha with a ridge every row and 4 t/ha with no ridges (Pinczés 1980; data are conversions assuming a bulk density of 1 Mg/m³). The broad bed and furrow practices described in Section 9.2.3 for use on vertisols can also be operated on the contour with furrow grades ranging between 1:125 and 1:250. Steeper grades result in too much erosion whilst gentler grades do not provide sufficient drainage under wet conditions (Kampen, Hari Krishna and Pathak 1981).

Greater storage of water and more effective erosion control can be achieved by connecting the ridges with cross-ties over the intervening furrows, thereby forming a series of rectangular depressions which fill with water during rain. Because crop damage can occur if the water cannot soak into the soil within 48 h, this practice, known as tied ridging, should only be used on well-drained soils. If it is applied to clay soils, waterlogging is likely to occur. When the practice was tried in Israel on wheat lands in Gaza, the soil loss over the 1980–81 winter was 0.16 t/ha with 1.6-m wide ridges, 0.07 t/ha with 0.6-m wide ridges and 0.29 t/ha with flat planting. The wheat yield was highest on the 1.6-m wide ridges (Morin et al. 1984). Tied ridging increased yields of millet, maize and cotton in years of average and below average rainfall in Burkina Faso compared with open ridging or flat planting due to better water retention. In wet years, however, crops like cowpeas which are sensitive to waterlogging produced lower yields (Hulugalle 1988). Tied ridging with no till gave soil losses over a three-year period of less than 0.5 t/ha compared with up to 9.5 t/ha for conventional ploughing with a mouldboard under maize cultivation on erodible sandy soils in Zimbabwe (Vogel 1992). Maize yields increased with the system in the semi-humid region of the country but decreased in the semi-arid region (Vogel 1991).

A similar technique to tied ridging, called range pitting, is sometimes used on grazing land whereby a series of pits is dug, these being about 50 cm by 50 cm, 7.5 cm deep and 40 cm apart. About 50 per cent of the storage capacity is lost after 8 years and about 90 per cent after 28 years (Neff 1973). The Katumani pits, described in Section 8.7.4, perform similar functions of storing water and trapping sediment.

10.2 Contour Bunds

Contour bunds are earth banks, 1.5 to 2 m wide, thrown across the slope to act as a barrier to runoff, to form a water storage area on their upslope side and to break up a slope into segments shorter in length than is required to generate overland flow. They are suitable for slopes of 1° to 7° and are frequently used on smallholdings in the tropics where they form permanent buffers in a strip-cropping system, being planted with grasses or trees. The banks, spaced at 10 to 20 m intervals, are generally hand-constructed. There are no precise specifications for their design and deviations in their alignment of up to 10 per cent from the contour are permissible.

Hurni (1984) calculated the effectiveness of contour bunds to control erosion in Wallo Province, Ethiopia, and showed that they would only reduce soil loss sufficiently on the lowest of the slopes examined (Table 10.1). Contour bunding of an alfisol soil at the ICRISAT Research Centre near Hyderabad, India, reduced annual soil loss from sorghum and pearl millet, both intercropped with pigeon pea, to 0.97 t/ha compared with 4.79 t/ha for flat cultivation with field bunds (Pathak, Singh and Sudi 1987).

Although contour earth bunds were found to reduce erosion in Burkina Faso, they required a large input of labour to construct and maintain and did not lead to increased crop yields (Roose and Cavalie 1988). In Mali, they were frequently damaged by heavy rain and sometimes failed completely (Rands 1992). As an alternative technique, experiments in Mali are being conducted with contour stone bunds (small stone structures, 25–30 cm high, set in a shallow trench)

and permeable rock dams (long low structures, often over 100 m in length and 50–70 cm high, made from loose stones laid across valley floors). These are easier to build and maintain than earth bunds and allow runoff to be slowly filtered through the structure rather than ponded behind.

10.3 Terraces

Terraces are earth embankments constructed across the slope to intercept surface runoff and convey it to a stable outlet at a non-erosive velocity, and to shorten slope length. They thus perform similar functions to contour bunds. They differ from them by being larger and designed to more stringent specifications. Decisions are required on the spacing and length of the terraces, the location of terrace outlets, the gradient and dimensions of the terrace channel and the layout of the terrace system (Tables 10.2 and 10.3).

Terraces can be classified into three main types: diversion, retention and bench (Table 10.4). The primary aim of diversion terraces is to intercept runoff and channel it across the slope to a suitable outlet. They therefore run at a slight grade, usually 1:250, to the contour. There are several varieties of diversion terrace. The Mangum terrace, formed by taking soil from both sides of the embankment, and the Nichols terrace, constructed by moving soil from the upslope side only, are broad-based with the embankment and channel occupying a width of about 15 m. Narrow-based terraces are only 3 to 4 m wide and consequently have steeper banks which cannot be cultivated. For cultivation to be possible, the banks should not exceed 14° slope if small machinery is used or 8.5° if large reaping machines are operated. Diversion terraces are not suitable for ground slopes greater than

Table 10.1 Capacity of contour bunds to reduce erosion in Wallo Province, Ethiopia (after Hurni 1984)

Bund form				
Slope	6°	14°	27°	33°
Height (m)	0.20	0.20	0.20	0.20
Width (m)	1.90	0.70	0.40	0.30
Storage capacity (m²)	0.19	0.07	0.04	0.03
Spacing (m)	20	15	10	8
Predicted annual soil loss (t/ha)	30	115	115	125
Capacity of bunds to store soil (t/ha)	114	59	48	45
Percentage of soil loss stored behind bunds in first and second year	100 (1) 100 (2)	51 (1) 0 (2)	42 (1) 0 (2)	36 (1) 0 (2)

Table 10.2 Design lengths and grades for terrace channels (after Hudson 1981)

Maximum length:	normal	250 m (sandy soils) to 400 m (clay soils)
	absolute	400 m (sandy soils) to 450 m (clay soils)
Maximum grade:	first 100 m	1:1,000
	second 100 m	1:500
	third 100 m	1:330
	fourth 100 m	1:250
	constant grade	1:250
Ground slopes:	diversion terraces	usable on slopes up to 7°; on steeper slopes the cost of construction is too great and the spacing too close to allow mechanized farming
	retention terraces	recommended only on slopes up to 4.5°
	bench terraces	recommended on slopes of 7 to 30°

7° because of the expense of construction and the close spacing that would be required.

Retention terraces are used where it is necessary to conserve water by storing it on the hillside. They are therefore ungraded or level and generally designed with the capacity to store runoff volume with a 10-year return period without overtopping. These terraces are normally recommended only for permeable soils on slopes of less than 4.5°.

Bench terraces consist of a series of alternating shelves and risers and are employed where steep slopes, up to 30°, need to be cultivated. The riser is vulnerable to erosion and is protected by a vegetation cover and sometimes faced with stones or concrete. There is no channel as such but a storage area is created by sloping the shelf into the hillside (Fig. 10.1). The basic bench terrace system can be modified according to the nature and value of the crops grown. Two kinds of system are used in Malaysia. Where tree crops are grown, the terraces are widely-spaced, the shelves being wide enough for one row of plants, usually rubber or oil palm, and the long, relatively gentle riser banks being planted with grass or a ground creeper. With more valuable crops such as temperate vegetables grown in the highlands, the shelves are closely-spaced and the steeply sloping risers frequently protected by masonry. Level bench terraces are used where water conservation is also a requirement, as in the loess areas of China (Fang *et al.* 1981).

Bench terraces appear to be reasonably satisfactory as a conservation measure under a wide range of conditions, provided sufficient labour is available for their construction and maintenance. On a clay loam soil with a 17° slope cropped to yams in Jamaica, annual soil loss was 17 t/ha with bench terracing compared with 133 t/ha without terraces. For a similar soil and slope, the soil loss under bananas was 17 t/ha with bench terracing and 183 t/ha without (Sheng 1981). In the Uluguru Mountains in Tanzania, bench terraces were found to be unsuitable on shallow soils (Temple 1972b). Their construction exposed the infertile subsoil, they required too high a labour input for construction and maintenance, and they held back so much water on the hillsides that the soils became saturated and landsliding was induced. As an alternative conservation measure, *fanya juu* terraces were recommended.

Fanya juu terraces consist of narrow shelves constructed by digging a ditch on the contour and throwing the soil upslope to form an embankment which is later stabilized by planting grass (Thomas and Biamah 1989). During cultivation, vegetation and crop residues are spread over the shelves. Over time, redistribution of soil within the inter-terrace area causes the inter-terrace slope to decline in angle and bench-like features to develop. Since this decreases the storage area for runoff behind the embankment, maintenance is required to raise the height of the bank to prevent overtopping. Some degree of safety from overtopping is provided, however, because water flowing over the embankment is trapped by the ditch. Experiments with graded *fanya juu* in Ethiopia show that the technique reduces soil loss to levels similar to those achieved with contour grass strips (Grunder 1988; Yohannes 1992). Despite the apparent simplicity of grass strips compared with *fanya juu*, the latter have been widely adopted voluntarily by farmers in the Machakos District of Kenya (Tiffen, Mortimore and Gichuki 1994). Thomas and Biamah (1989) recommend their use on slopes up to 17° although Hurni (1986) suggests that they can be used on slopes up to 26°.

10.4 Waterways

The purpose of waterways in a conservation system is to convey runoff at a non-erosive velocity to a suitable disposal point. A waterway must therefore be carefully designed. Normally its dimensions must provide sufficient capacity to confine the peak runoff from a storm with a ten-year return period. Three types of waterway can be incorporated in a complete surface water disposal system: diversion channels, terrace channels and grass waterways (Fig. 10.2). Diversions are placed upslope of areas of farmland to intercept water running off the slope above and divert it across the slope to a grass waterway. Terrace channels collect runoff

Table 10.3 Formulae for determining spacing of terraces

Approach

Many formulae have been developed for determining the difference in height between two successive terraces; this height difference is known as the vertical interval (*VI*).

Theoretical formula

For steady-state conditions, the runoff (*Qw*) at slope length (*L*) on a hillside can be expressed as:

$$Qw = (R - i) \, L \cos \theta$$

where *R* is the rainfall intensity, *i* is the infiltration capacity and θ is the slope angle.

From the Manning equation of flow velocity:

$$Qw = (R - i) \, L \cos \theta = \frac{r^{5/3} \sin^{1/2} \theta}{n}$$

The hydraulic radius (*r*) is expressed by:

$$r = \left(\frac{v \, n}{\sin^{1/2} \theta} \right)^{3/2}$$

Therefore:

$$(R - i) \, L \cos \theta = \left[\left(\frac{v \, n}{\sin^{1/2} \theta} \right)^{3/2} \right]^{5/3} \frac{\sin^{1/2} \theta}{n}$$

Rearranging for given values of *R* and *i*, say those for the 1 hour rainfall with a 10-year return period, and for a preselected value of *v*, say the maximum permissible velocity for the soil (Table 10.6), gives a slope distance *L* which can be used as the distance between the terraces down the slope:

$$L = \frac{v^{5/2} \, n^{3/2}}{(R - i) \sin^{3/4} \theta \cos \theta}$$

A value of *n* = 0.01 is recommended for bare soil.

For example, if the peak rainfall excess (*R – i*) on a sandy loam soil is 0.02 m/s, and the selected value for *v* = 0.75 m/s, then for a slope of 3°:

$$L = \frac{0.75^{5/2} \times 0.20^{3/2}}{0.02 \times 0.1094 \times 0.9886}$$

$$L = 19.91 \, \text{m}$$

$$VI = 1.04 \, \text{m}$$

Empirical formulae

United States Soil Conservation Service	VI (m) $= aS + b$	where *a* varies from 0.12 in the south to 0.24 in the north and *b* varies between 0.3 and 1.2 according to the erodibility of the soil
Zimbabwe	VI (ft) $= \dfrac{S + f}{2}$	where *f* varies from 3 to 6 according to the erodibility of the soil
South Africa	VI (ft) $= \dfrac{S}{a} + b$	where *a* varies from 1.5 for low rainfall areas to 4 for high rainfall areas and *b* varies from 1 to 3 according to the erodibility of the soil
Algeria	VI (m) $= \dfrac{S}{10} + 2$	
Israel	VI (m) $= XS + Y$	where *X* varies from 0.25 to 0.3 according to the rainfall and *Y* is 1.5 or 2.0 according to the erodibility of the soil
Kenya	VI (m) $= \dfrac{0.3(S + 2)}{4}$	
Kenya (*fanya juu*)	VI (m) $= aS + b$	where *a* = 0.075 and *b* = 0.6
New South Wales	HI (m) $= K \, S^{-0.5}$	where *K* varies from 1.0 to 1.4 according to the erodibility of the soil

Note: It is recommended that 3 or more of the formulae be used and that design spacing be based on a consensus of the results.

Special formulae for bench terraces

Algeria/Morocco	VI (m) $= (260S)^{-0.3}$ for slopes of 10–25%	
	VI (m) $= (64S)^{-0.5}$ for slopes over 25%	
India	VI (m) $= 2(D - 0.15)$	where *D* is the depth of productive soil (m)
Taiwan/Jamaica	VI (m) $= \dfrac{S.Wb}{100 - (S.U)}$	where *Wb* is the width of the shelf (m) and *U* is the slope of the riser (expressed as a ratio of horizontal distance to vertical rise and usually taken as 1.0 or 0.75)
China	VI (m) $= \dfrac{Wb}{(\cos S - \cos \beta)}$	where β is the angle of slope of the riser (normally 70–75°)
Taiwan	VI (m) $= \dfrac{(Wb.S) + (0.1S - U)}{100 - (S.U)}$	for inward sloping bench terraces

VI = vertical interval between terraces
HI = horizontal interval between terraces
S = slope (per cent)

After Lakshmipathy and Narayanswamy (1956); Gichungwa (1970); Charman (1978); Hudson (1981); Sheng (1972b); Bensalem (1977); Fang *et al.* (1981); Chan (1981b); Thomas and Biamah (1989).

from the inter-terrace areas and also convey it across the slope to a grass waterway. Grass waterways are therefore designed to transport downslope the runoff from these sources to empty into the natural river system (Table 10.5); they are located in natural depressions wherever possible but, occasionally, natural channels are reshaped to serve as grass waterways. Design procedures for waterways are described in Table 10.6. A method of predicting the design runoff is given in Table 10.7. Grass waterways are recommended for slopes up to 11°; on steeper slopes the channels should be lined with stones, acceptable for slopes up to 15°, or concrete. On hillsides with alternating gentle and steep sections, a grass waterway with drop structures on the steeper slopes should be used. The selection of grasses for planting should take account of the local soil and climatic environment and the need to establish dense cover very rapidly. Commonly used grasses are *Cynodon dactylon* (Bermuda grass), *Poa pratensis* (Kentucky bluegrass), *Bromis inermis* (smooth bromegrass) and *Pennisetum purpureum* (Napier grass). It is recommended, however, to seek local agronomic or ecological advice before making the final selection.

Table 10.4 Types of terraces

Diversion terraces	Used to intercept overland flow on a hillside and channel it across slope to a suitable outlet, e.g. grass waterway or soak away to tile drain; built at slight downslope grade from contour.
• Mangum type	Formed by taking soil from both sides of embankment.
• Nichols type	Formed by taking soil from upslope side of embankment only.
• Broad-based type	Bank and channel occupy width of 15 m.
• Narrow-based type	Bank and channel occupy width of 3–4 m.
Retention terraces	Level terraces; used where water must be conserved by storage on the hillside.
Bench terraces	Alternating series of shelves and risers used to cultivate steep slopes. Riser often faced with stones or concrete. Various modifications to permit inward-sloping shelves for greater water storage or protection on very steep slopes or to allow cultivation of tree crops and market-garden crops.
Fanya juu terraces	Terraces formed by digging a ditch on the contour and throwing the soil on the upslope side to form a bank.

Grass waterways can be replaced in the water disposal system by tile drains. Diversion and terrace channels are graded to a soak-away, normally located in a natural depression, which provides the intake to the drain. The tile system is designed to remove surface water over a period not exceeding 48 h so that crop damage does not occur. Soil loss from tile-outlet terraces is much reduced because less than 5 per cent of the sediment delivered to the soak-aways passes into the drainage system (Laflen, Johnson and Reeve 1972). The tile outlet consists of four parts: the inlet tube, the orifice plate, the conducting pipe and the outlet. The inlet tube is usually made of plastic and rises from a pipe below ground to a height which is 70 to 100 mm higher than the adjacent terrace bank; the tube has holes or slots at regular intervals above ground level and a removable cap to prevent entry of debris and allow access. The orifice plate is positioned at the base of the tube where it connects with the conducting pipe; it regulates the downward flow of water. The conducting pipe, also of plastic, carries water from one or more inlet tubes to the outlet which is normally in a natural waterway. The terrace bank adjacent to the inlet must be level to reduce the risk of overtopping by ponded water. Although tile outlets are becoming more popular than grass waterways in the USA because they take up less crop land, they are more expensive. On a world-wide basis, grass waterways remain the cheapest and most effective form of terrace outlet. The design procedures for tile outlets are not described in this text, therefore, but they are similar in principle to those for tile drainage systems (Griessel and Beasley 1971).

The main reason why terrace and waterway systems reduce erosion is the way they manage the runoff. The terraces divide the hillside into inter-terrace areas which should be small enough in size to generate only small quantities of runoff. The grass waterway reduces the speed of flow because of the retardance effects of the vegetation. The arrangement of the waterway network gives a high tributary (diversion and terrace channels) to main channel (grass waterway) ratio and a catchment which is elongate in shape rather than square or circular; both attributes contribute to a reduction in peak flow.

A terrace and waterway system must be designed to give the most efficient layout possible in terms of farming operations. This can be achieved by following a systematic design procedure (Table 10.8) and then making adjustments within certain tolerance limits to take account of local topography. Once a system has been constructed, regular maintenance is required to prevent it from deteriorating. This includes cutting the grass in the waterways to maintain it at the height on which the channel design is based; regular applications of fertilizers to promote grass growth; closure of the waterways to animals and vehicles, especially when the soil is wet and damage could occur; and regular inspection and repair of breaks in the terrace banks. Although a terrace and waterway system will fail with the occurrence of a storm of much higher magnitude than that for which it is designed, by far the most common cause of failure is inadequate maintenance. Once a failure in vegetation cover occurs, the reduction in flow resistance is immediate and it may take two or more years for natural recovery to take effect (Temple and Alspach 1992).

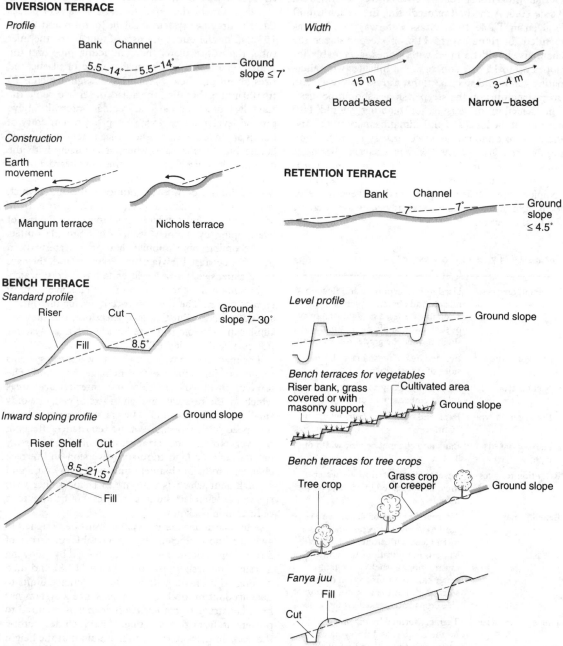

Figure 10.1 Terraces.

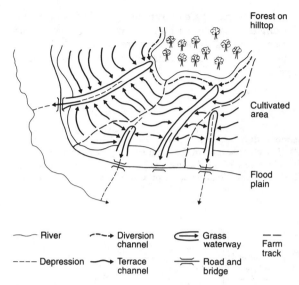

—— River	✏→ Diversion channel	⊂ Grass waterway	— — Farm track
---- Depression	→ Terrace channel	≋ Road and bridge	

Figure 10.2 Typical layout of waterways in a soil conservation scheme.

10.5 Stabilization Structures

Stabilization structures play an important role in gully reclamation and gully erosion control. Small dams, usually 0.4 to 2.0 m in height, made from locally available materials such as earth, wooden planks, brushwood or loose rock, are built across gullies to trap sediment and thereby reduce channel depth and slope. The structures have a high risk of failure but provide temporary stability and are therefore used in association with agronomic treatment of the sur-

Table 10.5 Types of waterway used in soil conservation systems

Diversion ditches	Placed upslope of areas where protection is required to intercept water from top of hillside; built across slope at slight grade so as to convey the intercepted runoff to a suitable outlet.
Terrace channels	Placed upslope of terrace bank to collect runoff from inter-terraced area; built across slope at slight grade so as to convey the runoff to a suitable outlet.
Grass waterways	Used as the outlet for diversions and terrace channels; run downslope, at grade of the sloping surface; empty into river system or other outlet; located in natural depressions on hillside.

rounding land where grasses, trees and shrubs are planted. If the agronomic measures successfully hold the soil and reduce runoff, the dams can be allowed to fall into disrepair. Even though they are temporary, the dams have to be carefully designed. They must be provided with a spillway to deal with overtopping during high flows and installed at a spacing appropriate to the slope of the channel.

The spacing of the dams can be determined from the formula of Heede (1976):

$$SPACING = \frac{HE}{K \tan \theta \cos \theta} \qquad (10.1)$$

where HE is the dam height, θ is the slope angle of the gully and K is a constant. $K = 0.3$ for $\tan \theta \leqslant 0.2$; $K = 0.5$ for $\tan \theta > 0.2$. The dam height is measured from the crest of the spillway to the gully floor. Based on the costs of construction, loose-rock dams are only economical with heights up to 0.45 m; single-fence dams are the most economical for dam heights of 0.45 to 0.75 m; and double-fence dams for heights of 0.75 to 1.7 m (Heede and Mufich 1973). The gully depths for which these dam heights are recommended are less than 1.2, 1.2 to 1.5 and 1.5 to 2.1 m respectively. Keying a check dam into the sides and floor of the gully greatly improves its stability. This entails digging a trench, usually 0.6 m deep and wide, across the channel, but if the channel walls are deeply cracked and fissured, the trench should be increased in depth to 1.2 or even 1.8 m. Aprons must be installed on the gully floor downstream of the check dam to prevent flows from undercutting the structure. Where the slope of the gully floor is less than 8.5°, the length of the apron should be 1.5 times the height of the structure; for steeper slopes it should be 1.75 times its height. At the downstream end of the apron, a loose rock sill about 0.15 m high should be built to create a pool to help absorb the energy of the water falling over the spillway. The spillway should be designed to convey peak flows with a given return period, usually 25 years. It is recommended that the spillway be trapezoidal in cross-section with a bottom length (L) which is equal to the bottom width of the gully. A longer spillway is not desirable because water flowing over the spillway will strike the gully sides where protection against erosion is less. The depth of the spillway (D) is given by the equation:

$$D = \left(\frac{Q}{1.65 L}\right)^{2/3} \qquad (10.2)$$

and, assuming the spillway sides are sloped at 1:1, the top length (Lt) is obtained from (Heede 1976):

$$Lt = L + D \qquad (10.3)$$

This approach assumes that the spillways approximate to a broad-crested weir.

Construction of a loose-rock dam (Fig. 10.3) begins by sloping back the tops of the banks. A trench is then dug across the floor of the gully and into the banks into which the large rocks are placed to form the toe of the structure. The dam is built upwards from the toe, using the flatter rocks on the downstream face. Rocks smaller than 100 mm in diameter should not be used because they will be quickly washed out. A dam made

Table 10.6 Procedures for waterway design

Approach

The design procedures are based on the principles of open-channel hydraulics. The method presented here represents an application of the Manning equation of flow velocity (Eq. 2.13). The cross-section of the waterway may be triangular, trapezoidal or parabolic. Triangular sections are not recommended because of the risk of scour at the lowest point. Since channels which are excavated as a trapezoidal section tend to become parabolic in time, the procedure described here is for a parabolic section.

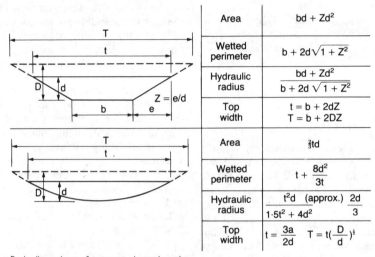

Area	$bd + Zd^2$
Wetted perimeter	$b + 2d\sqrt{1 + Z^2}$
Hydraulic radius	$\dfrac{bd + Zd^2}{b + 2d\sqrt{1 + Z^2}}$
Top width	$t = b + 2dZ$ $T = b + 2DZ$
Area	$\tfrac{2}{3}td$
Wetted perimeter	$t + \dfrac{8d^2}{3t}$
Hydraulic radius	$\dfrac{t^2 d}{1\cdot 5t^2 + 4d^2}$ (approx.) $\dfrac{2d}{3}$
Top width	$t = \dfrac{3a}{2d}$ $T = t\left(\dfrac{D}{d}\right)^{\frac{1}{2}}$

Basic dimensions of common channel sections

The waterways used in soil conservation works on agricultural land are normally designed to convey the peak runoff expected with a 10-year return period without causing scour or fill. The peak runoff can be estimated using the method shown in Table 10.7.

The design is based on the conditions expected to prevail 2 years after installation. Although the waterway is therefore vulnerable to erosion during this period, providing a stable design for this time would result in overdesign for the rest of its life. It would also lead to an unnecessary reduction in the area which can be devoted to arable farming.

Designs are based on the concept of a maximum allowable or safe velocity of flow in the channel. Temple (1991) has shown that the influence of flow duration on the stability of the channel is small in the first 100 hours of flow unless the vegetation is destroyed or has a clumpy growth habit, in which case erosion of the bare areas can undermine the vegetation cover. Since, in these channels, flow durations of 100 hours are unlikely, no allowance is made in the channel design for the duration of the flow. The design, therefore, is based on peak flows not exceeding the maximum permitted value.

Design problem

Design a parabolic grass waterway to convey a peak flow of 6 m³/s on a 1 per cent slope over an erodible sandy soil with Bermuda grass vegetation in a good stand cut to a height of 60 mm.

Procedure

Discharge (Q)	= 6 m³/s (given)
Slope (s)	= 0.01 (given)
Velocity (v)	select a maximum permissible velocity according to proposed vegetation cover and local soil
	= 1.5 m/s

Table 10.6 Continued

Maximum safe velocities (m/s) in channels based on covers expected after two seasons

Material	Bare	Medium grass cover	Very good grass cover
very light silty sand	0.3	0.75	1.5
light loose sand	0.5	0.9	1.5
coarse sand	0.75	1.25	1.7
sandy soil	0.75	1.5	2.0
firm clay loam	1.0	1.7	2.3
stiff clay or stiff gravelly soil	1.5	1.8	2.5
coarse gravels	1.5	1.8	⋆
shale, hardpan, soft rock etc	1.8	2.1	⋆
hard cemented conglomerates	2.5	⋆	⋆

⋆ Material is unlikely to yield a very good grass cover.
Intermediate values may be selected.

Roughness (n) select a suitable value according to the vegetation retardance class (CI); the value of CI can be estimated knowing the length of the plant stems (m) and the density of the stems per unit area (M); the latter can be estimated in turn from the grass type, a qualitative description of the stand and the percentage cover.

$m = 0.06$ (given)

$M = 5,380$ stems/m^2 (from table)

$CI = 2.5 \, (m \, \sqrt{M})^{1/3}$ (Temple 1982)

$CI = 2.5 \, (0.06 \times \sqrt{5380})^{1/3}$

$CI = 6.75$

Interpolating between values in table relating n to CI, select a value for n

$= 0.034$

Properties of grass channel linings for good uniform stands⋆

Cover group	Estimated cover factor, CF	Covers tested	Reference stem density (stems/m^2)
Creeping grasses	0.90	Bermuda grass	5,380
		Centipede grass	5,380
Sod-forming grasses	0.87	Buffalo grass	4,300
		Kentucky bluegrass	3,770
		Blue grama	3,770
Bunch grasses	0.50	Weeping love grass	3,770
		Yellow blue stem	2,690
Legumes†	0.50	Alfalfa	5,380
		Lespedeza sericea	3,230
Annuals	0.50	Common *Lespedeza*	1,610
		Sudan grass	538

⋆ Multiply the stem densities given by 1/3, 2/3, 1, 4/3 and 5/3 for 'poor, fair, good, very good and excellent' covers respectively. The equivalent adjustment to CF remains a matter of engineering judgement until more data are obtained or a more analytical model is developed.
† For the legumes tested, the effective stem count for resistance (given) is approximately five times the actual count very close to the bed. Similar adjustment may be needed for other unusually large-stemmed and/or woody vegetation.

Table 10.6 Continued

Values of Manning's n for vegetated channels

CI	Description	n
10.0	very long (over 600 mm) dense grass	0.06–0.20
7.6	long (250–500 mm) grass	0.04–0.15
5.6	medium (150–250 mm) grass	0.03–0.08
4.4	short (50–150 mm) grass	0.03–0.06
2.9	very short (less than 50 mm) grass	0.02–0.04

Calculate the hydraulic radius (r) from the Manning equation

$$r = \left(\frac{v\,n}{s^{0.5}}\right)^{1.5}$$

$$r = \left(\frac{1.5 \times 0.034}{0.01^{0.5}}\right)^{1.5}$$

$$r = 0.364 \text{ m}$$

Calculate the required cross-sectional area (A) of the channel

$$A = \frac{Q}{v}$$

$$A = 6 \,/\, 1.5$$

$$A = 4 \,\text{m}^2$$

Calculate the design depth which for a parabolic section can be approximated by:

$$d = 1.5\,r$$

$$d = 1.5 \times 0.364$$

$$d = 0.55 \,\text{m}$$

Calculate the top width which for a parabolic section is expressed by:

$$t = \frac{A}{0.67\,d}$$

$$t = \frac{4}{0.67 \times 0.55}$$

$$t = 10.86 \,\text{m}$$

Check that the capacity given by the design criteria is adequate. For a parabolic section:

$$Q = A\,v = 0.67\,t\,d\,v$$

$$Q = 0.67 \times 10.86 \times 0.55 \times 1.5$$

$$Q = 6 \,\text{m}^3/\text{s} \text{ which is adequate}$$

Add 20 per cent freeboard to design depth

$$d = 0.55 + 0.11 = 0.66 \,\text{m}$$

Final design criteria:

depth	= 0.66 m
top width	= 10.86 m

After Schwab *et al.* (1966); Hudson (1981); Temple (1982).

Notes: This procedure can be used for all waterways in a terrace and waterway system. With terrace channels and diversion channels, however, the slope is not predetermined by the ground slope but should be selected from guidelines given in Table 10.2. Terrace channels are usually unvegetated except with broad-based terraces where they may be cropped as part of the inter-terrace area. A value of $n = 0.02$ should be used for bare soil. When the above procedure is applied to small discharges, the design depths are sometimes greater than the design widths; since the channel then resembles a gully, it is undesirable. Also, the dimensions are too small for the channel to be constructed easily. To avoid these problems, a minimum size of 2.0 m wide and 0.5 m deep is recommended for terrace channels.

of large rocks will leave large voids in the structure through which water jets may flow, weakening the dam. These jets will also carry sediment through the

Cross-section

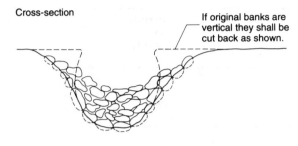

If original banks are vertical they shall be cut back as shown.

Side-section

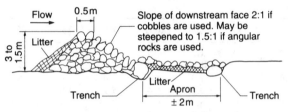

Flow 0.5m

3 to 1.5 m

Litter

Slope of downstream face 2:1 if cobbles are used. May be steepened to 1.5:1 if angular rocks are used.

Trench

Litter

Apron
±2m

Trench

Figure 10.3 Construction of a loose-rock dam (after Gray and Leiser 1982).

dam instead of allowing it to accumulate on the upstream side. To avoid these effects, the dam should be made with a graded rock structure. An effective composition is 25 per cent of the rocks between 100 and 140 mm in diameter, 20 per cent between 150 and 190 mm, 25 per cent between 200 and 300 mm, and 30 per cent between 310 and 450 mm (Heede 1976). A second trench should be made to mark the downstream end of the apron and filled with heavy rocks. A 100-mm thick layer of litter, such as leaves, straw or fine twigs, is laid on the floor of the apron and covered with a solid pavement of rock. A thick layer of litter is also placed on the upstream face of the dam.

When building a single-fence brush dam, the gully banks are first sloped back and stout posts are then driven into the floor and banks of the gully to a depth of about 1 m below the surface and about 0.5 m apart (Fig. 10.4); willow is the recommended material. A 150-mm thick layer of litter is placed on the floor of the gully between the posts extending upstream to the proposed base of the dam and downstream to the end of the apron. Green tree branches or brush are laid on top of the litter, the longer ones at the bottom, with the butt ends upstream. Usually the gully is filled with brush which is trampled to compress it into a compact mass. Cross poles are fixed on the upstream side of the posts and the brush is tied to the structure with

Table 10.7 Estimating the volume of peak runoff

Approach

Several methods have been developed for estimating the volume of peak runoff from small areas where no measured data exist. These include the rational formula and the United States Soil Conservation Service Curve Number. Both these require meteorological information which is not always readily obtainable. The procedure described here can be operated with the minimum of data. It has been developed by Hudson (1981) for use in tropical Africa.

Problem

Estimate the volume of peak runoff for a roughly circular catchment of 50 ha of which (A) 10 ha comprises steeply sloping land with shallow rocky soils, (B) 15 ha is cultivated land with loamy soils and slopes of 6–9° and (C) 25 ha is flat land devoted to pasture on clay soils.

Procedure

The runoff generating characteristics for any catchment can be represented by an area-weighted score based on the vegetation, soil and slope conditions. Typical scores for each of these factors are shown in the table below.

Catchment characteristics

Cover		Soil type and drainage		Slope	
Heavy grass or forest	10	Deep, well drained soils, sands	10	Very flat to gentle (0–3°)	5
Scrub or medium grass	15	Deep, moderately pervious soil, silts	20	Moderate (3–6°)	10
Cultivated lands	20	Soils of fair permeability and depth, loams	25	Rolling (6–9°)	15
Bare or eroded	25	Shallow soils with impeded drainage	30	Hilly or steep	20
		Medium heavy clays or rocky surface	40	Mountainous	25
		Impervious surfaces and waterlogged soils	50		

Table 10.7 Continued

The value of the catchment characteristic (*CC*) for the problem catchment is calculated as follows:

Region	Factor values								Percentage area weighting	Total
	Cover		Soil		Slope					
A	25	+	40	+	20	×	0.20			17.0
B	20	+	25	+	10	×	0.30			18.0
C	10	+	40	+	5	×	0.50			27.5
Catchment characteristic (*CC*)						=				62.5

From table, read peak runoff with a 10-year return period for area (*A*) = 50 ha and *CC* = 62.5. Interpolating gives peak runoff = 9.25 m³/s.

Peak runoff as a function of catchment characteristics and area

A\CC	25	30	35	40	45	50	55	60	65	70	75	80
5	0.2	0.3	0.4	0.5	0.7	0.9	1.1	1.3	1.5	1.7	1.9	2.1
10	0.3	0.5	0.7	0.9	1.1	1.4	1.7	2.0	2.4	2.8	3.2	3.7
15	0.5	0.8	1.1	1.4	1.7	2.0	2.4	2.9	3.4	4.0	4.6	5.2
20	0.6	1.0	1.4	1.8	2.2	2.7	3.2	3.8	4.4	5.1	5.8	6.5
30	0.8	1.3	1.8	2.3	2.9	3.6	4.4	5.3	6.3	7.3	8.4	9.5
40	1.1	1.5	2.1	2.8	3.5	4.5	5.5	6.6	7.8	9.1	10.5	12.3
50	1.2	1.8	2.5	3.5	4.6	5.8	7.1	8.5	10.0	11.6	13.3	15.1
75	1.6	2.4	3.6	4.9	6.3	8.0	9.9	11.9	14.0	16.4	18.9	21.7
100	1.8	3.2	4.7	6.4	8.3	10.4	12.7	15.4	18.2	21.2	24.5	28.0
150	2.1	4.1	6.3	8.8	11.6	14.7	18.2	21.8	25.6	29.9	35.0	40.6
200	2.8	5.5	8.4	11.7	15.3	19.1	23.3	28.0	33.1	38.5	45.0	52.5
250	3.5	6.5	9.7	13.2	17.2	21.7	27.0	32.9	39.6	46.9	55.0	63.7
300	4.2	7.0	10.5	14.7	19.6	25.2	31.5	38.5	46.2	54.6	63.7	73.5
350	4.9	8.4	12.6	17.2	23.2	30.2	37.8	46.3	53.8	62.5	71.5	81.0
400	5.6	10.0	14.4	19.4	25.6	33.6	42.2	51.0	60.0	69.3	79.5	90.0
450	6.3	10.5	15.5	21.5	28.5	36.5	45.5	55.5	65.5	76.0	86.5	97.5
500	7.0	11.0	17.0	23.5	31.0	40.5	51.0	62.0	73.0	84.0	95.0	106.5

A is the area of the catchment in hectares, *CC* is the catchment characteristics from previous table, and the runoff (in cubic metres per second) is for a 10-year frequency.

Notes:

Rainfall intensity:	tropical (high)	multiply by 1.0
	temperate (low)	multiply by 0.75
Catchment shape:	long, narrow	multiply by 0.8
	square/circular	multiply by 1.0
	broad, short	multiply by 1.25
Return period:	2 years	multiply by 0.90
	5 years	multiply by 0.95
	10 years	multiply by 1.0
	25 years	multiply by 1.25
	50 years	multiply by 1.5

galvanized wire. A layer of litter is placed on the upstream face of the dam and packed into the openings between the butt ends of the brush.

For the double-fence brush dam, the gully banks are sloped back and two rows of stout posts are erected. A 150-mm thick litter layer is placed on the floor of the gully, again extending upstream to the proposed base of the dam and downstream to the end of the apron (Fig. 10.5). A 0.3 m layer of brush is positioned on the apron and attached to the lower row of posts. A row of stakes is driven through the middle of the apron into the gully floor and the brush tied to it to form a dense mat. The space between the two rows of posts is filled with brush laid across the gully; this is compressed tightly and held in position with wire. Litter is placed on the upstream face of the dam.

More permanent structures are sometimes required on large gullies to control the overfall of water on the headwall. These are designed to deal safely with the peak runoff with a ten-year return period. They must therefore dissipate the energy of the flow in a manner which protects both the structure and the channel downstream. The structures comprise three compo-

nents: an inlet, a conduit and an outlet. The various types of each component are outlined in Schwab et al. (1966). Where the drop is less than 3 m, the structure should incorporate a drop spillway. For drops between 3 and 6 m, a chute is used and for greater drops a pipe spillway is required. These structures are expensive and, since they are built in adverse conditions with unstable soils subject to extreme fluctuations in moisture, their failure rate is also high. Their foundations may be undermined by animals and the structure may be circumvented if the gully cuts a new channel in the next major flood. Thus they cannot be generally recommended as an economic investment. If the gully erosion is severe enough to require them, a cheaper alternative is to take the land out of cultivation and allow it to revegetate naturally or by reseeding. Their greatest value is where agricultural land in

Table 10.8 Procedures for laying out terrace and waterway systems

1. Using aerial photographs, topographic maps and reconnaissance field surveys, determine the preliminary positions of the grass waterways. Locate the waterways in the natural depressions or drainage lines of the ground surface.

2. Locate the main breaks of slope and any badly eroded or gullied areas. Terrace banks should wherever possible be located to incorporate slope breaks and be positioned upslope of eroded lands.

3. Determine the spacing of the terraces (Table 10.3). The computed spacings may be varied by 25–30 per cent to allow for adjustments in position of the terraces to conform with slope breaks and avoid eroded areas.

4. Determine terrace lengths (Table 10.2). Terraces must be limited in length to avoid dangerous accumulations of runoff and large cross-sectional areas to the terrace channels.

5. Adjust the positions of the grass waterways if necessary to avoid excessive terrace lengths.

6. Locate paths and farm roads along the divides between separate terrace and waterways systems (Fig. 10.2). The use of crest locations minimizes the catchment area contributing runoff to the road. Runoff may then be discharged into the surrounding land without the need for side drains. Crest locations also dispense with the need for bridges and culverts, avoid breaking up terraces to allow roads to cross them and keep vehicles away from grass waterways.

7. Using contour maps and aerial photographs, plan the layout of the system. Locate the grass waterways and diversion channels on the maps and photographs. Locate the top or key terrace and position the others in relation to it in accordance with the design spacings, lengths and gradients and in keeping with the location of slope breaks and eroded areas. Note that terraces often begin at a ridge or high point on a spur and run away from it before turning approximately parallel to it and running across the slope.

8. Examine the layout to see if it is practical for farming. Check whether the smallest inter-terrace areas can be worked with the proposed machinery, assuming contour cultivation and allowing room for turning at each end; if they cannot, the terrace network may need adjustment or the land involved will have to be taken out of cultivation. The terrace spacing should be adjusted to the nearest multiple of the width of the equipment to be used on the inter-terrace area.

9. On land of irregular topography, the terraces will converge in steeper sloping areas, resulting in many point rows during farming operations. This inconvenience can be minimized by the greater expense of installing a parallel terrace system. This involves levelling the land by removing material from the spurs and filling in the depressions. With parallel layouts, the recommended terrace spacing is 0.67 *VI*.

Locate the top or key terrace and, using this new spacing, draw in successive terraces parallel to it. Check each terrace in turn to see that it always has a downhill grade and that the grade is nowhere excessive. When a terrace line is unsatisfactory because it has an uphill section or the downhill grade is too steep, remove this line and replace it with a new key line. Locate a new set of parallel terraces in relation to the second keyline.

10. Calculate the design dimensions of the grass waterways, terrace channels and diversion channels. It may be desirable to design the grass waterways in sections. This will allow adjustments to be made to the width and depth as slope steepness changes and avoid excessively wide channels at the top of a slope where the runoff is less. Where the slope exceeds 11°, drop structures should be used.

11. Stake out the waterway in the field and construct it by excavating soil from the centre and throwing it to each side to form the banks; check first, however, that the natural depression does not already meet the design depth and top width requirements. Seed the grass cover along with compost, fertilizer and mulch.

Table 10.8 **Continued**

12. Stake out the positions of the diversion channels and terrace channels along the outlet or grass waterway to the appropriate vertical interval. From these positions, the terraces are pegged out across the slope according to the selected grade. The vertical intervals should be checked regularly. Begin with the diversion channel, then the top terrace and work downslope. Minor adjustments to the positions of the terrace channels can be made within a tolerance of ± 50 mm *VI*. Mark the lines of the channels with a plough furrow.

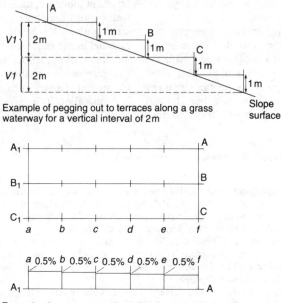

Example of pegging out to terraces along a grass waterway for a vertical interval of 2 m

Example of pegging out across the slope at a grade of 1:200 or 0.5 per cent

13. Construct the diversion channel and then the terraces, beginning at the top and working downslope. This method of working is necessary because each terrace channel is designed to convey only the runoff from its inter-terrace area; it cannot cope with runoff from further upslope and so the protection works must be in place prior to construction. If the top soil is thin, it should be scraped off first, stored and then returned to the inter-terrace area after the terraces are complete.

After Schwab *et al.* (1966); Troeh, Hobbs and Donahue (1980), Hudson (1981); Wenner (1981).

flatter areas needs to be protected from channel erosion and where water needs to be conserved. Advice should be sought from civil engineers on the design and construction of the structures.

Stabilization structures are also used to control erosion on steep slopes such as gully sidewalls, embankments and cuttings. Stability for a short period until a dense vegetation cover has had a chance to grow can be achieved by brush matting. Live willow stakes are driven into the soil to a depth of 200 mm. The stakes are placed about 1 m apart to form a row along the contour with a distance of 10 to 20 m between the rows depending upon the steepness of the slope. Sprouting brushwood is spread over the top of the slope, butt ends downslope, compressed to ensure no area is left uncovered, and tied to the stakes with galvanized wire. A similar technique is wattling which consists of placing fascines or bundles of woody plant stems, using species which will root easily such as *Salix, Leucaena, Baccharis* and *Tamarix*, in shallow trenches on the contour with rows spaced 1 to 6 m apart. The wattling is tied to stakes on the downslope side of the trench and secured by further stakes driven through the material (Fig. 10.6). After installation, the wattling is covered with soil until only about 10 per cent of the fascine is exposed. Grasses and trees can be planted between the wattles.

More permanent protection of banks and embankments can be achieved by facing them with a resistant material such as concrete. Better, however, is to construct retaining walls with gabions. These are rectangular steel wire-mesh baskets, packed tightly with stones. They have the advantages of allowing seepage of water through the facing and of deforming by bending without loss of structural efficiency rather than by cracking. Gabions are supplied flat and then folded into their rectangular shape on site. They are placed in position and anchored before being filled with stones 125 to 200 mm in diameter. The gabion is filled to one-third of its depth after which two connecting wires or braces are inserted front to back to prevent bulging of the wire basket on further filling. The bracing is repeated when the basket is two-thirds full. The gabion is slightly overfilled to allow for settlement, the hinged lid is closed and wired to the sides. The simplest structure consists of one tier of gabions, 1 m high. A second tier can be positioned on top of the first, set back about 0.5 m. The addition of further tiers, however, requires bracing the structure against overturning and should not be attempted without civil engineering advice. Similar advice should be sought for other types of retaining walls and on the benching of long, steep slopes.

Control of surface and seepage water is often needed on steep slopes to minimize landslides and slumps. This can be achieved by drainage which will help to prevent the build-up of soil water. The princi-

ples are to construct a diversion channel to run at grade across the hillslope, upslope of the area of risk, and thereby reduce the amount of water coming into the area. Subsurface drains of 75 mm in diameter PVC perforated pipe wrapped in a filter cloth to prevent blockage can also be installed to intercept subsurface flow. The area itself can be drained by rubble drains, excavated to a depth of 300 mm and a width of 500 mm and filled with rocks. Safe outlets must be provided for all components of the drainage system.

Stabilization methods are also used in the construction and maintenance of footpaths, especially in recreational areas on sloping land. The path itself must be

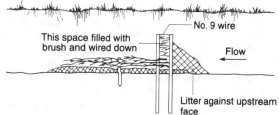

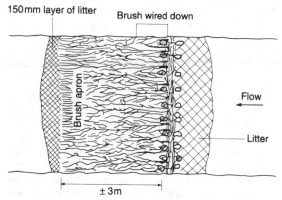

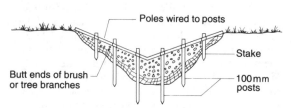

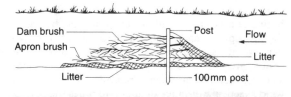

Figure 10.4 Construction of a single-fence brush dam (after Gray and Leiser 1982).

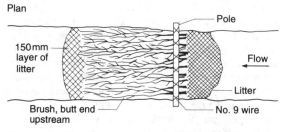

Figure 10.5 Construction of a double-fence brush dam (after Gray and Leiser 1982).

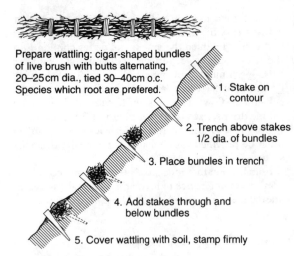

Prepare wattling: cigar-shaped bundles
of live brush with butts alternating,
20–25cm dia., tied 30–40cm o.c.
Species which root are prefered.

1. Stake on contour

2. Trench above stakes 1/2 dia. of bundles

3. Place bundles in trench

4. Add stakes through and below bundles

5. Cover wattling with soil, stamp firmly

NOTE: Work starts at bottom of cut or fill and proceeds
from step 1 through step 5

Figure 10.6 Installation of wattles for slope stabilization
(after Gray and Leiser 1982).

constructed with either a camber or cross-fall (Fig.
10.7) to shed runoff to a suitable outlet. On gently-
sloping land with porous soil on the slope above, the
path can be outward-sloping and the runoff allowed to
discharge on to the vegetated hillside below but, in
other situations, it is preferable for the path to be
inward-sloping with a fall of about 1:12 to a side drain
which can collect runoff from both the path and the
land above. The side drain should be graded across
the slope to a safe outlet, either a vegetated waterway
or a soak away area. Culverts should be used to take
water beneath the path. Where the path runs downslope, cut-off drains formed by making a small trench,
reinforced with a wooden plank, should be placed at
regular intervals, to collect and drain runoff from the
path. The plank must be high enough to divert the
flow but small enough to merge into the profile of the
path and not form a barrier to walking. The cut-offs
should have an angle of 30–45° (Fig. 10.7). A shallower angle will cause water to pond, leading to
siltation in the drain and the risk of overtopping. A
steeper angle will lead to scouring of the path. The
cut-off should also extend some 300 mm beyond the
path to avoid runoff by-passing the structure. Bends
or turns in the path, for example hair-pins on a steep
slope, are vulnerable areas for damage and erosion.
They should be made at zero slope by locally increasing the steepness of the path above and below the
turn. Alternatively, but more expensively, the turn can
be made using log or stone steps. Logs and stones may
also be used to form revetments along the side of a
path.

10.6 Windbreaks

Windbreaks are placed at right-angles to erosive winds
to reduce wind velocity and, by spacing them at
regular intervals, break up the length of open wind
blow. Windbreaks may be inert structures, such as
stone walls, slat and brush fences and cloth screens or
living vegetation. Living windbreaks are known as
shelterbelts. In addition to reducing wind speed, shelterbelts result in lower evapotranspiration, higher soil
temperatures in winter and lower in summer, and
higher soil moisture; in many instances, these effects
can lead to increases in crop yield.

A shelterbelt is designed so that it rises abruptly on
the windward side and provides both a barrier and a
filter to wind movement. A complete belt can vary
from a single line of trees to one of two or three tree
rows and up to three shrub rows, one of which is
placed on the windward side. Belt widths vary from
about 9 m for a two-row tree belt with associated
shrubs to about 3 m for single-row hedge belts. These
widths mean that belts can occupy about 3 per cent of
the land area they are protecting. The density of the
belt should not be so great as to form an impermeable
barrier nor so sparse that the belt is transparent. The
correct density is equivalent to a porosity of 40 to 50
per cent. More open barriers do not reduce wind
velocity sufficiently. Where only a single row of trees is
used, it is important that the branches and foliage
extend to ground level to give the required level of
porosity in the lower metre where most of the sediment movement by saltation takes place (Section 2.7).
With denser barriers there is a much greater reduction
in wind speed initially but, since the velocity increases
more rapidly with distance downwind than is the case
for more porous barriers, they are effective for only
very short distances.

The reduction in wind velocity by a shelterbelt
begins at a distance of about five times the height of
the belt upwind and reaches a maximum of about 40
per cent of the original wind velocity at a distance of
about three times the height of the belt downwind.
Velocity then increases again returning to the original
wind speed at a distance of about 30 times the barrier
height (Marshall 1967). Shelterbelts are designed to
maintain the wind velocity at about 80 per cent of the
open wind velocity. Wind tunnel studies by Woodruff
and Zingg (1952) showed that tree belts at right-
angles to the wind afford this level of protection for
distances up to 17 times their height for open wind
velocities up to 44 km/h. Allowing for variations in
wind speed and deviations in wind direction, they
developed the following equation for determining
shelterbelt spacing:

$$L = 17H \, (V_t/V) \cos \alpha \qquad (10.4)$$

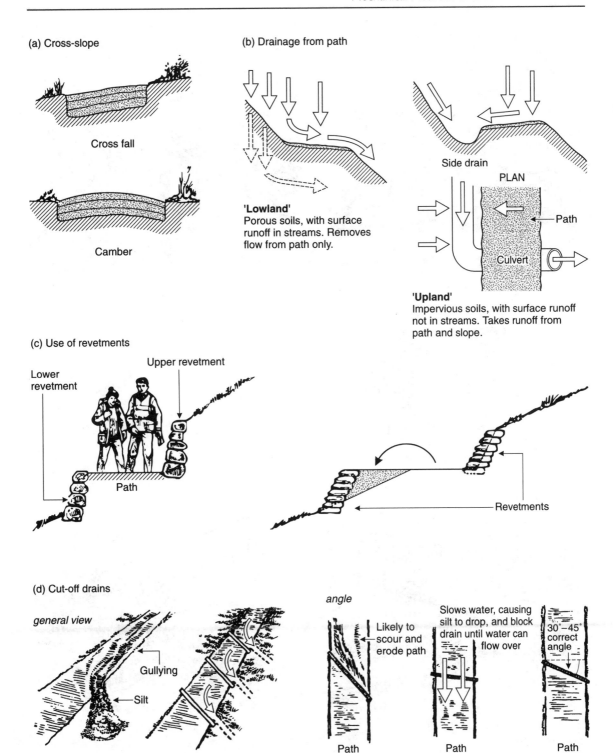

(a) Cross-slope

Cross fall

Camber

(b) Drainage from path

'Lowland'
Porous soils, with surface runoff in streams. Removes flow from path only.

Side drain

PLAN

Path

Culvert

'Upland'
Impervious soils, with surface runoff not in streams. Takes runoff from path and slope.

(c) Use of revetments

Upper revetment

Lower revetment

Path

Revetments

(d) Cut-off drains

general view

Gullying

Silt

angle

Likely to scour and erode path

Slows water, causing silt to drop, and block drain until water can flow over

30°–45° correct angle

Path

Path

Path

Figure 10.7 Measures for erosion control on footpaths (after Agate 1983).

(d) Cut-off drains *(continued)*

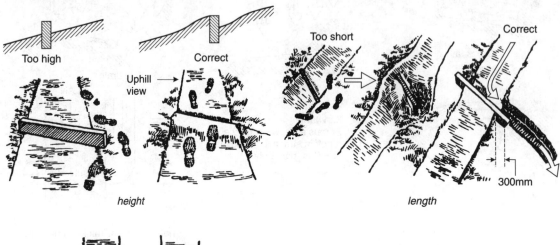

height

length

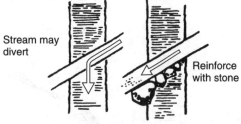

stone reinforcement

(e) Making a turn

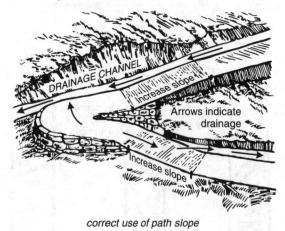

correct use of path slope

use of steps

Figure 10.7 Continued

where L is the spacing or distance apart of the belts, H is the height of the belt, V is the actual or design wind velocity measured at a height of 15 m above the ground surface, V_t is the threshold wind velocity for particle movement, taken as 34 km/h, and α is the angle of deviation of the prevailing wind from a line perpendicular to the belt. Effective protection in the field rarely reaches this theoretical level of $17H$, however, being reduced in unstable air and by variable growth and poor maintenance of the trees. A distance of 10 or 12 times the height of the belt is more realistic. Where 5- to 7-m high hedges are used, the effective distance protected increases to about 30 times the barrier height but, because of their lower height, the absolute distance protected is much less and more frequent spacing is required. Belts of 6.7-m high trees gave protection for a distance of $20H$ near Youyu, Shaanxi, China (Li 1984).

Belt lengths should be a minimum of 12 times the belt height provided that the belt is at right-angles to the wind. To allow for deviations in wind direction, a longer length is desirable and a length of $24H$ is generally recommended (Bates 1924). For winds ranging $\pm$ 45° from the perpendicular, the effective area protected by the belt increases rapidly with belt length. Figure 10.8 (Olesen 1979) shows that for 2-m tall hedges, the area protected is 156 m² when the belt is 25 m long but increases to 2,500 m² when the belt is 100 m long,

Where there is a dominant erosive wind from a single direction, the best protection is obtained by aligning the shelterbelts in parallel rows at right-angles to it. Where erosive winds come from several directions, grid or herringbone layouts may be necessary. The requirement is to provide maximum protection averaged over all wind directions and all wind velocities above the threshold level. This may be achieved by a scheme in which complete protection is not obtained for any single wind direction. The effectiveness of shelterbelt layouts can be evaluated using Eq. 3.7. This is first applied to obtain a measure of wind erosivity with no protection. A measure is then obtained for the shelterbelt layout by reducing the values of $\bar{V}_t$ by the ratio V_x/V_o, where V_x is the wind velocity at distance x from the belt, with x measured in units of barrier height, and V_o is the wind velocity in the open field. Values of V_x/V_o can be determined for a belt with 40 per cent porosity by the equation:

$$V_x/V_o = 0.85 - 4\,e^{-0.2H'} + e^{-0.3H'} + 0.0002H'^2$$
$$(10.5)$$

where $H' = x/sin\ \beta$ when β is the acute angle of incident wind (Skidmore and Hagen 1977).

The greatest effect of shelterbelts is found where, as

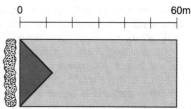

Length: 25m

Maximum distance protected: 12.5 H
Area protected: 156m²

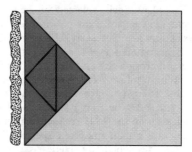

Length: 50m

Maximum distance protected: 25 H
Area protected: 625m²

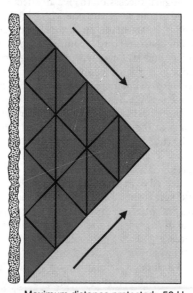

Length: 100m

Maximum distance protected: 50 H
Area protected: 2500m²

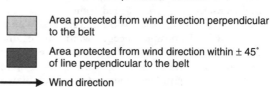

Area protected from wind direction perpendicular to the belt

Area protected from wind direction within ± 45° of line perpendicular to the belt

Wind direction

Figure 10.8 Effect of shelterbelt length on size and shape of area protected by a 2-m tall hedge belt (after Olesen 1979).

a result of farmer collaboration in a collective belt planting scheme, a regional framework exists of belts placed along property boundaries in a coordinated way so that they form a parallel series of main line barriers, 200 to 400 m apart. Within this framework, individual farmers are free to plant additional hedges. Collective shelterbelt schemes are encouraged by Hedeselskabet (Danish Land Development Service) to control erosion on the sandy soils of Jylland (Olesen 1979).

The plant species selected for shelterbelts should be rapid growing, tolerant of wind and light and frost resistant where necessary. Their growth habit should give the required level of porosity at the time of year of greatest erosion risk and a conical or cylindrical shape avoiding top-heavy crowns. The branches should be pliable so that they bend with the wind instead of breaking off. The root system should provide a firm anchorage to the soil. Preference should be given to local rather than imported species. The shelterbelt system developed by Hedeselskabet meets these requirements and, at the same time, provides for an ecological succession to give an effective barrier within three to four years and one with a life of 50 to 80 years. The belts are made up of three parallel rows, 1.25 to 1.5 m apart, each row comprising nurse trees, such as alders and willows, which are fast-growing and provide the early protection; durable trees, such as oaks, sycamore, elm, maple and rowan, which take longer to grow but provide the long-term protection; and shade-tolerant bushes to provide undergrowth in the lower levels of the belt. The mixture of species provides a belt which is less vulnerable to attack by diseases and pests, is visually more attractive and able to give a varied habitat for wildlife. The belt needs to be protected in its early years against damage by livestock and spray drift. After about five years and then at an interval of every three to four years, mechanical cutting of the sides of the belt is required to maintain the necessary shape, particularly the sharp rise from the ground on the windward side.

Windbreaks of brushwood or plastic meshes with about 50 per cent porosity are often used to help stabilize mobile sand dunes and, thereby provide a more suitable environment for vegetation growth. The windbreaks, sometimes termed sand fences, are placed at right-angles to the wind (Savage and Wood-house 1968). Deposition occurs windward of the barrier for a distance of 0.4 to 2.0 times the barrier height and in the lee of the barrier for a distance up to four times the height. Once the sand has accumulated and almost buried the fence, a second fence is built on top of the newly-formed dune. Further fences are added until the dune is reformed by the wind into a streamlined shape so that air flows over it without loss

of transport capacity. Where wind velocities exceed 18 m/s, double or triple fence systems are used, spaced at intervals of four times the barrier height.

10.7 Geotextiles

A geotextile is any permeable textile material used with foundation, soil, rock, earth or any geotechnical engineering-related material. It can be in the form of a mat, sheet, grid or web of either natural fibre, such as jute or coir, or artificial fibre, such as nylon. Several products are commercially available for use in erosion control where they interact as a composite with the soil and vegetation. They are supplied in rolls, unrolled over the hillslope from the top and anchored with large pins. The natural fibre types are biodegradable and are designed to be laid over the surface of the slope to provide temporary protection against erosion until a vegetation cover is established. The artificial fibre types, which include geowebs and geogrids as well as mats, are buried and designed to give permanent protection to a slope by reinforcing the soil; once a vegetation cover is established, the plant roots and the fibre act together to increase the cohesion of the soil and the fibre provides a back-up resistance should the vegetation fail.

Experimental data on the performance of some geotextiles are given in Table 10.9, based on work by Rickson (1988; 1990; 1992). She found that surface-laid mats of natural fibres are the most effective in

Table 10.9 Effectiveness of geotextile materials in controlling water erosion on a sandy loam soil

Material	Moisture absorption*	Soil detachment by raindrop impact (% of control)	Sediment removal by runoff (% of control)	Soil loss by rain and runoff combined (% of control)
Bare soil (control)	–	100	100	100
Jute-based net	570	14	13	22
Coir-based net	250	30	49	
Wood–wool mulch	320	25	67	32
Polyamide mat (buried)	120	130	n/d	72
Polyamide mat (on surface)	120	35	65	n/d

* Ratio of weight at saturation compared with dry weight expressed as a percentage.

n/d = no data

After Rickson (1988; 1990; 1992); Morgan and Rickson (1988).

controlling soil detachment by raindrop impact because they provide good surface cover, high water absorption, thick fibres able to intercept splashed particles from their point of ejection, and a rough surface in which water is ponded, thereby further inhibiting the splash action on the soil. In contrast, buried mats of artificial fibres do not effectively control the splash process. Their percentage cover and water absorbing capacities are low and problems with backfilling them mean that they tend to be filled with highly erodible unconsolidated material. Despite the ability of the natural fibres to hold water, no significant differences in runoff production were observed in laboratory experiments between an unprotected slope and slopes protected by natural or artificial fibres. However, erosion resulting from the runoff was significantly lower for slopes protected by surface-laid jute mats because of the higher roughness they imparted to the flow. Although surface laid mats of coir, wood-chip and artificial fibres also reduced soil loss by runoff, they were less effective than the jute because they did not adhere to the soil surface so well.

In small plot studies with both rainfall and runoff simulation, Morgan and Rickson (1988) found that surface-laid jute mats offered the best protection, reflecting their ability to control both soil detachment by raindrop impact and the transport capacity of the runoff. A mat of wood-chips held within a light-degradable plastic mesh also controlled the raindrop impact but was less effective in reducing the transport capacity of the runoff. Armstrong and Wall (1991) also found that a wood-chip based geotextile failed to reduce runoff compared with an unprotected slope, although it successfully controlled soil loss. Buried artificial fibre mats reduced erosion but to a lesser extent because they could not control the soil detachment by raindrop impact. Very similar results were obtained by Rustom and Weggel (1993) but rather different findings emerged in studies by Fifield and Malnor (1990) and Cazzuffi, Monti and Rimoldi (1991) who showed that surface-laid jute and buried artificial fibre mats gave very similar reductions in soil loss. However, it should be stressed that the above trials were carried out immediately after installation of the geotextile materials. Over time, the natural fibre mats will become less effective as they biodegrade whereas the performance of the artificial fibre mats is likely to remain constant.

CHAPTER 11

Implementation

The ultimate objective of research on soil erosion is to solve erosion problems by adopting suitable conservation measures. Suitability can be judged in various ways. The measures may reduce erosion to an acceptable level. They may be appropriate to the local farming system in terms of their level of technology and compatibility with farming practices. They may be economically justifiable. They must also be capable of implementation. The aim of this last chapter is to provide a background to some of the issues involved in implementing soil conservation proposals. For a more comprehensive coverage, the reader is referred to Halcrow, Heady and Cotner (1982), Blaikie (1985), Lovejoy and Napier (1986), Blaikie and Brookfield (1987) and Chisholm and Dumsday (1987).

11.1 Socio-economic Setting

As indicated in Chapter 6, soil erosion is frequently a response to the breakdown of farming systems. The three major periods in history associated with extensive soil erosion (Dregne 1982) all reflected the inability of existing farming systems to deal with population growth and the intensification of agriculture. These periods were the expansion of agriculture into China, the Middle East and the Mediterranean some 1,000 to 3,000 years ago; the migration of Europeans to develop colonies some 50 to 150 years ago; and the expansion in the last 30 years of people on to marginal lands in Latin America, Africa and Asia. In all cases, the outcome was land degradation and the movement of population to new areas. The migrants took with them agricultural practices which had worked for hundreds of years in their home areas but which were generally unsuited to their new environments. Unfortunately, knowledge of this unsuitability was only possible from experience. Today, we are seeing a further cause of erosion due to the lack of maintenance of soil conservation works in areas of rural depopulation. Although most pronounced in Mediterranean Europe, it is a trend which will become increasingly important world-wide as all countries over the next century experience a decline in the proportion of their population employed in agricul-

ture and a movement of people to the towns. The need to supply food for the urban market will also create an increasing strain on many existing farming systems. No longer is it appropriate to view soil erosion solely as a problem arising from population pressure; it can also be associated with decreasing numbers of people and a declining labour force.

The information provided in Chapter 6 suggests that farmers will adopt soil conservation practices if they have the necessary labour, capital and technological inputs to do so and if they perceive an immediate economic benefit. Unfortunately, the overall take-up of soil conservation remains poor. After six decades of voluntary soil conservation programmes in the USA, erosion is still at an unacceptably high level (Swanson, Camboni and Napier 1986). Farmers are unlikely to adopt conservation measures if there is no immediate threat to the productivity of their land or if the main justification for their use is to prevent pollution and other off-site damage (Napier 1990). These issues bring into question the role that governments should play.

11.2 Political Context

Although most governments have some form of soil protection policies that include erosion control, few are translated into effective action because of the lack of political will; soil conservation is not a vote-winning concern (Hudson 1981). Yet, under certain conditions, governments do respond quite rapidly. The United States Soil Conservation Service was created as a result of political pressure from all segments of society who were affected by severe wind erosion during the 1930s. Farmers lost their soil and often their farms, exacerbating the effects of the world-wide economic depression of the time; national food supplies were threatened; and off-site sedimentation created problems for the urban population and the non-agricultural population in rural areas (Rasmussen 1982). A much earlier but smaller-scale example occurred at the end of the last century in Iceland where drifting sand seriously affected the livelihood of farmers in the southern and north-eastern parts of the country. Between 1860 and 1890, the land surface was lowered at rates over 2 mm/y. Under pressure from the Agricultural Society of Iceland, the Icelandic Parliament (Althing) made available a small grant to bring in Danish specialists to investigate the problem. In 1895, the Althing passed the Act for Resolution on Sand Erosion and Reclamation giving District Commissions the authority to take action to solve the problem. Without any financial wherewithal or knowledge of the best measures to take, however, this legislation was of little value. Political pressure from farmers continued and in 1907 the Act of Forestry and Prevention of Erosion of Land was passed, effectively marking the foundation of the State Soil Conservation Service of Iceland (Runólfsson 1978; 1987). These two examples serve to indicate what can be achieved if the farmers in particular and society in general have sufficient political voice. Since this occurs only when disaster is imminent, the response of governments seems to be limited to crisis management.

A similar history applies to much of the soil conservation work carried out by European colonial administrations. Concern over land degradation in much of Africa was expressed by British administrators between 1900 and 1920 but it was not until droughts followed by heavy rains in the 1930s resulted in serious erosion that the problem was addressed. Aided by the example of the US Soil Conservation Service, similar organizations were founded in many countries, often under political pressure from European settlers who envisaged potential damage to their land from deposition of the sediment eroded from areas reserved for the native population. As seen in Chapter 6, the colonial approach to soil conservation was based on transferring to communal and smallholder farms the methods developed in the USA and tested on European-owned farms in the local region. These methods relied heavily on mechanical conservation works. The legacy of crisis management was therefore a top-down approach to soil conservation based on imposing often inappropriate technology with an inadequate understanding of the erosion processes and the social conditions in the local environment (Anderson 1984). In many instances, the same philosophy was adopted on independence by post-colonial governments although it was often subsumed beneath a policy of ignoring soil conservation because of its political associations with the former colonial power (Wood 1992).

Within the last 20 years many governments have endeavoured to set up conservation initiatives but have been constrained by lack of funds and the limited political value of being involved in conservation work. In some cases, these initiatives can be viewed as further crisis management in response to the droughts of the Sahelian and Soudanian regions of Africa in the 1970s. In Kenya, protection of the environment was formally endorsed by the President in 1977 and a permanent Presidential Commission on Soil Conservation and Afforestation was established in 1980 to develop strategy and policy, ensure coordination between interested bodies and monitor progress. This has encouraged external financial and technical assistance such as that provided by the Swedish International Development Authority (SIDA) to the National Soil and Water Conservation Programme. In many countries, so many organizations are now involved in promoting soil conservation that new political problems have arisen such as how the work might best be coordinated, how the responsibilities of each organization might be defined, and what should be the role of external organisations vis-à-vis those of the host country.

11.3 New Approaches

The last two decades have seen considerable changes in the approaches used to promote and implement soil conservation. Perhaps the most fundamental has been the move from a top-down towards a bottom-up approach involving the participation of the farmers. Once it became recognized that sending in technical experts to identify the erosion problem and design an erosion-control system was inadequate, erosion began to be regarded as a social as well as a technical problem. Although this led to recognizing that a sociologist might have a role to play, that role was initially limited to obtaining the consent of the farmers to the technical solution (Heusch 1981). The technical solutions being proposed still took no account of the social structure and economic condition of the society on which they were being imposed. The approach today is one of participatory development in which the farming family, not just the head male, is involved in defining priorities and the job of the technical experts, including physical scientists and socio-economists, is to work with the family in producing a site-specific scheme. In many cases, soil conservation is not promoted directly but is presented by stealth within a background of water conservation, improved soil fertility and overall promotion of wise land use and good land husbandry (Shaxson 1988). The new approach also depends on recognizing that many traditional agricultural systems relied on soundly-based soil protection practices and that more acceptable conservation schemes can be developed by building on these (Shaxson 1987).

Although the results of this new approach are promising, it is too early to know whether they can be sustained in the long term. Further, the implication that the top-down approach must always be unsuccessful cannot always be supported. Experience from a number of projects in India, involving different levels of initial consultation with farmers, shows that how farmers perceive the risk and the economic benefits is more important in the end. Projects where this is not recognized will fail no matter how much farmer participation is involved (Singh 1990). A further example which supports this view is the almost universal adoption of contour grass strips by farmers in Swaziland following the decree of King Sobhuza II in 1948 whereas later attempts to introduce bench terraces as part of a Rural Area Development Programme between 1977 and 1983 failed. Clearly, farmers will adopt whatever measures they feel are beneficial regardless of the approach used.

A second area where a change of attitude is taking place is the move from soil conservation projects towards conservation programmes. Soil conservation projects have been favoured by aid agencies and other donors because money can be made available for a specific purpose over a finite period and the outcome monitored. Traditionally, these projects have been influenced by American and colonial experience and been based on engineering structures. The results can be seen in the number of kilometres of terraces and waterways constructed. Present-day food-for-work programmes follow a similar philosophy. Since few of these projects have produced sustained conservation benefits, emphasis is now being placed on developing soil conservation programmes integrating research, extension work, education and training. A programme is based on a general mission statement and encompasses a range of activities targeted towards individual farmers, communities and the population in general. Although each activity has specific objectives and its performance is monitored, the programme is continually updated and modified. The Swedish Aid programme in Kenya, run since 1974, works with the Presidential Commission referred to above and through the existing government structure to train staff and build the appropriate institutions to liaise with farmers. The programme seeks to collaborate with the government instead of opposing it (Wenner 1988). An important outcome of the programme has been the training of soil conservation officers and technical assistants to maintain regular contact with and advise farmers through a Training and Visit System. Each technical assistant operates through a contact farmer and between four and eight follower farmers, visiting on a set day and time once a fortnight to discuss soil conservation, crops, mechanization, farm management and other issues. Where women's groups and other self-help groups exist, they are also used as contacts (Mbegera, Eriksson and Njoroge 1992).

International organizations such as the Food and Agricultural Organization of the United Nations now view the encouragement of national governments to establish conservation programmes as one of their major tasks and their future role as the promotion of conservation services, pilot demonstration schemes and watershed management within the context of such programmes (Sanders 1988). Their role will thus be similar to that of national government aid agencies and non-governmental organizations.

A third area of reappraisal is whether soil conservation should be approached on a watershed basis, at the level of the individual farmer or on some form of administrative district. The watershed has the advantage of being the natural geomorphological unit for water erosion. The risk of erosion at any point within a watershed can be understood in relation to its topographic position and the effect this has on local hydrology and sediment production. The off-site effects of erosion are also more easily appreciated within a watershed than by the study of an individual field. Emphasis on a watershed, however, can result in too much reliance on mechanical measures aimed at runoff control and on a top-down approach to erosion control. It also ignores the fact that the success or otherwise of soil conservation depends on the behaviour of the individual land user. At the most basic level, the effective watershed is a farmer's field and with the increasing emphasis on agronomic methods of erosion control this seems the most appropriate unit at which to operate.

Administrative units are required through which contact can be made with farmers. In the USA, these take the form of Conservation Districts set up under the 1937 Standard State Soil Conservation Districts Enabling Law. The Districts provide the forum through which farmers can approach the various federal, state and local agencies that provide technical and financial assistance for conservation work. They bring together farmers, public-spirited citizens in business, industry and education, and officers of the Soil Conservation Service. Although originally focusing only on soil erosion, the Districts are today concerned with all aspects of land use and the environment (Cook 1989). In Kenya, the Soil Conservation Districts are administrative units for the planning and implementation of soil conservation programmes and the participation of farmers is organized on watersheds within the District through catchment conservation committees. These comprise farmers, chiefs and assistant chiefs, as well as the soil conservation officer and technical assistants. All are involved in decision making on land use activities in the catchment and in their implementation, monitoring and evaluation (Mbegera, Eriksson and Njoroge 1992). In Burkina Faso, individual villages form the operating units with the farmers addressing the problems of land use planning for the whole village territory (Eger and Bado 1992). The programme, based on a participatory approach, begins with a request from the village and leads to the training of local extension agents and farmers and then to the development and implementation of a plan for rational land use and improved land management.

11.4 Responsible Bodies

In some countries, soil conservation is the responsibility of a specific agency, which has a clearly-defined mission and an administrative structure with a national core to define policy, and regional and local sub-divisions for implementation. The boundaries of its remit with respect to other agencies are also clear. This is generally the situation in countries with a formalized Soil Conservation Service. Elsewhere, soil conservation is administered alongside other activities through the advisory services to farmers. With the present focus on promoting soil conservation as an overall part of land husbandry, this might be considered a more appropriate model but, as seen above with respect to the US Soil Conservation Districts, many soil conservation services have taken wider issues on board. Since 1965, for example, the State Soil Conservation Service of Iceland has had an extended remit to cover protection of vegetation and soils in general.

In many countries, the implementation of soil conservation projects and programmes is the responsibility of a multiplicity of organizations, including national and local governments, often involving several ministries covering agriculture, forestry and public works; aid agencies, with several overseas governments and international organizations being involved; non-government organizations; and private companies. The result is ill-defined responsibilities, competition between the bodies for scarce resources, particularly skilled labour, and confusion on the part of the farmer on whom to contact. In a review of the situation in Java, Indonesia, in the 1980s, McCauley (1988) found six government organizations which included aspects of soil conservation in their responsibilities working alongside FAO, UNDP, USAID, the Asian Development Bank, the World Bank and the Dutch Government. There was little coordination of activities and often competition to work on different aspects within the same watershed project. It was to avoid this type of problem that the Kenyan Government set up the Presidential Commission on Soil Conservation and Afforestation. As long as there is an overall coordinating agency through which responsibilities can be defined and collaboration achieved, there is no reason why all these organizations cannot be usefully involved. Such coordination can also help enlist the assistance of others, ranging from individuals working in research organizations and universities, to school teachers, farming cooperatives and women's groups. Coordination, however, requires an understanding of the perspectives and objectives of each organization and why it is involved in soil conservation. The perspectives of national governments, donor agencies, non-governmental organizations, universities and farmer groups are all different. Coordination also requires ensuring that the farmer has a single, respected and well-defined point of contact. As seen above, this is best achieved using local institutions at the district or village level.

In addition to ensuring that the roles of responsible bodies are clear, they must have good leadership (Wenner 1988; Hudson 1991). The quality of leadership, from national down to village level, can influence whether or not a project succeeds. Indeed, a charismatic leader can often make a poorly-designed project work whilst a poor leader can mar the outcome of a well-designed one. The training of project managers, junior level managers and field technicians must therefore include organizational, team and personal management as well as the technical aspects of soil conservation. It must also be backed up by a suitable career structure in which promotion is based on performance and not solely on academic qualifications (Tejwani 1992). A post as a technical assistant or a field soil conservation officer must not in any way be an inferior position to one in an academic or research environment.

11.5 Requirements for Technology

Although, as stated earlier, the broad technology for controlling soil erosion is understood, it must satisfy a number of requirements as well as being scientifically sound. These include (Hudson 1991):

1. a high and quick financial return;
2. a reduction in risk;
3. no loss of existing benefits;
4. accessibility to the farmer in terms of extra inputs of labour and capital;
5. social acceptability, particularly in terms of gender issues;
6. being an extension or modification of an existing practice rather than something new.

In addition to meeting these requirements, the

farmers must be convinced that the technology will work. The early efforts of the State Soil Conservation Service of Iceland to reclaim degraded land by barbed-wire fencing to close areas from grazing and allow the vegetation to generate naturally were not taken up by farmers. Nor were the experiments with shelterbelts. Following severe erosion in the late 1940s, however, a major reclamation experiment was carried out near Gunnarsholt, using technology based on assisting the natural ecological succession of plant communities by planting nurse species to stabilize the land and adding fertilizer (Arnalds, Aradóttir and Thorsteinsson 1987; Section 8.7.4). Some 1,300 hectares were successfully reclaimed and this demonstration project provided the basis for all future work of the Service.

Ideally, the technology should be shown to work on farms as well as on research stations. In order to test a range of conservation measures for their transferability to farms, trial areas were established early in the current programme of soil conservation in Kenya. In 1974 some 160 target farms were selected in each of four districts. Within 6 years, trial farms were operative in 40 districts. A range of conservation measures was introduced on the farms and their performance monitored. The results were used to plan the national programme described earlier (Wenner 1988). In a scheme to introduce new technologies, including agroforestry and social forestry, to shifting cultivators in the uplands of Mindanao, Philippines, participatory programmes are being introduced on selected farms in target areas. To qualify, a target area must be accessible by land transport, have a minimum of 10° slope with visual evidence of land degradation, contain at least 30 farmers who have been awarded stewardship certificates by the Bureau of Forest Development, and be politically peaceful. Within each target area, some ten farmers are selected as programme participants according to their level of income, willingness to take part, leadership potential and possession of no previous criminal record. Conservation measures are introduced on each chosen farm after discussion with the farmer family about their aspirations and the constraints on their present farming systems (Pava *et al.* 1990). The success of the measures is monitored on each farm.

11.6 Legislative Instruments

There is still considerable disagreement among soil conservation workers on the importance of legislation. Many early soil and water conservation programmes relied on laws which required farmers to adopt certain techniques and to desist from using others. Laws were also passed to restrict activities on certain types of land. For example, cultivation should not be permitted on slopes above a certain steepness or forests in watershed areas should be protected. In some cases, legislation appears to work. As stated earlier, contour grass strips were introduced in Swaziland by royal decree and accepted by almost all farmers. In Iceland, the State Soil Conservation Service has the legal power to acquire degraded land from farmers. In reality, however, the Swazi farmers adopted the technology because it seemed to work. The Conservation Service in Iceland has never used its power because most farmers are only too willing to have the Service help them restore their land. In many other situations, legislation is ineffective because it cannot be enforced. Generally, if legislation is effective it is because the principles behind it have been accepted by the population in which case it is probably not required (Hudson 1981).

If legislation is unworkable, the alternative is for farmers to adopt soil conservation voluntarily. The low extent of farmer uptake, however, suggests that this does not work either, particularly where, as in the USA and Europe, the benefits of conservation are acquired by the community at large rather than the individual farmer. In this situation, the role of government is to provide the necessary framework within which soil conservation will work. In many cases, however, the framework is detrimental to soil protection. Policies designed to maximize agricultural production can result in greater erosion because highly-mechanized and specialized (monoculture) farming is often the most highly erosive. In recent years, the failure to contain erosion in the USA (Napier 1990) and the increasing erosion problem in parts of Europe (Chisci 1986; Boardman 1988) have been attributed to agricultural policies and have led to calls for more environmentally-friendly farming. Within this context, the following options are available as ways of encouraging farmers to adopt more conservation-oriented measures.

11.6.1 Regulatory instruments

Examples of regulatory instruments include prohibiting arable farming on land classified as unsuitable for arable crops, setting a maximum acceptable rate of erosion and making farmers liable to legal action if the rate is exceeded, and setting minimum requirements on farms for provision of wildlife habitats. As mentioned above, these measures are only effective if they can be monitored and enforced. Regulations based on rates of erosion would be almost impossible to apply because erosion is costly to measure and there is considerable uncertainty on the levels of accuracy that can be obtained (Section 6.1).

11.6.2 Advisory work

An effective advisory service is the vital element behind all participatory approaches to soil conservation aimed at arousing the interest of farmers and their families and building on their existing farming system. An effective service is, however, costly which means either a realistic charge must be made to farmers or the government must defray the expense. In many countries, the farmers are unable to pay and government finance is not available to run a staff of well-trained and motivated professionals. Even where farmers are able to pay, they are less likely to seek advice than if the service were free. Whatever advice is given, it is clear from much of the evidence presented in Section 11.1 and Section 7.3 that it will only be taken up if clear economic benefits ensue to the farmer. It is thus often necessary to support advisory work with other financial incentives.

11.6.3 Financial support

Financial support should only be provided if adopting particular soil conservation practices results in farming below the economic optimum. Ideally, it should be limited to stimulating the involvement of the farmers and not be aimed at buying their participation (IFAD 1992). The most suitable type of financial incentive is likely to be site specific. In some cases, it may consist of the provision of basic tools such as pickaxes, wheelbarrows, seeds and cuttings; in others it may include the construction of terraces, the costs of training courses for farmers, and the cost of land lost to cultivation by grassing valley floors or introducing contour grass strips or shelterbelts. Incentives may also be set within a broader environmental context, for example promotion of winter cover crops to control nitrate leaching.

Cost-sharing programmes between farmer and government were the basis of the early work of the US Soil Conservation Service (Napier 1990). Their disadvantages are that they lead to complacency that soil conservation is being adequately supported and they require a long-term financial commitment. Also, once the incentive is withdrawn, the conservation ceases. Programmes that rely on farmers taking land out of agricultural production suffer from the same short-term commitment. The Soil Bank Programme, introduced in the USA in 1956, encouraged farmers to enrol land in the Soil Bank in return for a payment not to cultivate. Although many farmers enrolled their most marginal and frequently most erodible land, no new entrants were permitted after 1958 and the scheme ended in the late 1960s, at which time much of the land was returned to production. The same

problem arose with the Conservation Reserve Provisions, introduced in the USA in 1985. Farmers entered a contractual arrangement with the government to retire highly-erodible land for ten years in return for a rent payment. Those who violated the contract within that time stood to lose federal farm benefits and be required to repay rent with interest. Although a significant reduction in soil erosion resulted, the cost was extremely high. Also, the eligibility rules, requiring fields to have erodible soils over at least two-thirds of their area and to have been cropped for two years between 1981 and 1985, meant that in Illinois, Iowa and Missouri some 25 per cent of the land classified as highly erodible was excluded from the scheme (Padgitt 1989).

More recently, the emphasis has been on policies which reward farmers for adopting Codes of Good Agricultural Practice which are aimed at protection of the environment. In the USA, the Conservation Title of the Food Security Act 1985 requires farmers of highly erodible land to develop a conservation farm plan that will reduce soil loss to a specified level. Those who do not comply with the programme will lose all other USDA-run farm benefits. According to Napier (1990), however, the specified level has been set so low that most farmers will qualify without any change in farming practice.

The main advantage of financial measures is that the cost is borne by the community which seems fair when it is the community that benefits from both protection of the environment from off-site damage and from the secure food supply associated with sustained agricultural production. However, a decision has to be made whether to provide the support to all farmers or to target farmers in designated areas of erosion risk or even to specific fields in those areas (Boardman 1988). Assistance in the USA has generally been targeted to farmers in areas with severe erosion problems. Although this has the advantage of allocating limited resources to areas of greatest concern, it meets resistance from those farmers who are unable to benefit. Where the assistance is clearly aimed at soil conservation this may be not be a problem but, where the benefit to soil conservation is indirect through schemes promoting creation of wildlife habitats, lowland forestry or agroforestry, the other farmers will feel that they should have equal access to the benefits.

Financial support must clearly be treated with care. Inappropriate support can either destroy the long-term prospects of a soil conservation programme or require a high level of long-term funding which cannot be sustained. Incentives ought not to be necessary to control the on-site effects of erosion since soil conservation measures should be designed so that farmers can see the economic benefits. Some support is likely to be required, however, where farmers are being asked to adopt practices which afford no economic advantage in order to control off-site damage. The situation here is similar in principle to using public finance to pay for erosion control on road banks or footpath restoration in national parks where the costs of erosion and the benefits of control accrue to the community. Under certain circumstances, where damage from erosion can be clearly related to inadequate on-site management and to a specific land user, financial costs can be reduced through legislation based on the principle of the 'polluter pays'. Land users need to be persuaded that the cost of conservation is preferable to any settlement that might be demanded. This is more likely to apply to land users responsible for preventing erosion on mining spoil, industrial land and construction sites than to farmers, where apportioning blame is more difficult. Erosion on agricultural land, particularly where it occurs infrequently, can often legally be considered as an 'act of God' and, in many instances in the recent past, can be related either directly or indirectly to government policies promoting more efficient agricultural production. In the future, it may be feasible to hold farmers responsible where it can be shown that they have not complied with a Code of Good Agricultural Practice.

11.7 Concluding Remarks

Soil conservation is an interdisciplinary subject. It requires assessments of erosion risk; an understanding of the geomorphological processes at work and the factors controlling them; an understanding of agricultural systems and the organizational structure of the society in which they are practised; the ability to design farming systems which are economically acceptable, satisfy the aspirations of the farmer and result in sustained production without environmental damage; and the ability to implement any proposals and advise on the legislative framework to support them. Although in many countries, soil conservation has been traditionally the preserve of agricultural engineers, with the shift towards vegetation-based control measures agriculturalists and agronomists are likely to have a greater role. Engineers, however, are still contributing through the development of better systems of soil management. With increasing emphasis being given to erosion as a socio-economic rather than a technical problem, the sociologist and the economist also have important roles to play. Overall, however, there is a need for people with a broad perspective of all these disciplines who can, at the same time, contribute specialist skills in at least one area. Thus there is a place and, indeed, an urgent need, for the geographer.

References

Abrahams A D, A J Parsons, P J Hirsch (1992), Field and laboratory studies of resistance to interrill overland flow on semi-arid hillslopes, southern Arizona. In A J Parsons and A D Abrahams (eds), *Overland flow: hydraulics and erosion mechanics*. London, UCL Press: 1–23.

Abrahams A D, A J Parsons, S H Luk (1991), The effect of spatial variability in overland flow on the downslope pattern of soil loss on a semiarid hillslope, southern Arizona. *Catena* **18**: 255–70.

Abrahim Y B, R J Rickson (1989), The effectiveness of stubble mulching in soil erosion control. In U Schwertmann, R J Rickson and K Auerswald (eds), *Soil erosion protection measures in Europe*. Soil Technology Series 1: 115–26.

Abtew W, J M Gregory, J Borrelli (1989), Wind profile: estimation of displacement height and aerodynamic roughness. *Transactions of the American Society of Agricultural Engineers* **32**: 521–7.

Agassi M, M Ben-Hur (1992), Stabilizing steep slopes with soil conditioners and plants. *Soil Technology* **5**: 249–56.

Agate E (1983), *Footpaths*. Wallingford, British Trust for Conservation Volunteers.

Aina P O (1979), Soil changes resulting from long-term management practices in western Nigeria. *Soil Science Society of America Journal* **43**: 173–7.

Aina P O, R Lal, G S Taylor (1977), Soil and crop management in relation to soil erosion in the rainforest of western Nigeria. In *Soil erosion: prediction and control*. Soil Conservation Society of America Special Publication 21: 75–82.

Aina P O, R Lal, G S Taylor (1979), Effects of vegetal cover on soil erosion on an alfisol. In R Lal and D J Greenland (eds), *Soil physical properties and crop production in the tropics*. London, Wiley: 501–8.

Alberts E E, W C Moldenhauer, G R Foster (1980), Soil aggregates and primary particles transported in rill and interrill flow. *Soil Science Society of America Journal* **44**: 590–5.

Aldon E F (1976), Soil ripping treatments for runoff and erosion control. In *Proceedings of the Federal Inter-Agency Sediment Conference, Denver, Colorado*, **3**: 2-24–2-29.

Al-Durrah M M, J M Bradford (1981), New methods of studying soil detachment due to water drop impact. *Soil Science Society of America Journal* **45**: 949–53.

Al-Durrah M M, J M Bradford (1982), Parameters for describing soil detachment due to single water drop impact. *Soil Science Society of America Journal* **46**: 836–40.

Alexander E B (1988), Rates of soil formation: implications for soil loss tolerance. *Soil Science* **145**: 37–45.

Alexander M J (1990) Reclamation after tin mining on the Jos Plateau, Nigeria. *Geographical Journal* **156**: 44–50.

Alfaro Moreno J (1988), Farmer income and soil conservation in the Peruvian Andes. In S Rimwanich (ed.), *Land conservation for future generations*. Bangkok, Department of Land Development: 711–26.

Alström K, A Bergman (1988), Sediment and nutrient losses by water erosion from arable land in south Sweden: a problem with nonpoint pollution? *Vatten* **44**: 193–204.

Alström K, A Bergman (1990), Water erosion on arable land in southern Sweden. In J Boardman, I D L Foster and J A Dearing (eds), *Soil erosion on agricultural land*. Chichester, Wiley: 107–17.

Altshul H J (1993), Comparison of soil erosion risk assessment techniques as applied to three sites in Swaziland on a land facet basis. MSc Thesis, Cranfield University.

Ambar S, K F Wiersum (1980), Comparison of different erodibility indices under various soil and land use conditions in west Java. *Indonesian Journal of Geography* **10**(39): 1–15.

Amphlett M B (1988), A nested catchment approach to sediment yield monitoring in the Magat catchment, central Luzon, The Philippines. In S Rimwanich (ed.), *Land conservation for future generations*. Bangkok, Department of Land Development: 283–98.

Anderson D (1984), Depression, dust bowl, demography and drought: the colonial state and soil conservation in East Africa during the 1930s. *African Affairs* **83**: 321–43.

Anderson M G, K S Richards (1987), *Slope stability: geotechnical engineering and geomorphology*. Chichester, Wiley.

André J E, H W Anderson (1961), Variation of soil erodibility with geology, geographic zone, elevation and vegetation type in northern California wildlands. *Journal of Geophysical Research* **66**: 3351–8.

Armstrong C L, J K Mitchell (1987), Transformations of rainfall by plant canopy. *Transactions of the American Society of Agricultural Engineers* **30**: 688–96.

Armstrong J J, G J Wall (1991), Quantitative evaluation of the effectiveness of erosion control materials. In *Erosion control: a global perspective*. Steamboat Springs CO, International Erosion Control Association: 165–79.

Arnalds A (1987), Ecosystem disturbance in Iceland. *Arctic and Alpine Research* **19**: 508–13.

Arnalds A (1988), Landgæði á Íslandi fyrr og nú. *Árbók Landgræðslu Ríkisins* **1**: 13–31.

Arnalds O, A L Aradóttir, I Thorsteinsson (1987), The nature and restoration of denuded areas in Iceland. *Arctic and Alpine Research* **19**: 518–25.

Arulanandan K, P Loganathan, R B Krone (1975), Pore and eroding fluid influences on the surface erosion of a soil. *Journal of the Geotechnical Engineering Division ASCE* **101**: 53–66.

Auerswald K, F Schmidt (1986), Atlas der Erosionsgefährdung in Bayern – Karten zum flächenhaften Bodenabtrag durch Regen. *GLA-Fachberichte* **1**, München, Geologisches Landesamt.

Awasthi R P, R N Prasad, B N Chatterjee (1987), Effects of cropping systems on runoff, soil losses and crop productivity from steep slope under slash and burn agriculture in the humid tropics. In T H Tay, A M Mokhtaruddin and A B Zahari (eds), *Steepland agriculture in the humid tropics*. Kuala Lumpur, MARDI/Malaysian Society of Soil Science: 713–31.

Babaji G A (1987), Some plant stem properties and overland flow hydraulics: a laboratory simulation. PhD Thesis, Cranfield Institute of Technology.

Bache D H (1986), Momentum transfer to plant canopies: influence of structure and variable drag. *Atmospheric Environment* **20**: 1369–78.

Bache D H, I A MacAskill (1984), *Vegetation in civil and landscape engineering*. London, Granada.

Baffoe-Bonnie E, C Quansah (1975), The effect of tillage on soil and water loss. *Ghana Journal of Agricultural Science* **8**: 191–5.

Bagnold R A (1937), The transport of sand by wind. *Geographical Journal* **89**: 409–38.

Bagnold R A (1941), *The physics of blown sand and desert dunes*. London, Chapman and Hall.

Bagnold R A (1951), The movement of a cohesionless granular bed by fluid flow over it. *British Journal of Applied Physics* **2**: 29–34.

Bagnold R A (1979), Sediment transport by wind and water. *Nordic Hydrology* **10**: 309–22.

Bamberger M (1991), Continuous sodding and its ecological effects on the landscape in steep slope vineyards in the Ruwer Valley area. In G Richter (ed.), *Combating soil erosion in vineyards of the Mosel Region*. Universität Trier Forschungsstelle Bodenerosion **10**: 125–34.

Banasik K, D Górski (1993), Evaluation of rainfall erosivity for east Poland. In K Banasik and A Żbikowski (eds), *Runoff and sediment yield modelling*. Warsaw, Warsaw Agricultural University Press: 129–40.

Barahona E, J Quirantes, J L Guardiola and A Iriarte (1990), Factors affecting the susceptibility of soils to interrill erosion in south-eastern Spain. In J L Rubio and R J Rickson (eds), *Strategies to combat desertification in Mediterranean Europe*. Commission of European Communities Report EUR 11175 EN/ES: 216–27.

Barrow E G C (1989), The value of traditional knowledge in present-day soil-conservation practice: the example of West Pokot and Turkana. In D B Thomas, E K Biamah, A M Kilewe, L Lundgren and B O Mochoge (eds), *Soil and water conservation in Kenya*. Nairobi, Department of Agricultural Engineering, University of Nairobi and Swedish International Development Authority: 471–85.

Barrow G, D I Brotherton and O C Maurice (1973), Tarn Hows experimental restoration project. *Countryside Recreation News Supplement* **9**: 13–18.

Bates C G (1924), The windbreak as a farm asset. *USDA Farmers' Bulletin* **1405**.

Bazzoffi P, D Torri and C Zanchi (1980), Stima dell'erodibilità dei suoli mediante simulazione di pioggia in laboratorio. Nota 1: simulatore di pioggia. *Annali Istituto Sperimentale per lo Studio e la Difesa del Suolo* **11**: 129–40.

Beasley D B, L F Huggins and E J Monke (1980), ANSWERS: A model for watershed planning. *Transactions of the American Society of Agricultural Engineers* **23**: 938–44.

Bekele M W, D B Thomas (1992), The influence of surface residue on soil loss and runoff. In H Hurni and Kebede Tato (eds), *Erosion, conservation and small-scale farming*. Bern, Geographica Bernensia: 439–52.

Bennett H H (1939), *Soil conservation.* New York, McGraw-Hill.

Bennett J P (1974), Concepts of mathematical modeling of sediment yield. *Water Resources Research* 10: 485-92.

Bensalem B (1977), Examples of soil and water conservation practices in North African countries. *FAO Soils Bulletin* 33: 151–60.

Bergsma E, C R Valenzuela (1981), Drop testing aggregate stability of some soils near Mérida, Spain. *Earth Surface Processes and Landforms* 6: 309–18.

Berry L, B P Ruxton (1960), The evolution of Hong Kong harbour basin. *Zeitschrift für Geomorphologie* 4: 97–115.

Berry M J (1956), Erosion control on Bukit Bakar, Kelantan. *Malayan Forester* 19: 3–11.

Beschta R L (1978), Long-term patterns of sediment production following road construction and logging in the Oregon Coast Range. *Water Resources Research* 14: 1011–16.

Besler H (1987), Slope properties, slope processes and soil erosion risk in the tropical rain forest of Kalimantan Timur (Indonesian Borneo). *Earth Surface Processes and Landforms* 12: 195–204.

Bharad G M, B C Bathkal (1991), Role of Vetiver grass in soil and moisture conservation. *Vetiver Newsletter* 6: 15–17.

Bhardwaj S P, M L Khybri, S Ram and S N Prasad (1985), Crop geometry – a nonmonetary input for reducing erosion in corn on four percent slope. In S A El-Swaify, W C Moldenhauer and A Lo (eds), *Soil erosion and conservation.* Ankeny IA, Soil Conservation Society of America: 644–8.

Biot Y (1990), THEPROM – an erosion productivity model. In J Boardman, I D L Foster and J A Dearing (eds), *Soil erosion on agricultural land.* Chichester, Wiley: 465–79.

Bisal F (1950), Calibration of splash cup for soil erosion studies. *Agricultural Engineering* 31: 621–2.

Bishop D M, M E Stevens (1964), Landslides on logged areas in southeast Alaska. *USDA Forest Research Service Paper* NOR-1.

Blaikie P M (1985), *The political economy of soil erosion in developing countries.* Harlow, Longman.

Blaikie P, H Brookfield (1987), *Land degradation and society.* London, Methuen.

Blong R J, O P Graham, J A Veness (1982), The role of sidewall processes in gully development: some N.S.W. examples. *Earth Surface Processes and Landforms* 7: 381–5.

Boardman J (1988), Public policy and soil erosion in Britain. In J M Hooke (ed.), *Geomorphology in environmental planning.* Chichester, Wiley: 33–50.

Boardman J (1990), Soil erosion on the South Downs: a review. In J Boardman, I D L Foster and J A Dearing (eds), *Soil erosion on agricultural land.* Chichester, Wiley: 87–105.

Boardman J, R Evans (1994), Soil erosion in Britain: a review. In R J Rickson (ed.), *Conserving soil resources: European perspectives.* Wallingford, CAB International: 3–12.

Bocharov A P (1984), *A description of devices used in the study of wind erosion of soils.* Rotterdam, Balkema.

Boiffin J (1985), Stage and time-dependency of soil crusting *in situ.* In F Callebaut, D Gabriels and M De Boodt (eds), *Assessment of soil surface crusting and sealing.* Flanders Research Centre for Soil Erosion and Conservation, State University of Gent: 91–8.

Boiffin J, G Monnier (1985), Infiltration rate as affected by soil surface crusting caused by rainfall. In F Callebaut, D Gabriels and M De Boodt (eds), *Assessment of soil surface crusting and sealing.* Flanders Research Centre for Soil Erosion and Conservation, State University of Gent: 210–17.

Bojö J (1992), Cost-benefit analysis of soil and water conservation projects: a review of 20 empirical studies. In Kebede Tato and H Hurni (eds), *Soil conservation for survival.* Ankeny IA, Soil and Water Conservation Society: 195–205.

Bollinne A (1975), La mesure de l'intensité du splash sur sol limoneux. Mise au point d'une technique de terrain et premiers résultats. *Pédologie* 25: 199–210.

Bollinne A (1977), La vitesse de l'érosion sous culture en région limoneuse. *Pédologie* 27: 191–206.

Bollinne A (1978), Study of the importance of splash and wash on cultivated loamy soils of Hesbaye (Belgium). *Earth Surface Processes* 3: 71–84.

Bollinne A, A Laurant, W Boon (1979), L'érosivité des précipitations à Florennes. Révision de la carte des isohyètes et de la carte d'érosivité de la Belgique. *Bulletin de la Société Géographique de Liège* 15: 77–99.

Bonell M, D A Gilmour (1978), The development of overland flow in a tropical rain forest catchment. *Journal of Hydrology* 39: 365–82.

Bonsu M (1981), Assessment of erosion under different cultural practices on a savanna soil in the Northern Region of Ghana. In R P C Morgan (ed.), *Soil conservation: problems and prospects.* Chichester, Wiley: 247–53.

Bonsu M (1985), Organic residues for less erosion and more grain in Ghana. In S A El-Swaify, W C Moldenhauer and A Lo (eds), *Soil erosion and conservation.* Ankeny IA, Soil Conservation Society of America: 615–20.

Bonsu M, H B Obeng (1979), Effects of cultural practices on soil erosion and maize production in the semi-deciduous rain forest and forest-savanna transition zones of Ghana. In R Lal and D J Greenland (eds), *Soil physical properties and crop production in the tropics*. Chichester, Wiley: 509–19.

Boon W, J Savat (1981), A nomogram for the prediction of rill erosion. In R P C Morgan (ed.), *Soil conservation: problems and prospects*. Chichester, Wiley: 303–19.

Boonchee S, S Peukrai, N Chinabutr (1988), Effects of land development and management on soil conservation in northern Thailand. In S Rimwanich (ed.), *Land conservation for future generations*. Bangkok, Department of Land Development: 1101–12.

Borghi C E, S M Giannoni, J P Martinez-Rica (1990), Soil removed by voles of the genus *Pitymys* in the Spanish Pyrenees. *Pirineos* **136**: 3–17.

Bork H R (1989), The history of soil erosion in southern Lower Saxony. *Landschaftgenese und Landschaftsökologie* **16**: 135–63.

Bork H R, H Rohdenburg (1981), Rainfall simulation in south-east Spain. In R P C Morgan (ed.), *Soil conservation: problems and prospects*. Chichester, Wiley: 293–302.

Borst H L, R Woodburn (1942), The effect of mulching and methods of cultivation on runoff and erosion from Muskingum silt loam. *Agricultural Engineering* **23**: 19–22.

Borsy Z (1972), Studies on wind erosion in the windblown sand areas of Hungary. *Acta Geographica Debrecina* **10**: 123–32.

Boubakari M (1992), Grass strip performance on steep lands. MSc Thesis, Silsoe College, Cranfield Institute of Technology.

Bouyoucos G J (1935), The clay ratio as a criterion of susceptibility of soils to erosion. *Journal of the American Society of Agronomy* **27**: 738–51.

Bowyer-Bower T A S (1993), Effects of rainfall intensity and antecedent moisture on the steady-state infiltration rate in a semi-arid region. *Soil Use and Management* **9**: 69–76.

Brabben T, J Bird, P Bolton (1988), Improving survey and computation of reservoir sedimentation. *ODU Bulletin* **12**: 4–7.

Bradford J M, R F Piest (1980), Erosional development of valley-bottom gullies in the Upper Midwestern United States. In D R Coates and J D Vitek (eds), *Thresholds in geomorphology*. London, Allen and Unwin: 75–101.

Bradford J M, C C Truman, C Huang (1992), Comparison of three measures of resistance of soil surface seals to raindrop splash. *Soil Technology* **5**: 47–56.

Brady N C (1990), *The nature and properties of soils*. New York, Macmillan.

Brandt C J (1989), The size distributions of through-fall drops under vegetation canopies. *Catena* **16**: 507–24.

Brandt C J (1990), Simulation of the size distribution and erosivity of raindrops and throughfall drops. *Earth Surface Processes and Landforms* **15**: 687–98.

Brierley J S (1976), Kitchen gardens in the West Indies with a contemporary study from Grenada. *Journal of Tropical Geography* **43**: 30–40.

Briggs D, A Giordano (1992), *CORINE soil erosion risk and important land resources in the southern regions of the European Community*. Commission of the European Communities Publication EUR 13233 EN.

Brown L R (1981), *Building a sustainable society*. New York, Norton.

Browning G M, R A Norton, A G McCall, F G Bell (1948), Investigation in erosion control and the reclamation of eroded land at the Missouri Valley Loess Conservation Experiment Station, Clarinda, Iowa. *USDA Technical Bulletin* 959.

Bruce-Okine E, R Lal (1975), Soil erodibility as determined by raindrop technique. *Soil Science* **119**: 149–57.

Brunori F, M C Penzo, D Torri (1989), Soil shear strength: its measurement and soil detachability. *Catena* **16**: 59–71.

Brunsden D, D B Prior (1984), *Slope instability*. Chichester, Wiley.

Bryan R B (1968), The development, use and efficiency of indices of soil erodibility. *Geoderma* **2**: 5–26.

Bryan R B (1979), The influence of slope angle on soil entrainment by sheetwash and rainsplash. *Earth Surface Processes* **4**: 43–58.

Bryan R B, J De Ploey (1983), Comparability of soil erosion measurements with different laboratory rainfall simulators. *Catena Supplement* **4**: 33–56.

Bryan R B, J W A Poesen (1989), Laboratory experiments on the influence of slope length on runoff, percolation and rill development. *Earth Surface Processes and Landforms* **14**: 211–31.

Bryan R, A Yair (1982), Perspectives on studies of badland geomorphology. In R Bryan and A Yair (eds), *Badland geomorphology and piping*. Norwich, Geo Books: 1–12.

Bubenzer G D (1979), Rainfall characteristics important for simulation. In *Proceedings, Rainfall simulator workshop, Tucson, Arizona*. USDA-SEA Agricultural Reviews and Manuals ARM-W-10: 22–34.

Bubenzer G D, B A Jones (1971), Drop size and impact velocity effects on the detachment of soil under simulated rainfall. *Transactions of the American Society of Agricultural Engineers* **14**: 625–8.

Buchner W (1988), Cultivation techniques for reducing soil erosion in the Rhineland. In R P C Morgan and R J Rickson (eds), *Erosion assessment and modelling*. Commission of the European Communities Report EUR 10860 EN: 283–97.

Bučko Š (1975), Investigation of normal and accelerated water erosion in Slovakia. *Studia Geomorphologica Carpatho-Balcanica* 4: 27–38.

Bunte K, J Poesen (1994), Effects of rock fragment size and cover on overland flow hydraulics, local turbulence and sediment yield on an erodible soil surface. *Earth Surface Processes and Landforms* 19: 115–35.

Buol S W, F D Hole, P J McCracken (1973), *Soil genesis and classification*. Ames IA, Iowa State University Press.

Burgess P F (1971), The effect of logging on hill dipterocarp forest. *Malayan Nature Journal* 24: 231–7.

Carson M A, M J Kirkby (1972), *Hillslope form and process*. Cambridge, Cambridge University Press.

Carte A (1971), Raindrop spectra in Pretoria. *South African Geographical Journal* 53: 100–3.

Carter C E, J D Greer, H J Braud, J M Floyd (1974), Raindrop characteristics in south central United States. *Transactions of the American Society of Agricultural Engineers* 17: 1033–7.

Cazzuffi D, R Monti, P Rimoldi (1991), Geosynthetics subjected to different conditions of rain and runoff in erosion control applications: a laboratory investigation. In *Erosion control: a global perspective*. Steamboat Springs CO, International Erosion Control Association: 191–208.

Chakela Q, M Stocking (1988), An improved methodology for erosion hazard mapping. Part II: Application to Lesotho. *Geografiska Annaler* 70-A: 181–9.

Chan C C (1981a), Evaluation of soil loss factors on cultivated slopelands of Taiwan. *Food and Fertilizer Technology Center Taipei, Technical Bulletin* 55.

Chan C C (1981b), Conservation measures on the cultivated slopelands of Taiwan. *Food and Fertilizer Technology Center Taipei, Extension Bulletin* 137.

Chapman D M (1990), Aeolian sand transport – an optimized model. *Earth Surface Processes and Landforms* 15: 751–60.

Chapman G (1948), Size of raindrops and their striking force at the soil surface in a red pine plantation. *Transactions of the American Geophysical Union* 29: 664–70.

Charman P E V (1978), Soils of New South Wales: their characterisation, classification and conservation. *NSW Soil Conservation Service Technical Handbook* 1.

Cheatle R J, S N J Njoroge (1993), Smallholder adoption of some land husbandry practices in Kenya. In N Hudson and R J Cheatle (eds), *Working with farmers for better land husbandry*. Southampton, IT Publications: 130–41.

Chepil W S (1945), Dynamics of wind erosion. III. Transport capacity of the wind. *Soil Science* 60: 475–80.

Chepil W S (1946), Dynamics of wind erosion. VI. Sorting of soil material by wind. *Soil Science* 61: 331–40.

Chepil W S (1950), Properties of soil which influence wind erosion. II. Dry aggregate structure as an index of erodibility. *Soil Science* 69: 403–14.

Chepil W S (1960), Conversion of relative field erodibility to annual soil loss by wind. *Soil Science Society of America Proceedings* 24: 143–5.

Chepil W S, N P Woodruff (1963), The physics of wind erosion and its control. *Advances in Agronomy* 15: 211–302.

Chepil W S, F H Siddoway, D V Armbrust (1964), Wind erodibility of knolly terrain. *Journal of Soil and Water Conservation* 19: 179–81.

Chepil WS, N P Woodruff, A W Zingg (1955), Field study of wind erosion in western Texas. *USDA Soil Conservation Service Technical Paper* SCS-TP-125.

Chisci G (1986), Influence of change in land use and management on the acceleration of land degradation phenomena in Apennines hilly areas. In G Chisci and R P C Morgan (eds), *Soil erosion in the European Community: impact of changing agriculture*. Rotterdam, Balkema: 3–18.

Chisci G, C Zanchi (1981), The influence of different tillage systems and different crops on soil losses on hilly silty-clayey soil. In R P C Morgan (ed.), *Soil conservation: problems and prospects*. Chichester, Wiley: 211–17.

Chisci G, P Bazzoffi, J S C Mbagwu (1989), Comparison of aggregate stability indices for soil classification and assessment of soil management practices. *Soil Technology* 2: 113–33.

Chischi G, C Zanchi, B Biagi (1978), Drenaggio profondo in un suolo limo-argilloso su sedimenti plioceni marini in funzione di differenti colture. *Annali Istituto Sperimentale per lo Studio e la Difesa del Suolo* 9: 257–72.

Chisholm A, R Dumsday (1987), *Land degradation*. Cambridge, Cambridge University Press.

Chorley R J (1959), The geomorphic significance of some Oxford soils. *American Journal of Science* 257: 503–15.

Cogo N P, W C Moldenhauer, G R Foster (1984), Soil loss reductions from conservation tillage practices. *Soil Science Society of America Journal* 48: 368–73.

Colclough J D (1965), Soil conservation and soil erosion control in Tasmania: tunnel erosion. *Tasmanian Journal of Agriculture* **36**: 7–12.

Coleman R (1981), Footpath erosion in the English Lake District. *Applied Geography* **1**: 121–31.

Combeau A, G Monnier (1961), A method for the study of structural stability: application to tropical soils. *African Soils* **6**: 33–52.

Cook M G (1989), Conservation Districts: a model for conservation planning and implementation in developing countries. In W C Moldenhauer, N W Hudson, T C Sheng and S W Lee (eds), *Development of conservation farming on hillslopes*. Ankeny IA, Soil and Water Conservation Society: 112–19.

Cooke R U, J C Doornkamp (1974), *Geomorphology in environmental management*. Oxford, Oxford University Press.

Coote D R, C A Malcolm-McGovern, G J Wall, W T Dickinson, R P Rudra (1988), Seasonal variation of erodibility indices based on shear strength and aggregate stability in some Ontario soils. *Canadian Journal of Soil Science* **68**: 405–16.

Coppin N J, I G Richards (1990), *Use of vegetation in civil engineering*. London, CIRIA/Butterworths.

Crouch R J (1976), Field tunnel erosion – a review. *Journal of the Soil Conservation Service NSW* **32**: 98–111.

Crouch R J (1978), Variation in the structural stability of soil in a tunnel-eroding area. In W W Emerson, R D Bond and A R Dexter (eds), *Modification of soil structure*. Chichester, Wiley: 267–74.

Crouch R J (1990a), Erosion processes and rates for gullies in granitic soils, Bathurst, New South Wales, Australia. *Earth Surface Processes and Landforms* **15**: 169–73.

Crouch R J (1990b), Rates and mechanisms of discontinuous gully erosion in a red-brown earth catchment, New South Wales, Australia. *Earth Surface Processes and Landforms* **15**: 277–82.

Crouch R J, J W McGarity, R R Storrier (1986), Tunnel formation processes in the Riverina area of N.S.W., Australia. *Earth Surface Processes and Landforms* **11**: 157–68.

Cruse R M, W E Larson (1977), Effect of soil shear strength on soil detachment due to raindrop impact. *Soil Science Society of America Journal* **41**: 777–81.

Davidson J L, J P Quirk (1961), The influence of dissolved gypsum on pasture establishment on irrigated sodic clays. *Australian Journal of Agricultural Research* **12**: 100–10.

De Bano L F, L D Mann, D A Hamilton (1970), Translocation of hydrophobic substances into soil by burning organic litter. *Proceedings of the Soil Science Society of America* **34**: 130–3.

De Jong E, H Villar, J R Bettany (1982), Prelimi-nary investigations on the use of Cs-137 to estimate erosion in Saskatchewan. *Canadian Journal of Soil Research* **82**: 673–83.

De Jong S M, H Th Riezebos (1992), *Assessment of erosion risk using multitemporal remote sensing data and an empirical erosion model.* Department of Physical Geography, University of Utrecht.

De Kesel M, D De Vleeschauwer (1981), Sand dune fixation in Tunisia by means of polyurea polyalkylene oxide (Uresol). In R Lal and E W Russell (eds), *Tropical agricultural hydrology.* Chichester, Wiley: 273–81.

De Meester T, P D Jungerius (1978), The relationship between the soil erodibility factor K (Universal Soil Loss Equation), aggregate stability and micromorphological properties of soils in the Hornos area, S. Spain. *Earth Surface Processes* **3**: 379–91.

De Meis M R M, J R S De Moura (1984), Upper Quaternary sedimentation and hillslope evolution: southeastern Brazilian plateau. *American Journal of Science* **284**: 241–54.

De Ploey J (1977), Some experimental data on slopewash and wind action with reference to Quaternary morphogenesis in Belgium. *Earth Surface Processes* **2**: 101–15.

De Ploey J (1981a), Crusting and time-dependent rainwash mechanisms on loamy soil. In R P C Morgan (ed.), *Soil conservation: problems and prospects.* Chichester, Wiley: 139–52.

De Ploey J (1981b), The ambivalent effects of some factors of erosion. *Mémoirs, Institute de Géologie, l'Université de Louvain* **31**: 171–81.

De Ploey J (1982), A stemflow equation for grasses and similar vegetation. *Catena* **9**: 139–52.

De Ploey J (1989), A model for headcut retreat in rills and gullies. *Catena Supplement* **14**: 81–6.

De Ploey J, O Cruz (1979), Landslides in the Serra do Mar, Brazil. *Catena* **6**: 111–22.

De Ploey J, D Gabriels (1980), Measuring soil loss and experimental studies. In M J Kirkby and R P C Morgan (eds), *Soil erosion.* Chichester, Wiley: 63–108.

De Ploey J, J Savat, J Moeyersons (1976), The differential impact of some soil factors on flow, runoff creep and rainwash. *Earth Surface Processes* **1**: 151–61.

De Roo A P J (1993), Modelling surface runoff and soil erosion in catchments using Geographical Information Systems. Validity and applicability of the ANSWERS model in two catchments in the loess area of south Limburg (The Netherlands) and one in Devon (UK). *Nederlandse Geografische Studies* No. 157.

Diaconu C (1969), Résultats de l'étude de l'écoulement des alluvions en suspension des rivières de la

Roumanie. *Bulletin of the International Association of Scientific Hydrology* 14: 51–89.

Dimitrakopoulos A P, R E Martin, N T Papamichos (1994), A simulation model of soil heating during wildland fires. In M Sala and J L Rubio (eds), *Soil erosion and degradation as a consequence of forest fires.* Logroño, Geoforma Ediciones: 199–206.

Disrud L A, R K Krauss (1971), Examining the process of soil detachment from clods exposed to wind-driven simulated rainfall. *Transactions of the American Society of Agricultural Engineers* 14: 90–2.

Dixon R M, J R Simanton (1980), Land imprinting for better watershed management. In *Symposium on Watershed Management.* St Giles MI, American Society of Civil Engineers: 809–26.

Dlamini P M, P S Maro (1988), The economic behaviour of the farming communities on conservation programmes: Swaziland and the SADCC region. Paper presented to Second Workshop, Economics of Conservation (Swaziland).

Dolgilevich M I, A A Sofronova, L L Mayevskaya (1973), Klassifikatsia pochv zapadnoy Sibiri, severnogo Kazakhstana po stepeniy podatlivosti k vetrovoy eroziy. *Bulletin VNIIA* 12.

Douglas I (1967), Man, vegetation and sediment yield of rivers. *Nature* 215: 925–8.

Douglas I (1978), The impact of urbanisation on fluvial geomorphology in the humid tropics. *Géo-Eco-Trop* 2: 239–42.

Douglas J T, M J Goss (1982), Stability and organic matter content of surface soil aggregates under different methods of cultivation and in grassland. *Soil and Tillage Research* 2: 155–76.

Downes R G (1946), Tunnelling erosion in northeastern Victoria. *Journal of the Council of Scientific and Industrial Research* 19: 283–92.

Drainage and Irrigation Department (1970), *Rainfall records for West Malaysia, 1959–1965.* Kuala Lumpur.

Dregne H E (1982), Desertification: man's abuse of the land. *Journal of Soil and Water Conservation* 33: 11–14.

D'Souza V P C, R P C Morgan (1976), A laboratory study of the effect of slope steepness and curvature on soil erosion. *Journal of Agricultural Engineering Research* 21: 21–31.

Duley F L, J C Russel (1943), Effect of stubble mulching on soil erosion and runoff. *Proceedings of the Soil Science Society of America* 7: 77–81.

Dunne T (1979) Sediment yield and land use in tropical catchments. *Journal of Hydrology* 42: 281–300.

Dunne T, B F Aubry (1986), Evaluation of Horton's theory of sheetwash and rill erosion on the basis of field experiments. In A D Abrahams (ed.), *Hillslope processes.* London, Allen and Unwin: 31–53.

Dunne T, R D Black (1970), Partial area contributions to storm runoff in a small New England watershed. *Water Resources Research* 6: 1296–311.

Dunne T, W E Dietrich, M J Brunengo (1978), Recent and past erosion rates in semi-arid Kenya. *Zeitschrift für Geomorphologie Supplementband* 29: 130–40.

Dyrness C T (1975), Grass–legume mixtures for erosion control along forest roads in western Oregon. *Journal of Soil and Water Conservation* 30: 169–73.

Dyrness C T (1976), Effect of wildfire on soil wettability in the High Cascade of Oregon. *USDA Forest Service Research Paper* PNW-202. Portland ORE, Pacific Northwest Forest and Range Experiment Station.

Edwards K (1993), Soil erosion and conservation in Australia. In D Pimental (ed.), *World soil erosion and conservation.* Cambridge, Cambridge University Press: 147–69.

Edwards W M, L B Owens (1991), Large storm effects on total soil loss. *Journal of Soil and Water Conservation* 46: 75–8.

Eger H, J Bado (1992), Village-level land management in the central plateau of Burkina Faso. In H Hurni and Kebede Tato (eds), *Erosion, conservation and small-scale farming.* Bern, Geographica Bernensia: 531–40.

Ekwue E I (1990), Effect of organic matter on splash detachment and the processes involved. *Earth Surface Processes and Landforms* 15: 175–81.

Ekwue E I (1992), Effect of organic and fertiliser treatments on soil physical properties and erodibility. *Soil and Tillage Research* 22: 199–209.

Ekwue E I, J O Ohu (1990), A model to describe soil detachment by rainfall. *Soil and Tillage Research* 16: 299–306.

Ekwue E I, J O Ohu, I H Wakawa (1993), Effects of incorporating two organic materials at varying levels on splash detachment of some soils from Borno State, Nigeria. *Earth Surface Processes and Landforms* 18: 399–406.

Ellison W D (1944), Two devices for measuring soil erosion. *Agricultural Engineering* 25: 53–5.

El-Swaify S A, E W Dangler, C L Armstrong (1982), *Soil erosion by water in the tropics.* College of Tropical Agriculture and Human Resources, University of Hawaii.

El-Swaify S A, P Pathak, T J Rego, S Singh (1985), Soil management for optimized productivity under rainfed conditions in the semi-arid tropics. *Advances in Soil Science* 1: 1–64.

El-Swaify S A, A Lo, R Joy, L Shinshiro, R S Yost (1988), Achieving conservation-effectiveness in the tropics using legume intercrops. *Soil Technology* **1**: 1–12.

Elwell H A (1978a), Modelling soil losses in Southern Africa. *Journal of Agricultural Engineering Research* **23**: 117–27.

Elwell H A (1978b), *Soil loss estimation. Compiled works of the Rhodesian multidisciplinary team on soil loss estimation.* Cyclo. Salisbury,

Elwell H A (1981), A soil loss estimation technique for southern Africa. In R P C Morgan (ed.), *Soil conservation: problems and prospects.* Chichester, Wiley: 281–92.

Elwell H A, M A Stocking (1976), Vegetal cover to estimate soil erosion hazard in Rhodesia. *Geoderma* **15**: 61–70.

Elwell H A, F E Wendelaar (1977), To initiate a vegetal cover data bank for soil loss estimation. *Department of Conservation and Extension Research Bulletin 23.*

Emama Ligdi, E. and Morgan R P C (1995), Contour grass strips: a laboratory simulation of their role in erosion control. Soil Technology **8**, 109-117.

Emmett W W (1970), The hydraulics of overland flow on hillslopes. *USGS Professional Paper 662-A.*

Engman E T (1986), Roughness coefficients for routing surface runoff. *Journal of the Irrigation and Drainage Division ASCE* **112**: 39–53.

Eppink L A A J, W P Spaan (1989), Agricultural wind erosion control measures in The Netherlands. In U Schwertmann, R J Rickson and K Auerswald (eds), *Soil erosion protection measures in Europe.* Soil Technology Series 1: 1–13.

Ervin D E, R Washburn (1981), Profitability of soil conservation practices in Missouri. *Journal of Soil and Water Conservation* **36**: 107–11.

Evans A C (1948), Studies on the relationships between earthworms and soil fertility. II. Some effects of earthworms and soil structure. *Applied Biology* **35**: 1–13.

Evans R (1977), Overgrazing and soil erosion on hill pastures with particular reference to the Peak District. *Journal of the British Grassland Society* **32**: 65–76.

Evans R (1980), Mechanics of water erosion and their spatial and temporal controls: an empirical viewpoint. In M J Kirkby and R P C Morgan (eds), *Soil erosion.* Chichester, Wiley: 109–28.

Evans R (1981), Potential soil and crop losses by erosion. In *Proceedings, SAWMA Conference on Soil and crop loss: developments in erosion control.* Stoneleigh, UK, National Agricultural Centre.

Evans R, S·Nortcliff (1978), Soil erosion in north Norfolk. *Journal of Agricultural Science Cambridge* **90**: 185-92.

Eyles R J (1967), Laterite at Kerdau, Pahang, Malaya. *Journal of Tropical Geography* **25**: 18–23.

Eyles R J (1968), Morphometric explanation: a case study. *Geographica (University of Malaya)* **4**: 17–23.

Fang Z S, P H Zhou, Q D Liu, B H Liu, L T Ren, H X Zhang (1981), Terraces in the loess plateau of China. In R P C Morgan (ed.), *Soil conservation: problems and prospects.* Chichester, Wiley: 481–513.

FAO (1965), Soil erosion by water. Rome.

FAO (1976), A framework for land evaluation. *FAO Soils Bulletin 32.*

Farres P (1978), The role of time and aggregate size in the crusting process. *Earth Surface Processes* **3**: 243–54.

Fenster C R, T M McCalla (1970), Tillage practices in western Nebraska with a wheat–fallow rotation. *Nebraska Agriculture Station, Lincoln, Bulletin 597.*

Fenster C T, G A Wicks (1977), Minimum tillage fallow systems for reducing wind erosion. *Transactions of the American Society of Agricultural Engineers* **20**: 906–10.

Fenton E W (1937), The influence of sheep on the vegetation of hill grazings in Scotland. *Journal of Ecology* **25**: 424–30.

Fifield J S, L K Malnor (1990), Erosion control materials vs. a semiarid environment: what has been learned from three years of testing. In *Erosion control: technology in transition.* Steamboat Springs CO, International Erosion Control Association: 233–48.

Finkel H J (1959), The barchans in southern Peru. *Journal of Geology* **67**: 614–47.

Finney H J (1984), The effect of crop covers on rainfall characteristics and splash detachment. *Journal of Agricultural Engineering Research* **29**: 337–43.

Floyd E J (1974), Tunnel erosion: a field study in the Riverina. *Journal of Soil Conservation Service NSW* **30**: 145–56.

Foster G R (1982), Modeling the erosion process. In C T Haan, H P Johnson and D L Brakensiek (eds), *Hydrologic modeling of small watersheds.* American Society of Agricultural Engineers Monograph 5: 297–380.

Foster G R, L J Lane (1981), Simulation of erosion and sediment yield from field-sized areas. In R Lal and E W Russell (eds), *Tropical agricultural hydrology.* Chichester, Wiley: 375–94.

Foster, G R, L D Meyer (1972), A closed-form soil erosion equation for upland areas. In H W Shen (ed.), *Sedimentation.* Dept. Civil Engng., Colorado State Univ., Fort Collins CO, 12-1–12.19.

Foster G R, L D Meyer (1975), Mathematical

simulation of upland erosion by fundamental erosion mechanics. In *Present and prospective technology for predicting sediment yields and sources*. USDA – ARS Publication ARS-S-40: 190–207.

Foster G R, C B Johnson, W C Moldenhauer (1982), Hydraulics of failure of unanchored cornstalk and wheat straw mulches for erosion control. *Transactions of the American Society of Agricultural Engineers* 25: 940–7.

Foster G R, L J Lane, J D Nowlin, J M Laflen, R A Young (1981), Estimating erosion and sediment yield on field-sized areas. *Transactions of the American Society of Agricultural Engineers* 24: 1253–63.

Foster R L, G L Martin (1969), Effect of unit weight and slope on erosion. *Journal of the Irrigation and Drainage Division ASCE* 95: 551–61.

Fournier F (1960), *Climat et érosion: la relation entre l'érosion du sol par l'eau et les précipitations atmosphériques*. Paris, Presses Universitaires de France.

Fournier F (1972), *Soil conservation*. Nature and Environment Series, Council of Europe.

Fox D, R B Bryan (1992), Influence of a polyacrylamide soil conditioner on runoff generation and soil erosion: field tests in Baringo District, Kenya. *Soil Technology* 5: 101–19.

Francis C F, J B Thornes (1990), Runoff hydrographs from three Mediterranean vegetation cover types. In J B Thornes (ed.), *Vegetation and erosion*. Chichester, Wiley: 363–84.

Francis I S, J A Taylor (1989), The effect of forestry drainage operations on upland sediment yields: a study of two peat-covered catchments. *Earth Surface Processes and Landforms* 14: 73–83.

Franke R, B H Chasin (1981), Peasants, peanuts, profits and pastoralists. *The Ecologist* 11: 156–68.

Fredén C, L Furuholm (1978), The Säterberget Gully at Brattforsheden, Värmland. *Geologiska Föreningens i Stockholm Förhandlingar* 100: 231–5.

Free G R (1960), Erosion characteristics of rainfall. *Agricultural Engineering* 41: 447–9, 455.

Freebairn D M, L D Ward, A L Clarke, G D Smith (1986), Research and development of reduced tillage systems for vertisols in Queensland, Australia. *Soil and Tillage Research* 8: 211–29.

Froehlich W, L Starkel (1993), The effects of deforestation on slope and channel evolution in the tectonically active Darjeeling Himalaya. *Earth Surface Processes and Landforms* 18: 285–90.

Fryrear D W (1984), Soil ridges-clods and wind erosion. *Transactions of the American Society of Agricultural Engineers* 27: 445–8.

Gabriels D, M De Boodt (1978), Evaluation of soil conditioners for water erosion control and sand stabilization. In W W Emerson, R D Bond and A R Dexter (eds), *Modification of soil structure*. Chichester, Wiley: 341–8.

Gabriels D, J M Pauwels, M De Boodt (1975), The slope gradient as it affects the amounts and size distribution of soil loss material from runoff on silt loam aggregates. *Mededelingen Fakulteit Landbouwwetenschappen. Rijksuniversiteit Gent* 40: 1333–8.

Gabriels D, L Maene, J Lenvain, M De Boodt (1977), Possibilities of using soil conditioners for soil erosion control. In D J Greenland and R Lal (eds), *Soil conservation and management in the humid tropics*. London, Wiley: 99–108.

George H, M G Jarvis (1979), Land for camping and caravan sites, picnic sites and footpaths. In M G Jarvis and D Mackney (eds), *Soil survey applications*. Soil Survey of England and Wales, Technical Monograph 13: 166–83.

Gerlach T (1966), Współczesby rozwój stoków w dorzeczu górnego Grajcarka (Beskid Wysoki-Karpaty Zachodnie). *Prace Geograf. IG PAN* 52 (with French summary).

Ghadiri H, D Payne (1979), Raindrop impact and soil splash. In R Lal and D J Greenland (eds), *Soil physical properties and crop production in the tropics*. Chichester, Wiley: 95–104.

Gichangi E M, R K Jones, D M Njarui, J R Simpson, J M N Mututho, S K Kitheka (1992), Pitting practices for rehabilitating eroded grazing land in the semi-arid tropics of Eastern Kenya: a progress report. In H Hurni and Kebede Tato (eds), *Erosion, conservation and small-scale farming*. Bern, Geographica Bernensia: 313–27.

Gichungwa J K (1970), *Soil conservation in Central Province (Kenya)*. Nairobi, Ministry of Agriculture.

Gifford G F (1976), Applicability of some infiltration formulae to rangeland infiltrometer data. *Journal of Hydrology* 28: 1–11.

Gilley J E, D A Woolhiser, D B McWhorter (1985), Interrill soil erosion – Part II: Testing and use of model equations. *Transactions of the American Society of Agricultural Engineers* 28: 154–9.

Giovannini G (1994), The effect of fire on soil quality. In M Sala and J L Rubio (eds), *Soil erosion and degradation as a consequence of forest fires*. Logroño, Geoforma Ediciones: 15–27.

Giovannini G, S Lucchesi, M Giachetti (1988), Effects of heating on some physical and chemical parameters related to soil aggregation and erodibility. *Soil Science* 146: 255–61.

Glasstetter M, V Prasuhn (1992), The influence of earthworm activity on erodibility and soil losses in central European agricultural soils. In H Hurni and Kebede Tato (eds), *Erosion, conservation and small-scale farming*. Bern, Geographica Bernensia: 285–98.

Godwin R J, G Spoor (1977), Soil failure with narrow tines. *Journal of Agricultural Engineering Research* 22: 213–28.

Goeck J, G Geisler (1989), Erosion control in maize fields in Schleswig-Holstein (F.R.G.). In U Schwertmann, R J Rickson and K Auerswald (eds), *Soil erosion protection measures in Europe*. Soil Technology Series 1: 83–92.

Gong S Y, D Q Jiang (1977), Soil erosion and its control in small gully watersheds in the rolling loess area on the middle reaches of the Yellow River. *Paris symposium on erosion and soil matter transport in inland waters*, preprint.

Gosh R L, O N Kaul, B K Subba-Rao (1978), Some aspects of water relations and nutrition in *Eucalyptus* plantations. *Indian Forester* 104: 517–24.

Govers G (1985), Selectivity and transport capacity of thin flows in relation to rill erosion. *Catena* 12: 35–49.

Govers G (1987), Initiation of motion in overland flow. *Sedimentology* 34: 1157–64.

Govers G (1989), Grain velocities in overland flow: a laboratory study. *Earth Surface Processes and Landforms* 14: 481–98.

Govers G (1990), Empirical relationships for the transporting capacity of overland flow. *International Association of Hydrological Sciences Publication* 189: 45–63.

Govers G (1992), Relationship between discharge, velocity and flow area for rills eroding loose, non-layered materials. *Earth Surface Processes and Landforms* 17: 515–28.

Govers G, J Poesen (1985), A field-scale study of surface sealing and compaction on loam and sandy loam soils. Part I. Spatial variability of soil surface sealing and crusting. In F Callebaut, D Gabriels and M De Boodt (eds), *Assessment of soil surface crusting and sealing*. Flanders Research Centre for Soil Erosion and Conservation, State University of Gent: 171–82.

Govers G, J Poesen (1988), Assessment of the interrill and rill contributions to total soil loss from an upland field plot. *Geomorphology* 1: 343–54.

Govers G, G Rauws (1986), Transporting capacity of overland flow on plane and on irregular beds. *Earth Surface Processes and Landforms* 11: 515–24.

Govers G, W Everaert, J Poesen, G Rauws, J De Ploey (1987), Susceptibilité d'un sol limoneux à l'érosion par rigoles: essais dans le grand canal de Caen. *Bulletin du Centre de Géomorphologie CNRS Caen* 33: 83–106.

Gray D H, A T Leiser (1982), *Biotechnical slope protection and erosion control*. New York, Van Nostrand Reinhold.

Green W H, G A Ampt (1911), Studies on soil physics. 1: The flow of air and water through soils. *Journal of Agricultural Science* 4: 1–24.

Greenway D R (1987), Vegetation and slope stability. In M G Anderson and K S Richards (eds), *Slope stability: geotechnical engineering and geomorphology*. Chichester, Wiley: 187–230.

Gregory K J, D E Walling (1973), *Drainage basin form and process*. London, Edward Arnold.

Grierson I T (1978), Gypsum and red-brown earths. In W W Emerson, R D Bond and A R Dexter (eds), *Modification of soil structure*. Chichester, Wiley: 315–24.

Gril J J, J P Canler, J Carsoulle (1989), The benefit of permanent grass and mulching for limiting runoff and erosion in vineyards. Experimentations using rainfall simulations in the Beaujolais. In U Schwertmann, R J Rickson and K Auerswald (eds), *Soil erosion protection measures in Europe*. Soil Technology Series 1: 157–66.

Griessel O, R P Beasley (1971), Design criteria for underground terrace outlets. *University of Missouri Extension Division, Science and Technology Guide* 1525.

Grissinger E H (1966), Resistance of selected clay systems to erosion by water. *Water Resources Research* 2: 131–8.

Grissinger E H, L E Asmussen (1963), Discussion of channel stability in undisturbed cohesive soils by E M Flaxman. *Journal of the Hydraulics Division ASCE* 89: 529–64.

Grunder M (1988), Soil conservation research in Ethiopia. In S Rimwanich (ed.), *Land conservation for future generations*. Bangkok, Department of Land Development: 901–11.

Guðmundsson Ó (1980), Approaches, materials and methods. In Ó Guðmundsson and A Arnalds (eds), *Consultancy reports for the project on utilization and conservation of grassland resources in Iceland*. Rannsóknastofnun Landbúnaðarins Report 61: 3–33.

Gupta A (1982), Observations on the effects of urbanization on runoff and sediment production in Singapore. *Singapore Journal of Tropical Geography* 3: 137–46.

Gupta J P (1979), Some observations on the periodic variations in moisture in stabilized and unstabilized dunes of the Indian desert. *Journal of Hydrology* 41: 153–6.

Guy B T, W T Dickinson (1990), Inception of sediment transport in shallow overland flow. *Catena Supplement* 17: 91–109.

Guy B T, W T Dickinson, R P Rudra, G J Wall (1990), Hydraulics of sediment-laden sheetflow and the influence of simulated rainfall. *Earth Surface Processes and Landforms* 15: 101–18.

Hagen L J (1991), A wind erosion prediction system to meet user needs. *Journal of Soil and Water Conservation* **46**: 106–11.

Haigh M J (1979), Ground retreat and slope evolution on plateau-type colliery spoil mounds at Blaenavon, Gwent. *Transactions of the Institute of British Geographers New Series* **4**: 321–8.

Halcrow H G, H O Heady, M L Cotner (1982), *Soil conservation policies, institutions and incentives.* Ankeny IA, Soil Conservation Society of America.

Hall G F, R B Daniels, J E Foss (1979), Soil formation and renewal rates in the US. In *Symposium on Determinants of soil loss tolerance.* Soil Science Society of America Annual Meeting, Fort Collins CO.

Hall M J (1970), A critique of methods of simulating rainfall. *Water Resources Research* **6**: 1104–14.

Hallsworth E G (1987), *Anatomy, physiology and psychology of erosion.* Wiley, Chichester.

Hannam I D, R W Hicks (1980), Soil conservation and urban land use planning. *Journal of the Soil Conservation Service NSW* **36**: 134–45.

Hardin G (1968), The tragedy of the commons. *Science* **162**: 1243–8.

Harper D E, S A El-Swaify (1988), Sustainable agricultural development in north Thailand: conservation as a component of success in assistance projects. In W C Moldenhauer and N W Hudson (eds), *Conservation farming on steep lands.* Ankeny IA, Soil and Water Conservation Society: 77–92.

Hashim G M, N C Wong (1988), Erosion from steep land under various plant covers and terrains. In T H Tay, A M Mokhtaruddin and A B Zahari (eds), *Steepland agriculture in the humid tropics.* Kuala Lumpur, Malaysian Agricultural Research and Development Institute/Malaysian Society of Soil Science: 424–61.

Hasholt B (1991), Influence of erosion on the transport of suspended sediment and phosphorus. *International Association of Hydrological Sciences Publication* **203**: 329–38.

Hazelhoff L, P van Hoof, A C Imeson, F J P M Kwaad (1981), The exposure of forest soil to erosion by earthworms. *Earth Surface Processes and Landforms* **6**: 235–50.

Hedfors L (1983), Evaluation and economic appraisal of soil conservation in a pilot area: a summarised report. In D B Thomas and W M Senga (eds), *Soil and water conservation in Kenya.* Institute of Development Studies and Faculty of Agriculture, University of Nairobi Occasional Paper 42: 257–62.

Heede B H (1975a), Watershed indicators of landform development. In *Hydrology and water resources in Arizona and the Southwest, Volume 5.* Proceedings,

1975 Meeting, Arizona Section, American Water Resources Association and Hydrology Section, Arizona Academy of Science: 43–6.

Heede B H (1975b), Stages of development of gullies in the west. In *Present and prospective technology for predicting sediment yields and sources.* USDA-ARS Publication ARS-S-40: 155–61.

Heede B H (1976), Gully development and control: the status of our knowledge. *USDA Forest Service Research Paper* RM-169, Fort Collins CO, Rocky Mountain Forest and Range Experiment Station.

Heede B H, J G Mufich (1973), Functional relationships and a computer program for structural gully control. *Journal of Environmental Management* **1**: 321–44.

Hénin S G, G Monnier, A Combeau (1958), Méthode pour l'étude de la stabilité structural des sols. *Annales Agronomie* **9**: 73–92.

Herwitz S R (1986), Infiltration-excess caused by stemflow in a cyclone-prone tropical rain forest. *Earth Surface Processes and Landforms* **11**: 401–12.

Hesp P (1979), Sand trapping ability of culms of marram grass (*Ammophila arenaria*). *Journal of the Soil Conservation Service NSW* **35**: 156–60.

Heusch B (1970), L'érosion du Pré-Rif. Une étude quantitative de l'érosion hydraulique dans les collines marneuses du Pré-Rif occidental. *Annales de Recherches Forestières de Maroc* **12**: 9–176.

Heusch B (1981), Sociological constraints in soil conservation: a case study, the Rif Mountains, Morocco. In R P C Morgan (ed.), *Soil conservation: problems and prospects.* Wiley, Chichester: 419–24.

Higginson F R (1973), Soil erosion of land systems within the Hunter Valley. *Journal of the Soil Conservation Service NSW* **29**: 103–10.

Hijkoop J, P van der Poel, B Kaya (1991), *Une lutte de longue haleine. Aménagements anti-érosifs et gestion de terroir.* IER Bamako/KIT Amsterdam.

Hillel D (1960), Crust formation in loessial soils. *Transactions VIIth International Soil Science Congress.* Madison, WI: 330–7.

Hills R C (1970), The determination of the infiltration capacity of field soils using the cylinder infiltrometer. *British Geomorphological Research Group Technical Bulletin* 5.

Holý M, V Svetlosanov, Z Handová, Z Kos, J Váska, K Vrána (1982), Procedures, numerical parameters and coefficients of the CREAMS model: application and verification in Czechoslovakia. *International Institute for Applied Systems Analysis Collaborative Paper* CP-82-23.

Hoogerbrugge I D, L O Fresco (1993), *Homegarden systems: agricultural characteristics and challenges.* International Institute for Environment and Development Gatekeeper Series No. 39.

Hoogmoed W B, L Stroosnijder (1984), Crust formation on sandy soils in the Sahel. I. Rainfall and infiltration. *Soil and Tillage Research* 4: 5–24.

Horikawa K, H W Shen (1960), Sand movement by wind (on the characteristics of sand traps). *US Beach Erosion Board Technical Memoir* 119.

Horton R E (1945), Erosional development of streams and their drainage basins: a hydrophysical approach to quantitative morphology. *Bulletin of the Geological Society of America* 56: 275–370.

Horváth V, B Erődi (1962), Determination of natural slope category limits by functional identity of erosion intensity. *International Association of Scientific Hydrology Publication* 59: 131–43.

Houze R A, P V Hobbs, D B Parsons, P H Herzeg (1979), Size distribution of precipitation particles in frontal clouds. *Journal of Atmospheric Science* 36: 156–62.

Howell J H, J E Clark, C J Lawrance, I Sunwar (1991), *Vegetation structures for stabilising highway slopes. A manual for Nepal*. Department of Roads, H M Government of Nepal, Kathmandu.

Hsu S A (1973), Computing eolian sand transport from shear velocity measurements. *Journal of Geology* 81: 739–43.

Huang C, J M Bradford, J H Cushman (1982), A numerical study of raindrop impact phenomena: the rigid case. *Soil Science Society of America Journal* 46: 14–19.

Hudson N W (1957), The design of field experiments on soil erosion. *Journal of Agricultural Engineering Research* 2: 56–65.

Hudson N W (1963), Raindrop size distribution in high intensity storms. *Rhodesian Journal of Agricultural Research* 1: 6–11.

Hudson N W (1965), The influence of rainfall on the mechanics of soil erosion with particular reference to Southern Rhodesia. MSc Thesis, University of Cape Town.

Hudson N W (1981), *Soil conservation*. London, Batsford.

Hudson N W (1988), Soil conservation strategies for the future. In S Rimwanich (ed.), *Land conservation for future generations*. Bangkok, Department of Land Development: 117–30.

Hudson N W (1991), Reasons for success and failure of soil conservation projects. *FAO Soils Bulletin* 64.

Hudson N (1993), *Land husbandry*. London, Batsford.

Hudson N, R J Cheatle (1993), *Working with farmers for better land husbandry*. Southampton, IT Publications.

Hudson N W, D C Jackson (1959), Results achieved in the measurement of erosion and runoff in Southern Rhodesia. *Proceedings of the Third Inter-African Soils Conference*, Dalaba: 575–83.

Hulugalle N R (1988), Properties of tied ridges in the Sudan savannah of the West African Semi-Arid Tropics. In S Rimwanich (ed.), *Land conservation for future generations*. Bangkok, Department of Land Development: 693–709.

Hulugalle N R, R Lal, C H H Terkuile (1984), Soil physical changes and crop root growth following different methods of land clearing in western Nigeria. *Soil Science* 138: 172–9.

Hurni H (1984), *Soil conservation research project Ethiopia. Volume 4. Third Progress Report (Year 1983)*. University of Bern and United Nations University.

Hurni H (1986), *Guidelines for development agents on soil conservation in Ethiopia*. Addis Ababa, Ministry of Agriculture.

Hurni H (1987), Erosion–productivity–conservation systems in Ethiopia. In I Pla Sentis (ed.), *Soil conservation and productivity*. Maracay, Sociedad Venezolana de la Ciencia del Suelo: 654–74.

Hurni H (1993), Land degradation, famine and land resource scenarios in Ethiopia. In D Pimental (ed.), *World soil erosion and conservation*. Cambridge, Cambridge University Press: 27–61.

Husse B (1991), Soil conservation in vineyards. *Universität Trier Forschungsstelle Bodenerosion* 10: 89–96.

Hussein M H, J M Laflen (1982), Effects of crop canopy and residue on rill and interrill soil erosion. *Transactions of the American Society of Agricultural Engineers* 25: 1310–15.

IFAD (1992), *Soil and water conservation in sub-Saharan Africa. Towards sustainable development by the rural poor*. Rome.

Imeson A C (1971), Heather burning and soil erosion on the North Yorkshire Moors. *Journal of Applied Ecology* 8: 537–42.

Imeson A C (1976), Some effects of burrowing animals on slope processes in the Luxembourg Ardennes: Part 1. The excavation of animal mounds in experimental plots. *Geografiska Annaler* 58–A: 115–25.

Imeson A C, F J P M Kwaad (1990), The response of tilled soils to wetting by rainfall and the dynamic character of soil erodibility. In J Boardman, I D L Foster and J A Dearing (eds), *Soil erosion on agricultural land*. Chichester, Wiley: 3–14.

Ireland H A, C F S Sharpe, D H Eargle (1939), Principles of gully erosion in the piedmont of South Carolina. *USDA Technical Bulletin* 633, Washington DC.

Iversen J D (1985), Aeolian threshold: effect of

density ratio. In O E Barndorff-Nielsen, J T Møller, K R Rasmussen and B B Willetts (eds), *Proceedings of international workshop on the physics of blown sand.* Memoirs No. 8, Dept. Theoretical Statistics, University Aarhus: 67–81.

Jackson S J (1984), The role of slope form on soil erosion at different scales as a possible link between soil erosion surveys and soil erosion models. PhD Thesis, Cranfield Institute of Technology.

Janssen W (1991), A new equation to determine saltation profiles for different friction velocities. In J Karácsony and G Szalai (eds), *Proceedings of the international wind erosion workshop of CIGR.* University of Agricultural Sciencies, Gödöllő, Hungary.

Janssen W, G Tetzlaff (1991), Construction and calibration of a registrating sediment trap. In J Karácsony and G Szalai (eds), *Proceedings of the international wind erosion workshop of CIGR.* University of Agricultural Sciencies, Gödöllő, Hungary.

Jeffers J N R (1986), The role of ecosystem theory in upland land use and management. In G Chisci and R P C Morgan (eds), *Soil erosion in the European Community: impact of changing agriculture.* Rotterdam, Balkema: 51–65.

Jensen J L, M Sørensen (1986), Estimation of some aeolian saltation transport parameters: a reanalysis of Williams' data. *Sedimentology* 33: 547–58.

Jiang D, L Qi, J Tan (1981), Soil erosion and conservation in the Wuding River Valley, China. In R P C Morgan (ed.), *Soil conservation: problems and prospects.* Chichester, Wiley: 461–79.

Jóhannesson B (1960), *The soils of Iceland.* University Research Institute, Department of Agriculture, Reports Series B, No. 13, Reykjavík.

Johnson C B, W C Moldenhauer (1979), Effect of chisel versus moldboard plowing on soil erosion by water. *Soil Science Society of America Journal* 43: 177–9.

Johnston A E (1982), The effects of farming systems on the amount of soil organic matter and its effect on yield at Rothamsted and Woburn. In D Boels, D B Davies, and A E Johnston (eds), *Soil degradation.* Rotterdam, Balkema: 187–202.

Jones M J (1971), The maintenance of soil organic matter under continuous cultivation at Samaru, Nigeria. *Journal of Agricultural Science Cambridge* 77: 473–82.

Jones R G B, M A Keech (1966), Identifying and assessing problem areas in soil erosion surveys using aerial photographs. *Photogrammetric Record* 5/27: 189–97.

Jovanović S, M Vukčević (1958), Suspended sediment regimen on some watercourses in Yugoslavia and analysis of erosion processes. *International Association of Scientific Hydrology Publication* 43: 337–59.

Józefaciuk C, A Józefaciuk (1993), Gullies net density as a factor of water system deformation in Vistula River basin. In K Banasik and A Żbikowski (eds), *Runoff and sediment yield modelling.* Warsaw, Warsaw Agricultural University Press: 169–74.

Jusoff K, N M Majid (1988), Effects of site preparation methods on soil physical characteristics. In S Rimwanich (ed.), *Land conservation for future generations.* Bangkok, Department of Land Development: 745–54.

Kairiukstis L, G Golubev (1982), Application of the CREAMS model as part of an overall system for optimizing environmental management in Lithuania, USSR: first experiments. In V Svetlosanov and W G Knisel (eds), *European and United States case studies in application of the CREAMS model.* International Institute of Applied Systems Analysis Collaborative Proceedings Series CP-82–S11: 99–119.

Kamalu C (1993), Soil erosion on road shoulders. PhD Thesis, Cranfield University.

Kampen J, J Hari Krishna, P Pathak (1981), Rainy season cropping on deep vertisols in the semiarid tropics: effects on hydrology and soil erosion. In R Lal and E W Russell (eds), *Tropical agricultural hydrology.* Chichester, Wiley: 257–71.

Keech M A (1969) Mondaro Tribal Trust Land. Determination of trend using air photo analysis. *Rhodesian Agricultural Journal* 66: 3–10.

Kellman M C (1969), Some environmental components of shifting cultivation in upland Mindanao. *Journal of Tropical Geography* 28: 40–56.

Kemper B, R Derpsch (1981), Results of studies made in 1978 and 1979 to control erosion by cover crops and no-tillage techniques in Paraná, Brazil. *Soil and Tillage Research* 1: 253–68.

Kermack K A, J B S Haldane (1950), Organic correlation and allometry. *Biometrika* 37: 30–41.

Khybri M L (1989), Mulch effects on soil and water loss in maize in India. In W C Moldenhauer, N W Hudson, T C Sheng and S W Lee (eds), *Development of conservation farming on hillslopes.* Ankeny IA, Soil and Water Conservation Society: 195–8.

Kinnell P I A (1974), Splash erosion: some observations on the splash-cup technique. *Soil Science Society of America Proceedings* 38: 657–60.

Kinnell P I A (1981), Rainfall intensity–kinetic energy relationships for soil loss prediction. *Soil Science Society of America Journal* 45: 153–5.

Kinnell P I A (1987), Rainfall energy in eastern Australia: intensity-kinetic energy relationships for Canberra, A.C.T. *Australian Journal of Soil Research* 25: 547–53.

Kinnell P I A (1990a) The mechanics of raindrop-induced flow transport. *Australian Journal of Soil Research* 28: 497–516.

Kinnell P I A (1990b), Modelling erosion by rain-impacted flow. *Catena Supplement* 17: 55–66.

Kinnell P I A (1991), The effect of flow depth on sediment transport induced by raindrops impacting shallow flows. *Transactions of the American Society of Agricultural Engineers* 34: 161–8.

Kirby P C, G R Mehuys (1987), Seasonal variation of soil erodibilities in southwestern Quebec. *Journal of Soil and Water Conservation* 42: 211–15.

Kirkby M J (1969a), Infiltration, throughflow and overland flow. In R J Chorley (ed.), *Water, earth and man*. London, Methuen: 215–27.

Kirkby M J (1969b), Erosion by water on hillslopes. In R J Chorley (ed.), *Water, earth and man*. London, Methuen: 229–38.

Kirkby M J (1971), Hillslope process-response models based on the continuity equation. In D Brunsden (ed.), *Slopes: form and process*. Institute of British Geographers Special Publication 3: 15–30.

Kirkby M J (1980a), The problem. In M J Kirkby and R P C Morgan (eds), *Soil erosion*. Chichester, Wiley: 1–16.

Kirkby M J (1980b), Modelling water erosion processes. In M J Kirkby and R P C Morgan (eds), *Soil erosion*. Chichester, Wiley: 183–216.

Kleene P, B Sanogo, G Vierstra (1989), *A partir de Fonsébougou. Présentation, objectifs et méthodologie du «Volet Fonsébougou» (1977–1987)*. IER Bamako/IRT Amsterdam.

Klingebiel A A, P H Montgomery (1966), Land capability classification. *USDA Soil Conservation Service Agricultural Handbook* 210.

Knisel W G (1980) CREAMS: a field scale model for chemicals, runoff and erosion from agricultural management systems. *USDA Conservation Research Report* 26.

Knisel W G, V Svetlosanov (1982), Review of case studies of CREAMS model application. In V Svetlosanov and W G Knisel (eds), *European and United States case studies in application of the CREAMS model*. International Institute for Applied Systems Analysis Collaborative Proceedings Series CP-82–S11: 121–35.

Knottnerus D J C (1976), Stabilisation of wind erodible land by an intermediate crop. *Mededelingen Fakulteit Landbouwwetenschappen, Rijksuniversiteit Gent* 41: 73–9.

Knottnerus D J C (1979), *Wind erosion research by means of a wind tunnel. Measures to control wind erosion of soil and other materials for reasons of economy and health*. Instituut voor Bodemvruchtbaarheid, Haren (Gr).

Kolenbrander G J (1974), Efficiency of organic manure in increasing soil organic matter content. *Transactions, 10th International Congress of Soil Science* 2: 129–36.

König D (1992), The potential of agroforestry methods for erosion control in Rwanda. *Soil Technology* 5: 167–76.

Konuche P K A (1983), Effects of forest management practices on soil and water conservation in Kenya forests. In D B Thomas and W M Senga (eds), *Soil and water conservation in Kenya*. Institute of Development Studies and Faculty of Agriculture, University of Nairobi, Occasional Paper No. 42: 350–9.

Kowal J M, A H Kassam (1976), Energy load and instantaneous intensity of rainstorms at Samaru, northern Nigeria. *Tropical Agriculture* 53: 185–97.

Kronvang B (1990), Sediment-associated phosphorus transport from two intensively farmed catchment areas. In J Boardman, I D L Foster and J A Dearing (eds), *Soil erosion on agricultural land*. Chichester, Wiley: 313–30.

Kumar A, J S Samra, N T Singh, L Batra (1990), Tree canopy–intercrop relationships in an agroforestry system. In *Proceedings, International symposium on water erosion, sedimentation and resource conservation*. Dehra Dun, Central Soil and Water Conservation Research and Training Institute: 446–52.

Kutiel P (1994), Fire and ecosystem heterogeneity: a Mediterranean case study. *Earth Surface Processes and Landforms* 19: 187–94.

Kwaad F J P M (1977), Measurements of rainsplash erosion and the formation of colluvium beneath deciduous woodland in the Luxembourg Ardennes. *Earth Surface Processes* 2: 161–73.

Kwaad F J P M (1991), Summer and winter regimes of runoff generation and soil erosion on cultivated loess soils (The Netherlands). *Earth Surface Processes and Landforms* 16: 653–62.

Kwaad F J P M (1993), Characteristics of runoff generating rains on bare loess soil in south Limbourg (The Netherlands). In S Wicherek (ed.), *Farm land erosion in temperate plains environment and hills*. Amsterdam, Elsevier: 71–86.

Kwaad F J P M, E J van Mulligen (1991), Cropping system effects of maize on infiltration, runoff and erosion on loess soils in south Limbourg, The Netherlands: a comparison of two rainfall events. *Soil Technology* 4: 281–95.

Laflen J M, T S Colvin (1981), Effect of crop residue on soil loss from continuous row cropping. *Transactions of the American Society of Agricultural Engineers* 24: 605–9.

Laflen J M, W C Moldenhauer (1979), Soil and

water losses from corn–soybean rotations. *Soil Science Society of America Journal* **43**: 1213–15.

Laflen J M, H P Johnson, R C Reeve (1972), Soil loss from tile-outlet terraces. *Journal of Soil and Water Conservation* **27**: 74–7.

Laflen J M, W J Elliot, J R Simanton, C S Holzhey, K D Kohl (1991), Soil erodibility experiments for rangeland and cropland soils. *Journal of Soil and Water Conservation* **46**: 39–44.

Laing D (1992), From the International Center for Tropical Agriculture (CIAT), Cali, Colombia. *Vetiver Newsletter* **8**: 13–15.

Laing I A F (1978), Soil surface treatment for runoff inducement. In W W Emerson, R D Bond and A R Dexter (eds), *Modification of soil structure*. Chichester, Wiley: 249–56.

Lakew D T (1991), Use of contour grass strips for soil erosion control. MSc Thesis, Silsoe College, Cranfield Institute of Technology.

Lakshmipathy B M, S Narayanswamy (1956), Bench terracing in the Nilgiris. *Journal of Soil and Water Conservation in India* **4**: 161–8.

Lal R (1976), *Soil erosion problems on an alfisol in western Nigeria and their control*. IITA Monograph No. 1.

Lal R (1977a), Soil-conserving versus soil-degrading crops and soil management for erosion control. In D J Greenland and R Lal (eds), *Soil conservation and management in the humid tropics*. London, Wiley: 81–6.

Lal R (1977b), Soil management systems and erosion control. In D J Greenland and R Lal (eds), *Soil conservation and management in the humid tropics*. Chichester, Wiley: 93–7.

Lal R (1981), Deforestation of tropical rainforest and hydrological problems. In R Lal and E W Russell (eds), *Tropical agricultural hydrology*. Chichester, Wiley: 131–40.

Lal R (1987), Research achievements towards soil and water conservation in the tropics: potential and priorities. In I Pla Sentis (ed.), *Soil conservation and productivity*. Maracay, Sociedad Venezolana de la Ciencia del Suelo: 755–87.

Lal R (1988), Soil erosion control with alley cropping. In S Rimwanich (ed.), *Land conservation for future generations*. Bangkok, Department of Land Development: 237–45.

Lal R, D J Cummings (1979), Clearing a tropical forest. I. Effects on soil and microclimate. *Field Crops Research* **2**: 91–107.

Landsberg J J, G B James (1971), Wind profiles in plant canopies: studies on an analytical model. *J. Applied Ecology* **8**: 729–41.

Lang R D, L A H McCaffrey (1984), Ground cover: its effects on soil loss from grazed runoff plots,

Gunnedah. *Journal of the Soil Conservation Service NSW* **40**: 56–61.

Langbein W B, S A Schumm (1958), Yield of sediment in relation to mean annual precipitation. *Transactions of the American Geophysical Union* **39**: 1076–84.

Langdale G W, W C Mills, A W Thomas (1992), Conservation tillage development for soil erosion control in the southern Piedmont. In H Hurni and Kebede Tato (eds), *Erosion, conservation and small-scale farming*. Bern, Geographica Bernensia: 453–8.

Langdale G W, A W Thomas, W L Hargrove (1987), Multiple crop conservation tillage systems to control soil erosion on southern Piedmont lands. In I Pla Sentis (ed.), *Soil conservation and productivity*. Maracay, Sociedad Venezolana de la Ciencia del Suelo: 1123–32.

Langdale G W, A W Thomas, W C Mills (1988), Effect of conservation tillage systems on seasonal soil loss probabilities. In S Rimwanich (ed.), *Land conservation for future generations*. Bangkok, Department of Land Development: 841–9.

Larson W E, F J Pierce, R H Dowdy (1985), Loss in long-term productivity from soil erosion in the United States. In S A El-Swaify, W C Moldenhauer and A Lo (eds), *Soil erosion and conservation*. Ankeny IA, Soil Conservation Society of America: 262–71.

Laurant A, A Bollinne (1978), Caractérisation des pluies en Belgique du point de vue de leur intensité et de leur erosivité. *Pédologie* **28**: 214–32.

Laws J O, D A Parsons (1943), The relationship of raindrop size to intensity. *Transactions of the American Geophysical Union* **24**: 452–60.

Leaf C F (1970), Sediment yields from central Colorado snow zone. *Journal of the Hydraulics Division ASCE* **96**: 87–93.

Le Bissonnais Y (1990), Experimental study and modelling of soil surface crusting processes. *Catena Supplement* **17**: 13–28.

Leigh C (1982a), Sediment transport by surface wash and throughflow at the Pasoh Forest Reserve, Negri Sembilan, Peninsular Malaysia. *Geografiska Annaler* **64-A**: 171–80.

Leigh C H (1982b), Urban development and soil erosion in Kuala Lumpur, Malaysia. *Journal of Environmental Management* **15**: 35–45.

Lenvain J S, W K Sakala, P L L Pauwelyn (1988), Iso-erodent map of Zambia. Part II: Erosivity prediction and mapping. *Soil Technology* **1**: 251–62.

Leopold L B, M G Wolman, J P Miller (1964), *Fluvial processes in geomorphology*. San Francisco, Freeman.

Li J H (1984), Benefits drawn from network of shelter

belts in farmland of Pengwa Production Brigade, Youyu County. *Soil and Water Conservation in China* 22: 24–5 (in Chinese with English summary).

Liddle M J (1973), The effects of trampling and vehicles on natural vegetation. PhD Thesis, University College of North Wales, Bangor.

Lim K H (1988), A study on soil erosion control under mature oil palm in Malaysia. In S Rimwanich (ed.), *Land conservation for future generations*. Bangkok, Department of Land Development: 783–95.

Lindsay J I, F A Gumbs (1982), Erodibility indices compared to measured values of selected Trinidad soils. *Soil Science Society of America Journal* 46: 393–5.

Loch R J, E C Thomas, T E Donnollan (1987), Interflow in a tilled, cracking clay soil under simulated rain. *Soil and Tillage Research* 9: 45–63.

Logie M (1982), Influence of roughness elements and soil moisture on the resistance of sand to wind erosion. *Catena Supplement* 1: 161–73.

López-Bermúdez F, M A Romero-Díaz (1989), Piping erosion and badland development in southeast Spain. *Catena Supplement* 14: 59–73.

Loughran R J, B L Campbell, G L Elliott (1988), Determination of erosion and accretion rates using caesium-137. In P F Warner (ed.), *Fluvial geomorphology of Australia*. Academic Press Australia: 87–103.

Lovejoy S B, T L Napier (1986), *Conserving soil: insights from socioeconomic research*. Ankeny IA, Soil Conservation Society of America.

Low F K (1967), Estimating potential erosion in developing countries. *Journal of Soil and Water Conservation* 22: 147–8.

Lugo-Lopez M A (1969), Prediction of the erosiveness of Puerto Rican soils on a basis of the percentage of particles of silt and clay when aggregated. *Journal of Agriculture, University of Puerto Rico* 53: 187–90.

Luk S H (1981), Variability of rainwash erosion within small sample areas. In *Proceedings, Twelfth Binghampton Geomorphology Symposium*. London, Allen and Unwin: 243–68.

Luk S H, C Morgan (1981), Spatial variations of rainwash and runoff within apparently homogeneous areas. *Catena* 8: 383–402.

Lundgren B, P K R Nair (1985), Agroforestry for soil conservation. In S A El-Swaify, W C Moldenhauer and A Lo (eds), *Soil erosion and conservation*. Ankeny IA, Soil Conservation Society of America: 703–11.

Lundgren L (1978), Studies of soil and vegetation development on fresh landslide scars in the Mgeta Valley, western Uluguru Mountains, Tanzania. *Geografiska Annaler* 60-A: 91–127.

Lundgren L, A Rapp (1974), A complex landslide with destructive effects on the water supply of Morogoro town, Tanzania. *Geografiska Annaler* 56-A: 251–60.

Lyles L, B E Allison (1980), Range grasses and their small grain equivalents for wind erosion control. *Journal of Range Management* 33: 143–6.

Lyles L, B E Allison (1981), Equivalent wind erosion protection from selected crop residues. *Transactions of the American Society of Agricultural Engineers* 24: 405–9.

Lyles L, J D Dickerson, M F Schmeidler (1974), Soil detachment from clods by rainfall: effects of wind, mulch cover and initial soil moisture. *Transactions of the American Society of Agricultural Engineers* 17: 697–700.

Lyles L, R L Schrandt, N F Schmeidler (1974), How aerodynamic roughness elements control sand movement. *Transactions of the American Society of Agricultural Engineers* 17: 134–9.

McCauley D S (1988), Overcoming institutional and organizational constraints to watershed management for the densely populated island of Java. In S Rimwanich (ed.), *Land conservation for future generations*. Bangkok, Department of Land Development: 1039–60.

McCormack D E, K K Young (1981), Technical and societal implications of soil loss tolerance. In R P C Morgan (ed.), *Soil conservation: problems and prospects*. Chichester, Wiley: 365–76.

McCormack R J (1971), The Canada Land Use Inventory: a basis for land use planning. *Journal of Soil and Water Conservation* 26: 141–6.

McGregor J C, C K Mutchler (1978), The effect of crop canopy on raindrop size distribution. *USDA Sedimentation Laboratory Annual Report*, Oxford MS.

McIsaac G F (1990), Apparent geographic and atmospheric influences on raindrop sizes and rainfall kinetic energy. *Journal of Soil and Water Conservation* 45: 663–6.

Maene L M, K C Thong, T S Ong, A M Mokhtaruddin (1979), Surface wash under mature oil palm. In E Pushparajah (ed.), *Proceedings, Symposium on Water in Malaysian agriculture*. Kuala Lumpur, Malaysian Society of Soil Science: 203–16.

Marshall C J, A R de Fegely (1987), Erosion control for tree plantation establishment. *Journal of Soil Conservation NSW* 43: 32–5.

Marshall J K (1967), The effect of shelter on the productivity of grasslands and field crops. *Field Crops Abstracts* 20: 1–14.

Marshall J S, W M Palmer (1948), Relation of rain

drop size to intensity. *Journal of Meteorology* 5: 165–6.

Marston D, C Hird (1978), Effect of stubble management on the structure of black cracking clays. In W W Emerson, R D Bond and A R Dexter (eds), *Modification of soil structure*. Chichester, Wiley: 411–17.

Marston D, S J Perrens (1981), Effect of stubble on erosion of a black earth soil. In T Tingsanchali and H Eggers (eds), *Southeast Asian regional symposium on problems of soil erosion and sedimentation*. Bangkok, Asian Institute of Technology: 289–300.

Martin L, R P C Morgan (1980), Soil erosion in mid-Bedfordshire. In J C Doornkamp, K J Gregory and A S Burn (eds), *Atlas of drought in Britain, 1975–76*. London, Institute of British Geographers: 47.

Mason B J, J B Andrews (1960), Drop size distributions from various types of rain. *Quarterly Journal of the Royal Meteorological Society* 86: 346–53.

Mason B J, R Ramandham (1953), A photoelectric spectrometer. *Quarterly Journal of the Royal Meteorological Society* 79: 490–5.

Mazurak A P, P N Mosher (1968), Detachment of soil particles in simulated rainfall. *Soil Science Society of America Proceedings* 32: 716–19.

Mbegera M, A Eriksson, S N J Njoroge (1992), Soil and water conservation training and extension: the Kenyan experience. In Kebede Tato and H Hurni (eds), *Soil conservation for survival*. Ankeny IA, Soil and Water Conservation Society: 284–95.

Mead R, R N Curnow (1983), *Statistical methods in agriculture and experimental biology*. London, Chapman and Hall.

Megahan W F, D C Molitor (1975), Erosional effects of wildfire and logging in Idaho. *Proceedings, Symposium on watershed management*. St Giles MI, American Society of Civil Engineers: 423–44.

Mein R G, C L Larson (1973), Modeling infiltration during a steady rain. *Water Resources Research* 9: 384–94.

Melville N (1992), The influence of grass density on the effectiveness of contour grass strips. MSc Thesis, Silsoe College, Cranfield Institute of Technology.

Merritt E (1984), The identification of four stages during microrill development. *Earth Surface Processes and Landforms* 9: 493–6.

Merzouk A, G R Blake (1991), Indices for the estimation of interrill erodibility of Moroccan soils. *Catena* 18: 537–50.

Meyer L D (1965), Mathematical relationships governing soil erosion by water. *Journal of Soil and Water Conservation* 20: 149–50.

Meyer L D (1979), Methods for attaining desired rainfall characteristics in rainfall simulators. In *Proceedings, Rainfall simulator workshop, Tucson, Arizona*. USDA-SEA Agricultural Reviews and Manuals ARM-W-10: 35–44.

Meyer L D (1981), How rain intensity affects interrill erosion. *Transactions of the American Society of Agricultural Engineers* 24: 1472–5.

Meyer L D, W C Harmon (1979), Multiple-intensity rainfall simulator for erosion research on row sideslopes. *Transactions of the American Society of Agricultural Engineers* 22: 100–3.

Meyer L D, E J Monke (1965), Mechanics of soil erosion by rainfall and overland flow. *Transactions of the American Society of Agricultural Engineers* 8: 572–7.

Meyer L D, W H Wischmeier (1969), Mathematical simulation of the process of soil erosion by water. *Transactions of the American Society of Agricultural Engineers* 12: 754–8, 762.

Meyer L D, G R Foster, S Nikolov (1975), Effect of flow rate and canopy on rill erosion. *Transactions of the American Society of Agricultural Engineers* 18: 905–11.

Meyer L D, G R Foster, M J M Römkens (1975), Source of soil eroded by water from upland slopes. In *Present and perspective technology for predicting sediment yields and sources*. USDA-ARS Publication ARS-AS-40: 177–89.

Middleton H E (1930), Properties of soils which influence soil erosion. *USDA Technical Bulletin* 178.

Midriak R (1965), Poškodenie pôdy eróziou pri prietrži mračien v oblasti Kendic pri Prešove. *Polnohospodárstvo* 9: 696–707.

Mihara Y (1951), Raindrops and soil erosion. *Bulletin of the National Institute of Agricultural Science Series A*, No. 1 (in Japanese).

Miller W P, M E Sumner (1988), Dispersion processes affecting runoff and erosion in highly weathered soils. In S Rimwanich (ed.), *Land conservation for future generations*. Bangkok, Department of Land Development: 419–27.

Millington A (1987), Local farmer perceptions of soil erosion hazards and indigenous soil conservation strategies in Sierra Leone, West Africa. In I Pla Sentis (ed.), *Soil conservation and productivity*. Maracay, Sociedad Venezolana de la Ciencia del Suelo: 675–90.

Milner C, M G Douglas (1989), *Problems of land degradation in Commonwealth Africa*. London, Commonwealth Secretariat.

Mitchell J K, S Mostaghimi, D S Freeny, J R McHenry (1983), Sediment deposition estimation from caesium-137 measurements. *Water Resources Bulletin* 19: 549–55.

Moeyersons J (1983), Measurements of splash-saltation fluxes under oblique rain. *Catena Supplement* 4: 19–31.

Moeyersons J, J De Ploey (1976), Quantitative data on splash erosion simulated on unvegetated slopes. *Zeitschrift für Geomorphologie Supplementband* 25: 120–31.

Mohammed A, F A Gumbs (1982), The effect of plant spacing on water runoff, soil erosion and yield of maize (*Zea mays* L.) on a steep slope of an ultisol in Trinidad. *Journal of Agricultural Engineering Research* 27: 481–8.

Mokhtaruddin A M, L M Maene (1979), Soil erosion under different crops and management practices. *Proceedings, International conference on agricultural engineering in national development*. Universiti Pertanian, Malaysia, Paper No. 79–53.

Moldenhauer W C (1965), Procedure for studying soil characteristics using disturbed samples and simulated rainfall. *Transactions of the American Society of Agricultural Engineers* 8: 74–5.

Moldenhauer W C, C A Onstad (1975), Achieving specified soil loss levels. *Journal of Soil and Water Conservation* 30: 166–8.

Monnier G, J Boiffin (1986), Effect of the agricultural use of soils on water erosion: the case of cropping systems in western Europe. In G Chisci and R P C Morgan (eds), *Soil erosion in the European Community: impact of changing agriculture*. Rotterdam, Balkema: 17–32.

Morgan R P C (1974), Estimating regional variations in soil erosion hazard in Peninsular Malaysia. *Malayan Nature Journal* 28: 94–106.

Morgan R P C (1976), The role of climate in the denudation system: a case study from West Malaysia. In E Derbyshire (ed.), *Climate and geomorphology*. London, Wiley: 317–43.

Morgan R P C (1980a), Field studies of sediment transport by overland flow. *Earth Surface Processes* 3: 307–16.

Morgan R P C (1980b), Soil erosion and conservation in Britain. *Progress in Physical Geography* 4: 24–47.

Morgan R P C (1980c), Preliminary testing of the CREAMS erosion sub-model with field data from Silsoe, Bedfordshire, England. *International Institute for Applied Systems Analysis Collaborative Paper* CP-80–21.

Morgan R P C (1981), Field measurement of splash erosion. *International Association of Scientific Hydrology Publication* 133: 373–82.

Morgan R P C (1985a), Assessment of soil erosion risk in England and Wales. *Soil Use and Management* 1: 127–31.

Morgan R P C (1985b), Effect of corn and soybean canopy on soil detachment by rainfall. *Transactions of the American Society of Agricultural Engineers* 28: 1135–40.

Morgan R P C (1989), Design of in-field shelter systems for wind erosion control. In U Schwertmann, R J Rickson and K Auerswald (eds), *Soil erosion protection measures in Europe*. Soil Technology Series 1: 15–23.

Morgan R P C (1994), The European Soil Erosion Model: an up-date on its structure and research base. In R J Rickson (ed.), *Conserving soil resources: European perspectives*. Wallingford, CAB International: 286–99.

Morgan R P C 1993, Soil erosion assessment. In *Proceedings, Workshop on Soil erosion in semi-arid Mediterranean areas, Taormina, Italy, Consiglio Nazionale delle Ricerche, Roma*, 3-17.

Morgan R P C, H J Finney (1987), Drag coefficients of single crop rows and their implications for wind erosion control. In V Gardiner (ed.), *International geomorphology 1986, Part II*. Chichester, Wiley: 449–58.

Morgan R P C, R J Rickson (1988), Soil erosion control: importance of geomorphological information. In J M Hooke (ed.), *Geomorphology in environmental planning*. Chichester, Wiley: 51–60.

Morgan R P C, R J Rickson (1995), *Slope stabilization and erosion control: a bioengineering approach*. London, E and F N Spon.

Morgan R P C, L Martin, C A Noble (1986), *Soil erosion in the United Kingdom: a case study from mid-Bedfordshire*. Silsoe College Occasional Paper No. 14.

Morgan R P C, D D V Morgan, H J Finney (1982), Stability of agricultural ecosystems: documentation of a simple model for soil erosion assessment. *International Institute for Applied Systems Analysis* Collaborative Paper CP-82-50.

Morgan R P C, D D V Morgan, H J Finney (1984), A predictive model for the assessment of soil erosion risk. *Journal of Agricultural Engineering Research* 30: 245–53.

Morgan R P C, D D V Morgan, H J Finney (1987), Predicting hillslope runoff and erosion in the Silsoe area of Bedfordshire, England, using the CREAMS model. In I Pla Sentis (ed.), *Soil conservation and productivity*. Maracay, Sociedad Venezolana de la Ciencia del Suelo: 892–9.

Morgan R P C, J N Quinton, R J Rickson (1994), Modelling methodology for soil erosion assessment and soil conservation design: the EUROSEM approach. *Outlook in Agriculture* 23: 5–9.

Morgan R P C, R J Rickson, K McIntyre, T R Brewer (1994), Soil erosion survey of the central part of the Middle Veld, Swaziland. In H M Mushala, T Scholten, P Felix-Henningsen, R P C Morgan and R J Rickson (eds), *Soil erosion and river sedimentation in Swaziland*. Final Report, University of Swaziland: 28–42.

Morgan R P C, K McIntyre, A W Vickers, J N Quinton, R J Rickson (1994), A rainfall simulation study of soil erosion on rangeland in Swaziland. In H M Mushala, T Scholten, P Felix-Henningsen, R P C Morgan and R J Rickson (eds), *Soil erosion and river sedimentation in Swaziland*. Final Report, University of Swaziland: 92–99.

Morin J, Y Benyamini, A Michaeli (1981), The effect of raindrop impact on the dynamics of soil surface crusting and water movement in the profile. *Journal of Hydrology* 52: 321–6.

Morin J, D Goldberg, I Seginer (1967), A rainfall simulator with a rotating disk. *Transactions of the American Society of Agricultural Engineers* 10: 74–7, 79.

Morin J, E Rawitz, Y Benyamini, W B Hoogmoed, H Etkin (1984), Tillage practices for soil and water conservation in the semi-arid zone. II. Development of the basin tillage system in wheat fields. *Soil and Tillage Research* 4: 155–64.

Mosley M P (1973), Rainsplash and the convexity of badland divides. *Zeitschrift für Geomorphologie Supplementband* 18: 10–25.

Mosley M P (1982), The effect of a New Zealand beech forest canopy on the kinetic energy of water drops and on surface erosion. *Earth Surface Processes and Landforms* 7: 103–7.

Moss A J (1988), Effects of flow velocity on rain-driven transportation and the role of rain impact in water erosion. *Australian Journal of Soil Research* 26: 443–50.

Moss A J, P H Walker (1978), Particle transport by continental water flows in relation to erosion, deposition, soils and human activities. *Sedimentary Geology* 20: 81–139.

Moss A J, P Green, J Hutka (1982), Small channels: their formation, nature and significance. *Earth Surface Processes and Landforms* 7: 401–15.

Mueller D H, R M Klemme, T C Daniel (1985), Short- and long-term cost comparisons of conventional and conservation tillage systems in corn production. *Journal of Soil and Water Conservation* 40: 466–70.

Mulligan K R (1988), Velocity profiles measured on the windward slope of a transverse dune. *Earth Surface Processes and Landforms* 13: 573–82.

Murgatroyd A L, J L Ternan (1983), The impact of afforestation on stream bank erosion and channel form. *Earth Surface Processes and Landforms* 8: 357–69.

Musgrave G W (1947), The quantitative evaluation of factors in water erosion: a first approximation. *Journal of Soil and Water Conservation* 2: 133–8.

Mutchler C K, R A Young (1975), Soil detachment by raindrops. In *Present and prospective technology for predicting sediment yields and sources*. USDA-ARS Publication ARS-S-40: 113–17.

Mututho J M N (1989), Some aspects of soil conservation on grazing lands. In D B Thomas, E K Biamah, A M Kilewe, L Lundgren and B O Mochoge (eds), *Soil and water conservation in Kenya*. Department of Agricultural Engineering, University of Nairobi / SIDA: 315–22.

Mwichabe S (1992), The effect of range management on vegetation cover in the Narok area, Kenya. In H Hurni and Kebede Tato (eds), *Erosion, conservation and small-scale farming*. Bern, Geographica Bernensia: 407–13.

Napier T L (1988), Socio-economic factors influencing the adoption of soil erosion control practices in the United States. In R P C Morgan and R J Rickson (eds), *Erosion assessment and modelling*. Commission of the European Communities Report EUR 10860 EN: 299–327.

Napier T L (1990), The evolution of US soil-conservation policy: from voluntary adoption to coercion. In J Boardman, I D L Foster and J A Dearing (eds), *Soil erosion on agricultural land*. Chichester, Wiley: 627–44.

Nassif S H, E M Wilson (1975), The influence of slope and rain intensity on runoff and infiltration. *Hydrological Sciences Bulletin* 20: 539–53.

National Research Council (1993), *Vetiver grass: a thin green line against erosion*. Washington DC, National Academy Press.

Naveh Z (1975), The evolutionary sequence of fire in the Mediterranean region. *Vegetation* 9: 199–206.

Nearing M A, G R Foster, L J Lane, S C Finkner (1989), A process-based soil erosion model for USDA-Water Erosion Prediction Project technology. *Transactions of the American Society of Agricultural Engineers* 32: 1587–93.

Neff E L (1973), Water storage capacity of contour furrows in Montana. *Journal of Range Management* 26: 298–301.

Newbould P (1982), Losses and accumulation of organic matter in soils. In D Boels, D B Davies and A E Johnston (eds), *Soil degradation*. Rotterdam, Balkema: 107–31.

Newson M (1980), The erosion of drainage ditches and its effect on bed-load yields in mid-Wales: reconnaissance case studies. *Earth Surface Processes* 3: 275–90.

Nicks A D (1985), Generation of climate data. In D G DeCoursey (ed.), *Proceedings of the Natural* *resources modeling symposium.* USDA-ARS Publication ARS-S-30.

Noble C A, R P C Morgan (1983), Rainfall interception and splash detachment with a Brussels sprouts plant: a laboratory simulation. *Earth Surface Processes and Landforms* **8**: 569–77.

Norton L D, N P Cogo, W C Moldenhauer (1985), Effectiveness of mulch in controlling erosion. In S A El-Swaify, W C Moldenhauer and A Lo (eds), *Soil erosion and conservation.* Ankeny IA, Soil Conservation Society of America: 598–606.

Nossin J J (1964), Geomorphology of the surroundings of Kuantan (Eastern Malaya). *Geologie en Mijnbouw* **43**: 157–82.

Nsibandze S M (1987), The history of soil conservation in Swaziland. Paper presented at SADCC Seminar, Maputo.

Odemerho F O (1986), Variation in erosion-slope relationship on cut slopes along a tropical highway. *Singapore Journal of Tropical Geography* **7**: 98–107.

Odemerho F O, A Avwunudiogba (1993), The effects of changing cassava management practices on soil loss: a Nigerian example. *Geographical Journal* **159**: 63–9.

Okigbo B N (1977), Farming systems and soil erosion in West Africa. In D J Greenland and R Lal (eds), *Soil conservation and management in the humid tropics.* London, Wiley: 151–63.

Okigbo B N, R Lal (1977), Role of cover crops in soil and water conservation. *FAO Soils Bulletin* **33**: 97–108.

Olesen F (1979), *Læplantning.* Viborg, Det Danske Hedeselskab.

O'Loughlin C L (1974), A study of tree root strength deterioration following clearfelling. *Canadian Journal of Forest Research* **4**: 107–13.

O'Loughlin C L, A Watson (1979), Root-wood strength deterioration in radiata pine after clearfelling. *New Zealand Journal of Forestry Science* **9**: 284–93.

Onaga K, K Shirai, A Yoshinaga (1988), Rainfall erosion and how to control its effects on farmland in Okinawa. In S Rimwanich (ed.), *Land conservation for future generations.* Bangkok, Department of Land Development: 627–39.

Osuji G E (1989), Raindrop characteristics in the humid tropics. *Journal of Environmental Management* **28**: 227–33.

Osuji G E, O Babalola, F O Aboaba (1980), Rainfall erosivity and tillage practices affecting soil and water loss on a tropical soil in Nigeria. *Journal of Environmental Management* **10**: 207–17.

Othieno C O (1978), An assessment of soil erosion on a field of young tea under different soil management practices. In *Soil and water conservation in Kenya.* Institute of Development Studies University of Nairobi, Occasional Paper No. 27: 62–73.

Padgitt M (1989), Soil diversity and the effects of field eligibility rules in implementing soil conservation programs targeted to highly erodible land. *Journal of Soil and Water Conservation* **44**: 91–5.

Palmer R S (1964), The influence of a thin water layer on water-drop impact forces. *International Association of Scientific Hydrology Publication* **65**: 141–8.

Parsons A J, A D Abrahams, S H Luk (1990), Hydraulics of interrill overland flow on a semi-arid hillslope, southern Arizona. *Journal of Hydrology* **117**: 255–73.

Pathak P, S M Miranda, S A El-Swaify (1985), Improved rainfed farming for semiarid tropics: implications for soil and water conservation. In S A El-Swaify, W C Moldenhauer and A Lo (eds), *Soil erosion and conservation.* Ankeny IA, Soil Conservation Society of America: 338–54.

Pathak P, S Singh, R Sudi (1987), Soil and water management alternatives for increased productivity on semiarid tropical alfisols. In I Pla Sentis (ed.), *Soil conservation and productivity.* Maracay, Sociedad Venezolana de la Ciencia del Suelo: 533–50.

Pava H M, J B Arances, J M Magallanes, I O Mugot, J M Manubag, I S Sealza (1990), Participatory processes to upland development: the MUSUAN model. *MUSUAN Program Monograph Series No. 2,* Bukidnon, Central Mindanao University.

Pearce A J (1976), Magnitude and frequency of erosion by Hortonian overland flow. *Journal of Geology* **84**: 65–80.

Peden D G (1987), Livestock and wildlife population distributions in relation to aridity and human population in Kenya. *Journal of Range Management* **40**: 67–71.

Perrens S J, N A Trustrum (1984), *Assessment and evaluation for soil conservation policy.* East-West Environment and Policy Institute, Workshop Report, Honolulu, HI.

Petryk S, G Bosmajian (1975), Analysis of flow through vegetation. *Journal of the Hydraulics Division ASCE* **101**: 871–84.

Philip J R (1957), The theory of infiltration. I. The infiltration equation and its solution. *Soil Science* **83**: 345–57.

Pidgeon J D, B D Soane (1978), Soil structure and strength relations following tillage, zero tillage and wheel traffic in Scotland. In W W Emerson, R D Bond and A R Dexter (eds), *Modification of soil structure.* Chichester, Wiley: 371–8.

Pihan J (1979), Risques climatiques d'érosion hydrique des sols en France. In H Vogt and Th Vogt (eds), *Colloque sur l'érosion agricole des sols en milieu tempéré non Mediterranéen*. Strasbourg, l'Université Louis Pasteur: 13–18.

Pike A H (1938), Soil conservation among the Matengo tribe. *Tanganyika Notes and Records* **6**: 79–81.

Pilgrim D H, D D Huff (1983), Suspended sediment in rapid subsurface stormflow on a large field plot. *Earth Surface Processes and Landforms* **8**: 451–63.

Pinczés Z (1980), A művelési ágak és módok hatása a talajerózióa. *Földrajzi Közlemények* **28**: 357–79.

Pla I (1977), Aggregate size and erosion control on sloping land treated with hydrophobic bitumen emulsion. In D J Greenland and R Lal (eds), *Soil conservation and management in the humid tropics*. London, Wiley: 109–15.

Poesen J (1981), Rainwash experiments on the erodibility of loose sediments. *Earth Surface Processes and Landforms* **6**: 285–307.

Poesen J (1984), The influence of slope angle on infiltration rate and Hortonian overland flow volume. *Zeitschrift für Geomorphologie Supplementband* **49**: 117–31.

Poesen J (1985), An improved splash transport model. *Zeitschrift für Geomorphologie* **29**: 193–211.

Poesen J (1987), Transport of rock fragments by rill flow: a field study. *Catena Supplement* **8**: 35–54.

Poesen J (1989), Conditions for gully formation in the Belgian loam belt and some ways to control them. In U Schwertmann, R J Rickson and K Auerswald (eds), *Soil erosion protection measures in Europe*. Soil Technology Series No. 1: 39–52.

Poesen J W A (1992), Mechanisms of overland-flow generation and sediment production on loamy and sandy soils with and without rock fragments. In A J Parsons and A D Abrahams (eds), *Overland flow: hydraulics and erosion mechanics*. London, UCL Press: 275–305.

Poesen J, F Ingelmo-Sanchez (1992), Runoff and sediment yield from topsoils with different porosity as affected by rock fragment cover and position. *Catena* **19**: 451–74.

Poesen J, D Torri (1988), The effect of cup size on splash detachment and transport measurements. Part I: Field measurements. *Catena Supplement* **12**: 113–26.

Pope A, J J Harper (1966), *Low-speed wind tunnel testing*. London, Wiley.

Porta J, R M Poch, J Boixadera (1989), Land evaluation and erosion control practices on mined soils in N.E. Spain. In U Schwertmann, R J Rickson and K Auerswald (eds), *Soil erosion protection measures in Europe*. Soil Technology Series 1: 189–206.

Pratt D T, M D Gwynne (1977), *Rangeland manage-*

ment and ecology in East Africa. London, Hodder and Stoughton.

Price-Williams D, A Watson, A S Goudie (1982), Quaternary colluvial stratigraphy, archaeological sequences and paleoenvironments in Swaziland, southern Africa. *Geographical Journal* **148**: 50–67.

Prinz D (1992), Towards an optimal utilisation of the soil and water resources in a semi-humid environment: an example from Thailand. In Kebede Tato and H Hurni (eds), *Soil conservation for survival*. Ankeny IA, Soil and Water Conservation Society: 321–31.

Proffitt A, C W Rose (1992), Relative contributions to soil loss by rainfall detachment and runoff entrainment. In H Hurni and Kebede Tato (eds), *Erosion, conservation and small-scale farming*. Bern, Geographica Bernensia: 75–89.

Proffitt A P B, C W Rose, P B Hairsine (1991), Detachment and deposition: experiments with low slopes and significant water depths. *Soil Science Society of America Journal* **55**: 325–32.

Quansah C (1981), The effect of soil type, slope, rain intensity and their interactions on splash detachment and transport. *Journal of Soil Science* **32**: 215–24.

Quansah C (1982), Laboratory experimentation for the statistical derivation of equations for soil erosion modelling and soil conservation design. PhD Thesis, Cranfield Institute of Technology.

Quansah C (1985), Rate of soil detachment by overland flow, with and without rain, and its relationship with discharge, slope steepness and soil type. In S A El-Swaify, W C Moldenhauer and A Lo (eds), *Soil erosion and conservation*. Ankeny IA, Soil Conservation Society of America: 406–23.

Quansah C, E Baffoe-Bonnie (1981), The effect of soil management systems on soil loss, runoff and fertility erosion in Ghana. In T Tingsanchali and H Eggers (eds), *Southeast Asian regional symposium on problems of soil erosion and sedimentation*. Bangkok, Asian Institute of Technology: 207–17.

Quinn N W, R P C Morgan, A J Smith (1980), Simulation of soil erosion induced by human trampling. *Journal of Environmental Management* **10**: 155–65.

Ramaswamy S D, M A Aziz, N Narayanan (1981), Some methods of control of erosion on natural slopes in urbanized areas. In T Tingsanchali and H Eggers (eds), *Southeast Asian regional symposium on Problems of soil erosion and sedimentation*. Bangkok, Asian Institute of Technology: 327–39.

Randall J M (1969), Wind profiles in an orchard plantation. *Agricultural Meteorology* **6**: 439–52.

Rands B C (1992), Experiences in soil and water

conservation work amongst pastoral peoples of northeastern Mali. In Kebede Tato and H Hurni (eds), *Soil conservation for survival*. Ankeny IA, Soil and Water Conservation Society: 345–55.

Rapp A, V Axelsson, L Berry, D H Murray-Rust (1972a), Soil erosion and sediment transport in the Morogoro River catchment. *Geografiska Annaler* **54-A**: 125–55.

Rapp A, D H Murray-Rust, C Christiansson, L Berry (1972b), Soil erosion and sedimentation in four catchments near Dodoma, Tanzania. *Geografiska Annaler* **54-A**: 255–318.

Rasmussen K R, M Sørensen, B B Willetts (1985), Measurement of saltation and wind strength on beaches. In O E Barndorff-Nielsen, J T Møller, K R Rasmussen and B B Willetts (eds), *Proceedings of international workshop on the physics of blown sand*. Memoirs No. 8, Department of Theoretical Statistics, University of Aarhus, 301–25.

Rasmussen W D (1982), History of soil conservation, institutions and incentives. In H G Halcrow, E O Heady and M L Cotner (eds), *Soil conservation policies, institutions and incentives*. Ankeny IA, Soil Conservation Society of America: 3–18.

Rauws G (1987), The initiation of rills on plane beds of non-cohesive sediments. *Catena Supplement* **8**: 107–18.

Rauws G, G Govers (1988), Hydraulic and soil mechanical aspects of rill generation on agricultural soils. *Journal of Soil Science* **39**: 111–24.

Ree W O (1949), Hydraulic characteristics of vegetation for vegetated waterways. *Agricultural Engineering* **30**: 184–7, 189.

Reid I (1979), Seasonal changes in microtopography and surface depression storage of arable soils. In G E Hollis (ed.), *Man's impact on the hydrological cycle in the United Kingdom*. Norwich, Geo Books: 19–30.

Reid L M, T Dunne, C J Cederholm (1981), Application of sediment budget studies to the evaluation of logging road impact. *New Zealand Journal of Hydrology* **20**: 49–62.

Renard K G, G R Foster, G A Weesies, J P Porter (1991), RUSLE: Revised universal soil loss equation. *Journal of Soil and Water Conservation* **46**: 30–3.

Rice R M, J S Krammes (1970), Mass-wasting processes in watershed management. *Proceedings, Symposium on Interdisciplinary aspects of watershed management*. St Giles MI, American Society of Civil Engineers: 231–60.

Richter G, J F W Negendank (1977), Soil erosion processes and their measurement in the German area of the Moselle river. *Earth Surface Processes* **2**: 261–78.

Rickson R J (1987), Small plot field studies of soil erodibility using a rainfall simulator. In I Pla Sentis (ed.), *Soil conservation and productivity*. Maracay, Sociedad Venezolana de la Ciencia del Suelo: 339–48.

Rickson R J (1988), The use of geotextiles in soil erosion control: comparison of performance on two soils. In S Rimwanich (ed.), *Land conservation for future generations*. Bangkok, Department of Land Development: 961–70.

Rickson R J (1990), The role of simulated vegetation in soil erosion control. In J B Thornes (ed.), *Vegetation and erosion*. Chichester, Wiley: 99–111.

Rickson R J (1992), The application of geotextiles in the protection of grassed waterways. In H Hurni and Kebede Tato (eds), *Erosion, conservation and small-scale farming*. Bern, Geographica Bernensia: 415–21.

Riley S J (1988), Soil loss from road batters in the Karuah State Forest, eastern Australia. *Soil Technology* **1**: 313–32.

Ripley E A, R E Redman (1976), Grassland. In J L Monteith (ed.), *Vegetation and the atmosphere. Volume 2. Case studies*. London, Academic Press: 349–98.

Ritchie J C, J R McHenry (1975), Fallout Cs-137: a tool in conservation research. *Journal of Soil and Water Conservation* **30**: 283–6.

Robinson M, K Blyth (1982), The effect of forestry drainage operations on upland sediment yields: a case study. *Earth Surface Processes and Landforms* **7**: 85–90.

Rodriguez O S, N Fernandez de la Paz (1992), Conservation practices for horticulture production in the mountainous regions of Venezuela. In H Hurni and Kebede Tato (eds), *Erosion, conservation and small-scale farming*. Bern, Geographica Bernensia: 393–406.

Roels J M, P J Jonker (1985), Representativity and accuracy of measurements of soil loss from runoff plots. *Transactions of the American Society of Agricultural Engineers* **28**: 1458–66.

Rogers N W, M J Selby (1980), Mechanisms of shallow translational landsliding during summer rainstorms: North Island, New Zealand. *Geografiska Annaler* **62–A**: 11–21.

Rogler H, U Schwertmann (1981), Erosivität der Niederschläge und Isoerodentkarte von Bayern. *Zeitschrift für Kulturtechnik und Flurbereiniging* **22**: 99–112.

Rohdenburg H (1989), *Landscape ecology – geomorphology*. Cremlingen-Destedt, Catena Verlag.

Roose E J (1967), Dix années de mesure de l'érosion et du ruissellement au Sénégal. *L'Agronomie Tropicale* **22**: 123–52.

Roose E J (1971), *Influence des modifications du milieu naturel sur l'érosion: le bilan hydrique et chimique suite à la mise en culture sous climat tropical.* Cyclo., ORSTOM, Adiopodoumé, Ivory Coast.

Roose E J (1975), *Erosion et ruissellement en Afrique de l'ouest: vingt années de mesures en petites parcelles expérimentales.* Cyclo. ORSTOM, Adiopodoumé, Ivory Coast.

Roose E J (1976), Contribution à l'étude de l'influence de la mésofaune sur la pédogenèse actuelle en milieu tropical. *Rapports ORSTOM,* Abidjan: 2–23.

Roose E J (1977), Application of the Universal Soil Loss Equation of Wischmeier and Smith in West Africa. In D J Greenland and R Lal (eds), *Soil conservation and management in the humid tropics.* London, Wiley: 177–87.

Roose E (1992), Traditional and modern strategies for soil and water conservation in the Sudano-Sahelian areas of western Africa. In H Hurni and Kebede Tato (eds), *Erosion, conservation and small-scale farming.* Bern, Geographica Bernensia: 349–65.

Roose E, J Cavalie (1988), New strategy of water management and soil conservation application in developed and developing countries. In S Rimwanich (ed.), *Land conservation for future generations.* Bangkok, Department of Land Development: 913–24.

Roose E J, F X Masson (1985), Consequences of heavy mechanization and new rotation on runoff and on loessial soil degradation in northern France. In S A El-Swaify, W C Moldenhauer and A Lo (eds), *Soil erosion and conservation.* Ankeny IA, Soil Conservation Society of America: 24–33.

Rooseboom A, G W Annandale (1981), Techniques applied in determining sediment loads in South African rivers. *International Association of Scientific Hydrology Publication* **133**: 219–24.

Rose C W, P B Hairsine (1988), Processes of water erosion. In W L Stefan, Q T Denmead and I White (eds), *Flow and transport in the natural environment.* Berlin, Springer-Verlag: 312–26.

Rose C W, J R Williams, G C Sander, D A Barry (1983), A mathematical model of soil erosion and deposition process. I. Theory for a plane element. *Soil Science Society of America Journal* **47**: 991–5.

Rose S J C (1989), The Three Peaks Project: tackling footpath erosion. In *Erosion knows no boundaries.* Steamboat Springs CO, International Erosion Control Association: 369–78.

Rosewell C J (1970), Investigations into the control of earth-work tunnelling. *Journal of the Soil Conservation Service NSW* **26**: 188–203.

Rowntree K M (1983), Rainfall erosivity in Kenya: some preliminary considerations. In D B Thomas and W M Senga (eds), *Soil and water conservation in Kenya.* Institute of Development Studies and Faculty of Agriculture, University of Nairobi Occasional Paper 42: 1–19.

Ruangpanit N (1985) Percent crown cover related to water and soil losses in mountainous forest in Thailand. In S A El-Swaify, W C Moldenhauer and A Lo (eds), *Soil erosion and conservation.* Ankeny IA, Soil Conservation Society of America: 462–71.

Runólfsson S (1978), Soil conservation in Iceland. In M W Holdgate and M J Woodman (eds), *The breakdown and restoration of ecosystems.* New York, Plenum: 231–40.

Runólfsson S (1987), Land reclamation in Iceland. *Arctic and Alpine Research* **19**: 514–17.

Rustom R N, J R Weggel (1993), A study of erosion control systems: experimental results. In *Preserving our environment: the race is on.* Steamboat Springs CO, International Erosion Control Association: 253–75.

Rutin J (1992), Geomorphic activity of rabbits on a coastal sand dune, De Blink Dunes, The Netherlands. *Earth Surface Processes and Landforms* **17**: 85–94.

Rydgren B (1992), Soil erosion and nutrient loss studies in the southern Lesotho lowlands. In H Hurni and Kebede Tato (eds), *Erosion, conservation and small-scale farming.* Bern, Geographica Bernensia: 213–28.

Sanders D W (1988), Food and Agriculture Organization activities in soil conservation. In W C Moldenhauer and N W Hudson (eds), *Conservation farming on steep lands.* Ankeny IA, Soil and Water Conservation Society: 54–62.

Savage R P, W W Woodhouse (1968), Creation and stabilization of coastal barrier dunes. In *Proceedings, Coastal engineering conference (London),* No. 1: 671–700.

Savat J (1975), Discharge velocities and total erosion of a calcareous loess: a comparison between pluvial and terminal runoff. *Revue de Géomorphologie Dynamique* **24**: 113–22.

Savat J (1977), The hydraulics of sheet flow on a smooth surface and the effect of simulated rainfall. *Earth Surface Processes* **2**: 125–40.

Savat J (1979), Laboratory experiments on erosion and deposition of loess by laminar sheet flow and turbulent rill flow. In H Vogt and Th Vogt (eds), *Colloque sur l'érosion agricole des sols en milieu tempéré non Mediterranéen.* Strasbourg, l'Université Louis Pasteur: 139–43.

Savat J (1981), Work done by splash: laboratory experiments. *Earth Surface Processes and Landforms* **6**: 275–83.

Savat J (1982), Common and uncommon selectivity in the process of fluid transportation: field observations and laboratory experiments on bare surfaces. *Catena Supplement* 1: 139–60.

Savat J, J De Ploey (1982), Sheetwash and rill development by surface flow. In R Bryan and A Yair (eds), *Badland geomorphology and piping*. Norwich, Geo Books: 113–26.

Schertz D L (1983), The basis for soil loss tolerance. *Journal of Soil and Water Conservation* 38: 10–14.

Schumm S A (1979), Geomorphic thresholds: the concept and its applications. *Transactions of the Institute of British Geographers New Series* 4: 485–515.

Schwab G O, R K Frevert, T W Edminster, K K Barnes (1966), *Soil and water conservation engineering*. New York, Wiley.

Schwertmann U, W Vogl, M Kainz (1987), *Bodenerosion durch Wasser*. Stuttgart, Ulmer Verlag.

Scoging H M, J B Thornes (1979), Infiltration characteristics in a semi-arid environment. *International Association of Scientific Hydrology Publication* 128: 159–68.

Scoging H, A J Parsons, A D Abrahams (1992), Application of a dynamic overland-flow hydraulic model to a semi-arid hillslope, Walnut Gulch, Arizona. In A J Parsons and A D Abrahams (eds), *Overland flow: hydraulics and erosion mechanics*. London, UCL Press: 105–45.

Seubert C E, P A Sanchez, C Valverde (1977), Effects of land clearing methods on soil properties and soil performance in the Amazon jungle of Peru. *Tropical Agriculture* 54: 307–21.

Shainberg I, J Letey (1984), Response of soils to sodic and saline conditions. *Hilgardia* 52: 1–57.

Shakesby R A, R Whitlow (1991), Perspectives on prehistoric and recent gullying in central Zimbabwe. *GeoJournal* 23: 49–58.

Shakesby R A, C de O A Coelho, A D Ferreira, J P Terry, R P D Walsh (1993), Wildfire impacts on soil erosion and hydrology in west Mediterranean forest, Portugal. *International Journal of Wildland Fire* 3: 95–110.

Sharpe C F S (1938), *Landslides and related phenomena*. New York, Columbia University Press.

Sharpley A N, S J Smith (1990), Phosphorus transport in agricultural runoff: the role of soil erosion. In J Boardman, I D L Foster and J A Dearing (eds), *Soil erosion on agricultural land*. Chichester, Wiley: 351–66.

Shaxson T F (1981), Reconciling social and technical needs in conservation work on village farmlands. In R P C Morgan (ed.), *Soil conservation: problems and prospects*. Chichester, Wiley: 385–97.

Shaxson T F (1987), Changing approaches to soil conservation. In I Pla Sentis (ed.), *Soil conservation and productivity*. Maracay, Sociedad Venezolana de la Ciencia del Suelo: 11–27.

Shaxson T F (1988), Conserving soil by stealth. In W C Moldenhauer and N W Hudson (eds), *Conserving farming on steep lands*. Ankeny IA, Soil and Water Conservation Society: 9–17.

Sheng T C (1972a), A treatment-oriented land capability classification scheme for hilly marginal lands in the humid tropics. *Journal of the Scientific Research Council Jamaica* 3: 93–112.

Sheng T C (1972b), Bench terracing. *Journal of the Scientific Research Council Jamaica* 3: 113–27.

Sheng T C (1981), The need for soil conservation structures for steep cultivated slopes in the humid tropics. In R Lal and E W Russell (eds), *Tropical agricultural hydrology*. Chichester, Wiley: 357–72.

Sherchan D P, S P Chand, Y B Thapa, T P Tiwari, G B Gurung (1990), Soil and nutrient losses in runoff on selected crop husbandry practices on hill slope soil of the Eastern Nepal. In *Proceedings, International symposium on water erosion, sedimentation and resource conservation*. Dehra Dun, Central Soil and Water Conservation Research and Training Institute: 188–98.

Shields A (1936), Anwendung der Ähnlichkeitsmechanik und der Turbulenzforschung auf die Geschiebebewegung. *Mitteilungen der Preussischen Anstalt Wasserbau und Schiffbau* 26.

Siemens J C, W R Oschwald (1978), Corn–soybean tillage systems: erosion control, effects on crop production, costs. *Transactions of the American Society of Agricultural Engineers* 21: 293–302.

Simanton J R, K G Renard (1992), Upland erosion research on rangeland. In A J Parsons and A D Abrahams (eds), *Overland flow: hydraulics and erosion mechanics*. London, UCL Press: 335–75.

Sinajin J S (1987), Alternatives to shifting cultivation in Sabah, East Malaysia. In T H Tay, A M Mokhtaruddin and A B Zahari (eds), *Steepland agriculture in the humid tropics*. Kuala Lumpur, MARDI/Malaysian Society of Soil Science: 601–29.

Singh G, R Babu, S Chandra (1981), *Soil loss prediction research in India*. Central Soil and Water Conservation Research and Training Institute Bulletin No. T 12/D9, Dehra Dun.

Singh G, S P Bhardwaj, B P Singh (1979), Effect of row cropping of maize and soybean on erosion losses. *Indian Journal of Soil Conservation* 7: 43–6.

Singh K (1990), Enlisting people's participation in soil and water conservation programmes: lessons of India's experience. In *Proceedings, International symposium on water erosion, sedimentation and resource conservation*. Dehra Dun, Central Soil and Water

Conservation Research and Training Institute: 471–81.

Skidmore E L, L J Hagen (1977), Reducing wind erosion with barriers. *Transactions of the American Society of Agricultural Engineers* 20: 911–15.

Skidmore E L, J R Williams (1991), Modified EPIC wind erosion model. In *Modeling plant and soil systems*. ASA-CSSA-SSSA Agronomy Monograph No. 31, Madison WI: 457–69.

Skidmore E L, N P Woodruff (1968), Wind erosion forces in the United States and their use in predicting soil loss. *USDA Agricultural Research Service Handbook* 346.

Slattery M C, R B Bryan (1992), Hydraulic conditions for rill incision under simulated rainfall: a laboratory experiment. *Earth Surface Processes and Landforms* 17: 127–46 .

Smalley I J (1970), Cohesion of soil particles and the intrinsic resistance of simple soil systems to wind erosion. *Journal of Soil Science* 21: 154–61.

Smith D D (1958), Factors affecting rainfall erosion and their evaluation. *International Association of Scientific Hydrology Publication* 43: 97–107.

Smithen A A, R E Schulze (1982), The spatial distribution in southern Africa of rainfall erosivity for use in the Universal Soil Loss Equation. *Water SA* 8: 74–8, 165–7.

Soane B D, J D Pidgeon (1975), Tillage requirement in relation to soil physical properties. *Soil Science* 119: 376–84.

Soil Survey of England and Wales (1979), *Land use capability: England and Wales.* 1:1,000,000 map sheets.

Soil Survey of England and Wales (1983), *Soil map of England and Wales.* 1:250,000 map sheets.

Sørensen M (1985), Estimation of some aeolian saltation transport parameters from transport rate profiles. In O E Barndorff-Nielsen, J T Møller, K R Rasmussen and B B Willetts (eds), *Proceedings of international workshop on the physics of blown sand.* Memoirs No. 8, Department of Theoretical Statistics, University of Aarhus: 141–90.

Spaan W P, G D van den Abeele (1991), Wind borne particle measurements with acoustic sensors. *Soil Technology* 4: 51–63.

Spaan W P, P M van Dijk, W A Hollemans, L A A J Eppink (1990), *Wind erosion measurements on Schiermonnikoog. Report II. Report of the workshop from 20 to 23 May 1990.* Wageningen Agricultural University.

Spomer R G, A T Hjelmfelt (1983), Snowmelt runoff and erosion on Iowa loess soils. *Transactions of the American Society of Agricultural Engineers* 26: 1109–11, 1116.

Spoor G, R J Godwin (1979), Soil deformation and shear strength characteristics of some clay soils at different moisture contents. *Journal of Soil Science* 30: 483–98.

Spoor G, P B Leeds-Harrison, R J Godwin (1982), Potential role of soil density and clay mineralogy in assessing the suitability of soils for mole drainage. *Journal of Soil Science* 33: 427–41.

Sreenivas L, J R Johnston, H O Hill (1947), Some relationships of vegetation and soil detachment in the erosion process. *Soil Science Society of America Proceedings* 11: 471–4.

Srivastva A K, M S Rama Mohan Rao (1988), Crop responses under different agroforestry practices in semi arid black soil region. *Indian Journal of Soil Conservation* 16: 1–10.

Stallings J H (1957), *Soil conservation.* Englewood Cliffs NJ, Prentice Hall.

Stanley R J, L A Watt (1990), Liquid sprays to assist stabilisation and revegetation of coastal sand dunes. *Soil Technology* 3: 9–20.

Starkel L (1972), The role of catastrophic rainfall in the shaping of the relief of the Lower Himalaya (Darjeeling Hills). *Geographica Polonica* 21: 103–47.

Steindórsson S (1980) Flokkun gróðurs í gróðurfélög. *Íslenzkar Landbúnaðarrannsóknir* 12: 11–52.

Stevens J H (1974), Stabilization of eolian sands in Saudi Arabia's Al Hasa oasis. *Journal of Soil and Water Conservation* 29: 129–33.

Stocking M (1985), Development projects for the small farmer: lessons from eastern and central Africa in adapting conservation. In S A El-Swaify, W C Moldenhauer and A Lo (eds), *Soil erosion and conservation.* Ankeny IA, Soil Conservation Society of America: 747–58.

Stocking M, N Abel (1992), Labour costs: a critical element in soil conservation. In Kebede Tato and H Hurni (eds), *Soil conservation for survival.* Ankeny IA, Soil and Water Conservation Society: 206–18.

Stocking M A, H A Elwell (1973a), Prediction of subtropical storm soil losses from field plot studies. *Agricultural Meteorology* 12: 193–201.

Stocking M A, H A Elwell (1973b), Soil erosion hazard in Rhodesia. *Rhodesian Agricultural Journal* 70: 93–101.

Stocking M A, H A Elwell (1976), Rainfall erosivity over Rhodesia. *Transactions of the Institute of British Geographers New Series* 1: 231–45.

Stredňanský J (1977), Kritické rýchlosti vetra z hladiska erodavtelnosti pôd na južnom Slovensku. *Vysoká škola polnohospodárska*, Nitra.

Streeter D T (1977), Gully restoration on Box Hill. *Countryside Recreation Review* 2: 28–40.

Stuttard M J (1984), Effect of tillage on clod stability to rainfall: laboratory simulation. *Journal of Agricultural Engineering Research* 30: 141–7.

Styczen M, K Høgh-Schmidt (1988), A new description of splash erosion in relation to raindrop sizes and vegetation. In R P C Morgan and R J Rickson (eds), *Erosion assessment and modelling.* Commission of the European Communities Report No. EUR 10860 EN: 147–84.

Styczen M E, R P C Morgan (1995), Engineering properties of vegetation. In R P C Morgan and R J Rickson (eds), *Slope stabilization and erosion control: a bioengineering approach.* London, E and F N Spon: 5–58.

Styczen M, S A Nielsen (1989), A view of soil erosion theory, process research and model building: possible interactions and future developments. *Quaderni di Scienza del Suolo* 2: 27–45 .

Suddhapreda N, E P Paningbatan, W Chankong, B Piadang (1988), Prediction of soil erosion in northern Thailand using a physical model. In S Rimwanich (ed.), *Land conservation for future generations.* Bangkok, Department of Land Development: 489–502.

Sulaiman W, L M Maene, A M Mokhtaruddin (1981), Runoff, soil and nutrient losses from an ultisol under different legumes. In T Tingsanchali and H Eggers (eds), *Southeast Asian regional symposium on problems of soil erosion and sedimentation.* Bangkok, Asian Institute of Technology: 275–86.

Swan S B St C (1970), Piedmont slope studies in a humid tropical region, Johor, southern Malaya. *Zeitschrift für Geomorphologie Supplementband* 10: 30–9.

Swanson F J, C T Dyrness (1975), Impact of clearcutting and road construction on soil erosion by landslides in the West Cascade Range, Oregon. *Geology* 3: 393–6.

Swanson L E, S M Camboni, T L Napier (1986), Barriers to the adoption of soil conservation practices on farms. In S B Lovejoy and T L Napier (eds), *Conserving soil: insights from socioeconomic research.* Ankeny IA, Soil Conservation Society of America: 108–20.

Swanson N P (1965), Rotating-boom rainfall simulator. *Transactions of the American Society of Agricultural Engineers* 8: 71–2.

Tackett J L, R W Pearson (1965), Some characteristics of soil crust formed by simulated rainfall. *Soil Science* 99: 407–13.

Talsma T (1969), *In situ* measurement of sorptivity. *Australian Journal of Soil Research* 17: 269–76.

Tang K, H Zhou, X Hou, Y Liu (1987), The influence of destruction and reconstruction of vegetation on soil erosion and its control in the loess plateau in China. In I Pla Sentis (ed.), *Soil conservation and productivity.* Maracay, Sociedad Venezolana de la Ciencia del Suelo: 963–72.

Tejwani K G (1981), Watershed management as a basis for land development and management in India. In R Lal and E W Russell (eds), *Tropical agricultural hydrology.* Chichester, Wiley: 239–55.

Tejwani K G (1992), Training, education and demonstration in soil and water conservation. In Kebede Tato and H Hurni (eds), *Soil conservation for survival.* Ankeny IA, Soil and Water Conservation Society: 269–76.

Temple D M (1982), Flow retardance of submerged grass channel linings. *Transactions of the American Society of Agricultural Engineers* 25: 1300–3.

Temple D M (1991), Changes in vegetal flow retardance during long-duration flows. *Transactions of the American Society of Agricultural Engineers* 34: 1769–74.

Temple D M, D Alspach (1992), Failure and recovery of a grass-lined channel. *Transactions of the American Society of Agricultural Engineers* 35: 171–3.

Temple P H (1972a), Soil and water conservation policies in the Uluguru Mountains, Tanzania. *Geografiska Annaler* 54–A: 110–23.

Temple P H (1972b), Measurements of runoff and soil erosion at an erosion plot scale with particular reference to Tanzania. *Geografiska Annaler* 54–A: 203–20.

Temple P H, A Rapp (1972), Landslides in the Mgeta area, western Uluguru Mountains, Tanzania. *Geografiska Annaler* 54–A: 157–93.

Terry J P, R A Shakesby (1993), Soil hydrophobicity effects on rainsplash: simulated rainfall and photographic evidence. *Earth Surface Processes and Landforms* 18: 519–25.

Terwilliger V J (1990), Effects of vegetation on soil slippage by pore pressure modification. *Earth Surface Processes and Landforms* 15: 553–70.

Thomas D B, E K Biamah (1989), Origin, application and design of the fanya juu terrace. In W C Moldenhauer, N W Hudson, T C Sheng and S W Lee (eds), *Development of conservation farming on hillslopes.* Ankeny IA, Soil and Water Conservation Society: 185–94.

Thórarinsson S (1958), The Öraefajökull eruption of 1362. *Acta Naturalia Islandica II.* No. 2, Reykjavík.

Thornes J B (1976), *Semi-arid erosion systems: case studies from Spain.* London School of Economics Geographical Papers No. 7.

Thornes J B (1980), Erosional processes of running water and their spatial and temporal controls: a theoretical viewpoint. In M J Kirkby and R P C

Morgan (eds), *Soil erosion*. Chichester, Wiley: 129–82.

Thorp J (1949), Effect of certain animals that live in soils. *Science Monthly* **68**: 180–91.

Thorsteinsson I (1980a), Gróðurskilyrði, gróðurfar, uppskera gróðurlenda og plöntuval búfjár. *Íslenzkar Landbúnaðarrannsóknir* **12**(2): 85–99.

Thorsteinsson I (1980b), Nýting úthaga: beitarþungi. *Íslenzkar Landbúnaðarrannsóknir* **12**(2): 113–22.

Thorsteinsson I (1980c), Beitagildi gróðurlenda. *Íslenzkar Landbúnaðarrannsóknir* **12**(2): 123–5.

Thorsteinsson I, G Ólafsson, G M van Dyne (1971), Range resources of Iceland. *Journal of Range Management* **24**: 86–93.

Tiffen M, M Mortimore, F Gichuki (1994), *More people, less erosion. Environmental recovery in Kenya*. Chichester, Wiley.

Till R (1973), The use of linear regression in geomorphology. *Area* **5**: 303–8.

Tisdall J M, J M Oades (1982), Organic matter and water stable aggregates in soils. *Journal of Soil Science* **33**: 141–64.

Tjernström R (1992), Yields from terraced and non-terraced fields in the Machakos District of Kenya. In Kebede Tato and H Hurni (eds), *Soil conservation for survival*. Ankeny IA, Soil and Water Conservation Society: 251–65.

Torri D, L Borselli (1991), Overland flow and soil erosion: some processes and their interactions. *Catena Supplement* **19**: 129–37.

Torri D, J Poesen (1988), Incipient motion conditions for single rock fragments in simulated rill flow. *Earth Surface Processes and Landforms* **13**: 225–37.

Torri D, J Poesen (1992), The effect of soil surface slope on raindrop detachment. *Catena* **19**: 561–78.

Torri D, M Sfalanga (1986), Some problems on soil erosion modelling. In A Giorgini and F Zingales (eds), *Agricultural nonpoint source pollution: model selection and applications*. Amsterdam, Elsevier: 161–71.

Torri D, M Sfalanga, G Chisci (1987), Threshold conditions for incipient rilling. *Catena Supplement* **8**: 97–105.

Torri D, M Sfalanga, M Del Sette (1987), Splash detachment: runoff depth and soil cohesion. *Catena* **14**: 149–55.

Toy T J (1989), An assessment of surface-mine reclamation based upon sheetwash erosion rates at the Glenrock Coal Company, Glenrock, Wyoming. *Earth Surface Processes and Landforms* **14**: 289–302.

Trabaud L, J Lepart (1980), Diversity and stability in garrigue ecosystems after fire. *Vegetation* **43**: 49–57.

Tracy F C, K G Renard, M M Fogel (1984), Rainfall energy characteristics for southeastern Arizona. In *Water today and tomorrow*. New York, Irrigation and Drainage Division ASCE: 559–66.

Troeh F R, J A Hobbs, R L Donahue (1980), *Soil and water conservation for productivity and environmental protection*. Englewood Cliffs NJ, Prentice Hall.

Truman C C, J M Bradford, J E Ferris (1990), Antecedent water content and rainfall energy influence on soil aggregate breakdown. *Soil Science Society of America Journal* **54**: 1385–92.

Trustrum N A, V J Thomas, M G Lambert (1984), Soil slip erosion as a constraint to hill country pasture production. *Proceedings of the New Zealand Grassland Association* **45**: 66–76.

Uchijima Z (1976), Maize and rice. In J L Monteith (ed.), *Vegetation and the atmosphere. Volume 2. Case studies*. London, Academic Press: 33–64.

Ulén B (1988), Erosion of phosphorus from arable land in Sweden. In S Skøien and R Aaker (eds), *Jorderosjon*. Informasjon fra Statens Fagtjeneste for Landbruket **13**: 80–4.

United States Department of Agriculture (1979), Field manual for research in agricultural hydrology. *USDA Agricultural Handbook 224*.

van Asch Th W J (1983), Water erosion on slopes in some land units in a Mediterranean area. *Catena Supplement* **4**: 129–40.

van Dijk P M (1990), *Wind erosion measurements on Schiermonnikoog. Report I. A study with acoustic sensors and sediment catchers*. Wageningen Agricultural University.

Vis M (1986), Interception, drop size distributions and rainfall kinetic energy in four Colombian forest ecosystems. *Earth Surface Processes and Landforms* **11**: 591–603.

Vittorini S (1972), The effect of soil erosion in an experimental station in the Pliocene clay of the Val d'Era (Tuscany) and its influence on the evolution of the slopes. *Acta Geographica Debrecina* **10**: 71–81.

Vogel H (1990), Deterioration of a mountainous agro-ecosystem in the Third World due to emigration of rural labour. In B Messerli and H Hurni (eds), *African mountains and highlands: problems and perspectives*. African Mountains Association: 389–406.

Vogel H (1991), Sheet erosion from sandveld soils under conservation tillage in Zimbabwe. *Conservation tillage for sustainable crop production systems, Project Research Report* No. 2. Agritex, Soil and Water Conservation Branch, Harare: 25–39.

Vogel H (1992), Effects of conservation tillage on sheet erosion from sandy soils at two experimental sites in Zimbabwe. *Applied Geography* **12**: 229–42.

Voorhees W B, M J Lindstrom (1984), Long-term effects of tillage method on soil tilth independent of wheel traffic compaction. *Soil Science Society of America Journal* **48**: 152–5.

Voroney R P, J A van Veen, E A Paul (1981), Organic carbon dynamics in grassland soils. II. Model validation and simulation of the long-term effects of cultivation and rainfall erosion. *Canadian Journal of Soil Science* **61**: 211–24.

Voznesensky A S, A B Artsruui (1940), Laboratoriya metod opretseleniy protivoerozionnoy ustoychivosti pochv. In *Voprosi protivoerozionnoy ustoychivosti pochv*. Izd. Zakavk. NIIVKH, Tbilisi.

Wallace G A, A Wallace (1986), Control of soil erosion by polymeric soil conditioners. *Soil Science* **141**: 363–7.

Walling D E (1983), The sediment delivery problem. *Journal of Hydrology* **69**: 209–37.

Walling D E, T A Quine (1990), Use of caesium-137 to investigate patterns and rates of soil erosion on arable fields. In J Boardman, I D L Foster and J A Dearing (eds), *Soil erosion in agricultural land*. Chichester, Wiley: 33–53.

Walling D E, B W Webb (1983), Patterns of sediment yield. In K J Gregory (ed.), *Background to palaeohydrology*. Chichester, Wiley: 69–100.

Wan Yusoff W A (1988), Forest management and conservation practices for controlling soil erosion in the natural forest of Peninsular Malaysia. In S Rimwanich (ed.), *Land conservation for future generations*. Bangkok, Department of Land Development: 869–75.

Watson A (1990), The control of blowing sand and mobile desert dunes. In A S Goudie (ed.), *Techniques for desert reclamation*. Chichester, Wiley: 35–85.

Weaver T, D Dale (1978), Trampling effects of hikers, motor cycles and horses in meadows and forests. *Journal of Applied Ecology* **15**: 451–7.

Wells N A, B Andriamihaja (1991), Growth of gullies on laterite in Madagascar: non-intuitive implications for control. In *Erosion control: a global perspective*. Steamboat Springs CO, International Erosion Control Association: 379–401.

Wells N A, B Andriamihaja, H F S Rakotovololona (1991), Patterns of development of lavaka, Madagascar's unusual gullies. *Earth Surface Processes and Landforms* **16**: 189–206.

Wen D (1993), Soil erosion and conservation in China. In D Pimental (ed.), *World soil erosion and conservation*. Cambridge, Cambridge University Press: 63–85.

Wenner C G (1981), *Soil conservation in Kenya*. Nairobi, Ministry of Agriculture.

Wenner C G (1988), The Kenyan model of soil conservation. In W C Moldenhauer and N W Hudson (eds), *Conservation farming on steep lands*. Ankeny IA, Soil and Water Conservation Society: 197–206.

Whitlow R, A Bullock (1986), Rapid gully development in Mangwende Communal Land. *Zimbabwe Agricultural Journal* **83**: 149–60.

Whitmore T C, C P Burnham (1969), The altitudinal sequence of forests and soils on granite near Kuala Lumpur. *Malayan Nature Journal* **22**: 99–118.

Wicks G A, D E Smika (1973), Chemical fallow in a winter wheat-fallow rotation. *Weed Science* **21**: 97–102.

Wiersum K F (1985), Effects of various vegetation layers of an *Acacia auriculiformis* forest plantation on surface erosion in Java, Indonesia. In S A El-Swaify, W C Moldenhauer and A Lo (eds), *Soil erosion and conservation*. Ankeny IA, Soil Conservation Society of America: 79–89.

Wiersum K F, P Budirijanto, D Rhomdoni (1979), Influence of forests on erosion. Seminar on the erosion problem in the Jatiluhur area. *Institute of Ecology Padjadjaran University, Bandung*, Report 3.

Wiggins S L (1981), The economics of soil conservation in the Acelhuate River Basin, El Salvador. In R P C Morgan (ed.), *Soil conservation: problems and prospects*. Chichester, Wiley: 399–417.

Willetts B B, M A Rice (1985), Inter-saltation collisions. In O E Barndorff-Nielsen, J T Møller, K R Rasmussen and B B Willetts (eds), *Proceedings of international workshop on the physics of blown sand*. Memoirs No. 8, Department of Theoretical Statistics, University of Aarhus: 83–100.

Willetts B B, M A Rice (1988), Particle dislodgement from a flat sand bed by wind. *Earth Surface Processes and Landforms* **13**: 717–28.

Williams A R, R P C Morgan (1976), Geomorphological mapping applied to soil erosion evaluation. *Journal of Soil and Water Conservation* **31**: 164–8.

Williams C N, K T Joseph (1970), *Climate, soil and crop production in the humid tropics*. Kuala Lumpur, Oxford University Press.

Williams J R, C A Jones, P T Dyke (1984), A modeling approach to determining the relationship between erosion and soil productivity. *Transactions of the American Society of Agricultural Engineers* **27**: 129–44.

Wischmeier W H (1973), Conservation tillage to control water erosion. In *Proceedings, National conservation tillage conference*. Ankeny IA, Soil Conservation Society of America: 133–41.

Wischmeier W H (1975), Estimating the soil loss equation's cover and management factor for undisturbed area. In *Present and prospective technology for* *predicting sediment yields and sources*. USDA ARS Publication ARS-S-40: 118–24.

Wischmeier W H (1978), Use and misuse of the Universal Soil Loss Equation. *Journal of Soil and Water Conservation* 31: 5–9.

Wischmeier W H, J V Mannering (1969), Relation of soil properties to its erodibility. *Soil Science Society of America Proceedings* 23: 131–7.

Wischmeier W H, D D Smith (1958), Rainfall energy and its relationship to soil loss. *Transactions of the American Geophysical Union* 39: 285–91.

Wischmeier W H, D D Smith (1978), Predicting rainfall erosion losses. *USDA Agricultural Research Service Handbook* 537.

Wischmeier W H, C B Johnson, B V Cross (1971), A soil erodibility nomograph for farmland and construction sites. *Journal of Soil and Water Conservation* 26: 189–93.

Wise S M, J B Thornes, A Gilman (1982), How old are the badlands? A case study from southeast Spain. In R Bryan and A Yair (eds), *Badland geomorphology and piping*. Norwich, Geo Books: 259–77.

Withers B, S Vipond (1974), *Irrigation: design and practice*. London, Batsford.

W M S Associates (1988), Gully erosion in Swaziland. Final Report. Fredericton NB, Canada.

Wolde F T, D B Thomas (1989), The effect of narrow grass strips in reducing soil loss and runoff in a Kabete nitisol, Kenya. In D B Thomas, E K Biamah, A M Kilewe, L Lundgren and B O Mochoge (eds), *Soil and water conservation in Kenya*. Department of Agricultural Engineering University of Nairobi/SIDA: 176–94.

Wolman M G (1967), A cycle of sedimentation and erosion in urban river channels. *Geografiska Annaler* 49–A: 385–95.

Wolman M G, J P Miller (1960), Magnitude and frequency of forces in geomorphic processes. *Journal of Geology* 68: 54–74.

Wood A P (1992), Zambia's soil conservation heritage: a review of policies and attitudes towards soil conservation from colonial times to the present. In Kebede Tato and H Hurni (eds), *Soil conservation for survival*. Ankeny IA, Soil and Water Conservation Society: 156–71.

Woodburn R, J Kozachyn (1956), A study of relative erodibility of a group of Mississippi gully soils. *Transactions of the American Geophysical Union* 37: 749–53.

Woodruff N P, F H Siddoway (1965), A wind erosion equation. *Soil Science of America Proceedings* 29: 602–8.

Woodruff N P, A W Zingg (1952), Wind-tunnel studies of fundamental problems related to wind-

breaks. *US Soil Conservation Service Publication* SCS-TP-112.

Woodruff N P, A W Zingg (1955), A comparative analysis of wind-tunnel and atmospheric air-flow patterns about single and successive barriers. *Transactions of the American Geophysical Union* 36: 203–8.

Woodruff N P, W S Chepil, R D Lynch (1957), Emergency chiseling to control wind erosion. *Kansas Agricultural Experimental Station Technical Bulletin* 90.

Woodruff N P, D W Fryrear, L Lyles (1963), Engineering similitude and momentum transfer principles applied to shelterbelt studies. *Transactions of the American Society of Agricultural Engineers* 6: 41–7.

World Bank (1990), *Vetiver grass: the hedge against erosion*. Washington, DC.

Wright J L, K W Brown (1967), Comparison of momentum and energy balance methods of computing vertical transfer within a crop. *Agronomy Journal* 59: 427–32.

Wu Q, Y Liang, K Tang, X Hou (1988), Aerial sowing of trees and herbs and its benefits to soil and water conservation in the semi-arid gullied hilly loess region. *International Journal of Sediment Research* 3: 1–9.

Wu T H (1995), Slope stabilization. In R P C Morgan and R J Rickson (eds), *Slope stabilization and erosion control: a bioengineering approach*. London, E and F N Spon: 221–64.

Yair A, J Rutin (1981), Some aspects of the regional variation in the amount of available sediment produced by isopods and porcupines, northern Negev, Israel. *Earth Surface Processes and Landforms* 6: 221–34.

Yalin M S (1963), An expression for bedload transportation. *Journal of the Hydraulics Division ASCE* 89: 221–50.

Yalin M S, E Karahan (1979), Inception of sediment transport. *Journal of the Hydraulics Division ASCE* 105: 1433–43.

Yariv S (1976), Comments on the mechanism of soil detachment by rainfall. *Geoderma* 15: 393–9.

Yohannes G M (1992), The effects of conservation on production in the Andit-Tid area, Ethiopia. In Kebede Tato and H Hurni (eds), *Soil conservation for survival*. Ankeny IA, Soil and Water Conservation Society: 239–50.

Young A (1969), Present rate of land erosion. *Nature* 224: 851–2.

Young A (1976), *Tropical soils and soil survey*. Cambridge, Cambridge University Press.

Zachar D (1982), *Soil erosion*. Amsterdam, Elsevier.

Zanchi C (1983), Influenze dell'azione battente della

pioggia e del ruscellamento nel processo erosivo e variazioni dell'erodibilità del suolo nei diversi periodi stagionali. *Annali Istituto Sperimentale per lo Studio e la Difesa del Suolo* **14**: 347–58.

Zanchi C (1989), Drainage as a soil conservation and soil stabilizing practice on hilly slopes. In U Schwertmann, R J Rickson and K Auerswald (eds), *Soil erosion protection measures in Europe*. Soil Technology Series **1**: 73–82.

Zanchi C, D Torri (1980), Evaluation of rainfall energy in central Italy. In M De Boodt and D Gabriels (eds), *Assessment of erosion*. London, Wiley: 133–42.

Zanchi C, P Bazzoffi, G D'Egidio, L Nistri (1981), Field rainfall simulator. In *Field excursion guide, Proceedings of the Florence Symposium*. International Association of Scientific Hydrology: 46–9.

Zaruba Q, V Mencl (1969), *Landslides and their control*. Amsterdam, Elsevier.

Zingg A W (1940), Degree and length of land slope as it affects soil loss in runoff. *Agricultural Engineering* **21**: 59–64.

Zingg A W (1951), A portable wind tunnel and dust collector developed to evaluate the erodibility of field surfaces. *Agronomy Journal* **43**: 189–91.

Zingg A W (1953), Wind tunnel studies of the movement of sedimentary material. *Proceedings 5th Hydraulics Conference, Iowa State University Bulletin* **34**: 111–35.

Index